T0331637

Proceedings of the Fifth Latin American Symposium
HIGH ENERGY PHYSICS

editors

C J Solano Salinas
O Pereyra Ravinez
R Ochoa Jiménez
Universidad Nacional de Ingeniería, Peru

Proceedings of the Fifth Latin American Symposium
HIGH ENERGY PHYSICS

Lima, Peru   12 – 17 July 2004

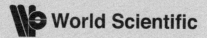

World Scientific

NEW JERSEY · LONDON · SINGAPORE · BEIJING · SHANGHAI · HONG KONG · TAIPEI · CHENNAI

Published by

World Scientific Publishing Co. Pte. Ltd.
5 Toh Tuck Link, Singapore 596224
USA office: 27 Warren Street, Suite 401-402, Hackensack, NJ 07601
UK office: 57 Shelton Street, Covent Garden, London WC2H 9HE

British Library Cataloguing-in-Publication Data
A catalogue record for this book is available from the British Library.

Proceedings of the Fifth Latin American Symposium
HIGH ENERGY PHYSICS

ISBN-13 978-981-256-731-4
ISBN-10 981-256-731-3

Printed in Singapore

PREFACE

The *Fifth Latin American Symposium on High Energy Physics* (SILAFAE V) was organized by the National Engineering University (UNI) in Lima, Perú, and took place during the period 12-17 July, 2004. The present volume contains the texts of most of the invited talks delivered at the conference and a selection from among the many contributed presentations including posters.

The general format and style of the conference followed the accepted and well-developed pattern for the series, focusing on the development, refinement and important applications of the techniques of Particle Physics. The intention of the series has always been to cover in a broad and balanced fashion both the entire spectrum of theoretical and experimental tools developed to tackle problems in particle physics and their major fields of application. One of the main aims of the series is to foster the exchange of ideas and techniques among physicists working in such diverse areas as high energy physics, nuclear and subnuclear physics, astrophysics, quantum field theory, relativity and gravitation.

The SILAFAE conferences are now firmly established as being one of the premier series of international meetings in the field. The history of the series itself provides a mirror in which to view the rapid development of this field in Latin America in recent years. The first official meeting in the SILAFAE series was held in Mérida, México in 1996. Later meetings in the series have been SILAFAE II in San Juan de Puerto Rico, Puerto Rico in 1998; SILAFAE III in Bogotá, Colombia in 2000; SILAFAE IV in Aguas de Lindoia, Brazil, in 2002; and, finally, the present meeting, SILAFAE V, Lima, Perú in 2004.

A special session was held the first moorning of SILAFAE V to honor Dr. Luis Masperi by dedicating the symposium to his memory. During his lifetime Masperi was an enthusiastic member of the Particle Physics community and also one of the prime instigators of the SILAFAE series, helping to shape it as a member of the Scientific Committee. He served as an inspiration to others throughout his career, especially during his final years when he served as Director of the Latin American Center for Physics (CLAF).

The organization of this SILAFAE in Peru was possible thanks to the financial support of the national agencies *Universidad Nacional de Ingeniería (UNI)*, *Instituto General de Investigación de la UNI (IGI)*, *Consejo Nacional de Ciencia y Tecnología (CONCYTEC)*, *Sociedad Peruana de Física (SOPERFI)*, and the foreign agencies *Centro Latinoamericano de Física (CLAF)*, *International Centre for Theoretical Physics (ICTP)*, and *National Science Foundation (NSF)*. Here, the support of some participants by the U. S. National Science Foundation through Award Number 0409377 is gratefully acknowledged. We must thank also to our friendly secretariat Margarita Mendoza.

In this event as well, the Organizing and Scientific Committees deserve great thanks in creating a well-run and productive meeting, with an exciting program of talks and poster presentations. It is a pleasure to thank all of them for their hard work, especially Orlando Pereyra who, as Chairman, has led and guided them throughout.

C. J. Solano Salinas   Lima, Perú
(Editor for the Fifth Latin American   31 December 2005
Symposium on High Energy Physics)

CONTENTS

Contributions

LOCAL ORGANIZING COMMITTEE

O. Pereyra Ravinez (Chairman) – Universidad Nacional de Ingeniería, Perú
M. Sheaff (Co-Chairman)   – University of Wisconsin, USA
C. J. Solano Salinas   – Universidad Nacional de Ingeniería, Perú
                               – Cinvestav Unidad Mérida, México
R. Ochoa Jimenez   – Universidad Nacional de Ingeniería, Perú
E. Salinas   – London South Bank University, UK
A. Bernui   – Universidad Nacional de Ingeniería, Perú
                               – CBPF, Brazil
A. M. Gago   – Pontificia Universidad Católica del Perú,
                               – Perú

FOREWORD BY THE EDITORS

The *Fifth Latin American Symposium on High Energy Physics* was organized by the Universidad Nacional de Ingeniería in Lima, Perú, over the period 12-17 July 2004. During the meeting all invited and some contributed papers were presented orally; other contributed papers were shown as posters.

All of the invited speakers were requested to submit a paper for these Proceedings. The Editors are grateful to the large majority who were able to do so by the tight deadline imposed. Since the quality of the contributed papers was so high, the Editors decided to also include a selection of them in these Proceedings. Accordingly, we invited some of the poster presenters to submit an article, and we have also incorporated these into the present volume.

In order to facilitate the usefulness of this book, related articles have been grouped together by the Editors. While these broadly reflect the themes of the conference sessions, we have not followed them exactly. In particular, the articles do not follow the same sequence as their presentation at the meeting. We are keenly aware that the classification scheme and the grouping of the articles within it that we have adopted are somewhat arbitrary. In particular, many articles would sit comfortably under more than one classification. The interested reader should not, therefore, follow the scheme too rigorously.

We must mention as well the wonderful trip to Cuzco that many of the participants did as a group following the symposium. We still miss those beautiful days.

Finally, we thank all of the contributors for the extremely high quality of the articles. We feel confident that the present volume will live up to the high standards set by earlier proceedings in this series.

Carlos Javier Solano Salinas   Lima, Perú
Orlando Pereyra Ravinez   31 December, 2005
Rosendo Ochoa Jimenez

GALLERY

HOMAGES

IN MEMORIAM LUIS MASPERI
(1940–2003)

1. Introduction

After six years as director of the Latin American Centre for Physics (CLAF), Luis Masperi, was planning to resettle in Bariloche in February 2004. Death truncated his plans on 2 December 2003 after a short illness.

2. His Academic Life

Born in Spoleto, Italy, in December 27th, 1940, Masperi grew up in Argentina and became an Argentinean citizen. He graduated as a Chemical engineer in 1961 and shifted to Physics afterward, obtaining his PhD at Instituto Balseiro in 1969. He became a theoretical physicist of international prestige and an outstanding teacher. He started, together with Andrs Garcia, the research in Particle Physics at Bariloche. He supervised several PhD theses and has published more than 70 articles in the areas of particle physics and field theory and more recently in Astroparticle Physics, after joining the Pierre Auger collaboration in 1995 on ultra high energy cosmic rays. He actually became one of the more active promoters of Argentinean participation in the project.

He made his scientific career at the Instituto Balseiro, starting as Professor Adjunto in 1969, Professor Associado in 1971, and Professor Titular in 1978. He was Vice Director of the Instituto Balseiro from 1972 to 1976 and Head of the Theory Division of the Centro Atomico de Bariloche from 1985 to 1991. From 1982 to 1984 he served as president of the Argentinean Physical Society.

3. His Political Life

Besides his enthusiastic contributions to research and academics, throughout his life, Luis showed commitment with world peace and a more just and equal society, at a local, national or international level. In 1972, at 32, Luis became Vice director of Instituto Balseiro. He resigned in1976 when the military coup struck Argentina, not before sanctioning, according to regulations, the Instituto Balseiro military students that became de facto authorities at the time. After the dictatorship he actively contributed to the reorganization of the Argentinian Physical Society (AFA), and became its president.

In 1992 he shared the American Physical Society Forum Award with fellow Latin American physicists Luis Pinguelli Rosa, Alberto Ridner and Fernando de Sousa Barros, for efforts in persuading Argentina and Brazil to abandon their nuclear weapons programmes. Luis was an engaged participant in INESAP (International Network of Engineers and Scientists Against Proliferation) activities. From 1999 to 2002 he, also, integrated the Pugwash Council. Pugwash was founded on the basis of the Einstein-Russell manifesto, aiming to bring scientific insight and reason to bear on threats to human security arising from science and technology in general, and above all from the catastrophic threat posed to humanity by nuclear and other weapons of mass destruction.

Luis Masperi was also involved in local politics. In a difficult context, by the end of the dictatorship in April 1982, he did not hesitate in opposing the adventure of the Malvinas war and its possible consequences when a very important fraction of the population and most of politicians, from left to right wing parties supported it. He was one of the founders of the Bariloche section of the Partido Intransigente and of the Bariloche local chapter of the national Asamblea Permanente por los Derecchos Humanos (Human Rights Forum). He was elected delegate in the debates that led to the establishment of the Municipal Constitution (Carta orgnica municipal) of the town of Bariloche.

Luis constantly stressed the importance of science in less developed countries and the crucial role of scientists to enhance dialog between peoples. He emphasized the urgent need of total nuclear disarmament as a necessary prerequisite to global peace and, above all, the defense of basic human rights.

It is probable that in these multiple engagements care of his own health was postponed. His body marked a limit.

As one of his friends mentioned, in particle physicist language, interactions with Luis were almost never weak but mostly strong. However, he was able to listen carefully and calmly debate on opinions neatly opposed to his.

Until his last days he kept his big brilliant eyes and expressive smile.

4. His Life in Brazil

First I would like to thank the organizers of the V-SILAFAE for inviting me to speak about Luis Masperi. It is a great honor and a big responsibility as I will certainly miss important parts of his work and his dedication to promote and support physics in Latin America.

Masperi was born in Spoleto, Italy, but was raised in Argentina and became an Argentinean citizen. He studied Physics at the Instituto Balseiro in Bariloche.

He made his scientific career at the Instituto Balseiro. After his PhD Masperi became very quickly a theoretical physicist of international. He was also a very good and loved teacher and contributed to the formation of many students at the Instituto Balseiro. He was thesis advisor of ten PhD students.

Masperi also maintained links with Italy through numerous visits and in particular a lifelong relationship with the International Centre for Theoretical Physics – ICTP, in Trieste, where he was Associated Senior Physicist from 1975 to 1992.

Masperi's international reputation rested not only on his research, but also on his dedication to world peace. In 1992 he shared with Alberto Ridner from Argentina and Luis Pinguelli Rosa and Fernando Souza Barros from Brazil the American Physical Society's Forum Award for his efforts in persuading Argentina and Brazil to resign nuclear weapons programs and to establish periodical mutual inspection of their nuclear facilities. His efforts to promote peace led him to be invited to be a member of the Pugwash Council from 1999 to 2002.

In 1998 he became a member of the Scientific Council of the Joint Institute for Nuclear Research – JINR in Dubna and has obtained several fellowships for Latin American students. Presently three Latin American students are doing their PhD at Dubna.

Personally I met Luis at a School organized by the Centro Internacional de Fisica – CIF in Bogota, Colombia in the late eighties. After that, when he was selected in 1998 to be Director of the Centro Latino Americano de Fisica – CLAF, he invited me to become the CLAF Vice Director. We had, during the six years we worked together, a very good relationship, and I would like to stress here his dedication and tireless work to CLAF.

Beside his very frequent visits to the 13 Latin American countries that are members of CLAF, Luis has consolidated an important collaboration with ICTP to finance a PhD of program known as "Doctorados Cooperativos" where students from one country can spend one year in a university of another Latin American country to do part of their research work. He has also created a joint program with the Instituto Balseiro in Bariloche known as "Diploma Bariloche" to support the formation of young physicists from relatively less developed countries. He has supported also the creation of the "Maestria Centro Americana", involving several countries of Central America.

To enumerate all the accomplishments of Luis Masperi as Director of CLAF is impossible in such a short time but I would like to stress his efforts to make CLAF play a coordinating role in large regional projects in the area of physics such as the Sincroton Radiation Source in Brazil, the Auger Observatory in Argentina, the Microton Accelerator in Cuba and the Cosmic Ray laboratory in Chacaltaya, Bolivia.

He also worked tirelessly to promote links between Latin American Physicists and the global physics community. A notable success was the organization of the first CERN-CLAF School of High Energy Physics in 2001.

Masperi also strongly supported meetings to promote Latin American collaboration. The realization of the V SILAFAE in Peru this year is a living example of his efforts.

Luis passed away on December 2, 2003 after a short but very aggressive illness. I am sure that Masperi's boundless energy and enthusiasm for what he believed in will be sorely missed by all those who worked with him and shared his ideals. Thank you.

J. C. dos Anjos and G. Aldazabal

ROUND TABLE: COLLABORATIONS IN PHYSICS IN LATIN AMERICA

1. Introduction

During the Fifth Latin American Symposium on High Energy Physics, from July 12–17, 2004, representatives of the different American countries, participating in the Symposium, gave their opinion about the situation of High Energy Physics in their countries and the collaborations in this field among the different countries. All agreed on the benefits of collaborations with physicists from other countries of the region. A wish for funds that would allow more visits by Latin American physicists to work with colleagues in other Latin American countries was universally stated.

At the start of the round table, Alberto Santoro of the State University of Rio de Janeiro put forth a proposal for a joint effort on a computer GRID linking all Latin American countries. While a number of the Latin American physicists in attendance were extremely enthusiastic about the idea, a plan for implementation would need to be worked out.

2. A. Melfo, U. de los Andes, Venezuela

The are a number of problems in trying to establish a program in high energy physics in Venezuela. The most pressing is the need for a stable source of support for both professors and students, especially funds for travel. Because the field is just getting started in Venezuela, travel to forefront facilities in other countries for collaborative work and to conferences is particularly important. While they have sent students abroad, they can't wait for them to come back before building up the program. They need help now. And, they need to be plugged in!!

3. J. Swain, Northeastern U., USA/Costa Rica

There is the problem of critical mass, as pointed out by Dr. Melfo. This is certainly the case in Costa Rica. There is the possibility of cooperative agreements with physicists from the U. S. or from countries in Europe. He has found that the funds available for this are quite limited both in the amount available and in the ways the money can be spent. E. g., he was not able to get funds from the U. S. funding agencies that could be put into the hands of his colleagues in Latin America. He believes a lack of initiative for this from the Latin American side is part of the problem.

4. M. Sheaff, University of Wisconsin, USA

There are funds available from the National Science Foundation for collaboration with physicists in Latin America. I have two such grants myself, one with colleagues in Mexico, the other with colleagues in Brazil. The cooperative work has had very positive results. NSF is most willing to support U. S. physicists to travel to Latin America with the understanding that the Latin American counterpart agency will provide travel support to the U. S. for its own nationals. There is an emphasis at NSF on sending younger physicists, who might otherwise not have support for foreign travel, to give them experience working in the international arena early in their careers. The young physicists from U. S. institutions that are participating in this symposium are supported by an award I received from NSF because of this policy.

5. J. dos Anjos, CBPF, Rio de Janeiro, Brazil

Experimental particle physics has come a long way from the early days in Brazil. What began in the early eighties as a two-year visit to Fermilab by three theorists to work on a very successful photoproduction of charm experiment has progressed along a path starting from work at Fermilab on hardware and software for that and follow-on experiments to formation of an analysis group and computer cluster at CBPF to allow students and professors to work at their home institution on analysis of experimental data and, finally, to building apparatus for two Fermilab experiments in Brazil. Because of this, the Brazilian groups now have credibility in the international community.

6. Jorge Alfaro, PUC, Chile

Postdocs A natural development of HEP is the implementation of a system of postdoctoral positions within our area. In this way, it will be easy to increase the number and quality of scientific collaborations.

Relation with the Press

Learning from other areas of Science, it would be useful to have a closer relationship with the media. In particular, to intensify our participation in programs to update the scientific knowledge of journalists, trough specialized seminars, workshops and joint projects.

Joint projects with astronomers

In the particular case of Chile, there has been a great development in Astronomy. HEP has many contact points with Astrophysics. So I can see many joint projects appearing in the near future.

7. Esteban Roulet, C. Bariloche, Argentina

One point raised by Esteban Roulet was the importance of attracting the good students so that they remain in their Latin-American countries for their graduate studies, at least in those countries which already invest a lot of effort in forming good undergraduate students (in particular, Argentina, Brasil and Mexico) and have already reached sufficient critical mass to perform advanced research in High Energy Physics. Graduate students are a key ingredient in the research groups and are hence vital for a healthy development of science in Latin-America. However, there is now for instance a difficulty to get the best students into frontier experimental projects like Auger. There is a widespread tendency for the good potential candidates to be attracted to the idea of doing their PhD's in the universities or in the large experimental facilities of the northern hemisphere. Although this is understandable for different reasons, it is something that Latin-American institutions, and in particular the SILAFAE, should try not to promote, and an effort should be made instead to attract the best students towards the local projects, promoting also the possibilities of exchanges between local groups. This change of attitude will just be consistent with the maturity now reached by the field in Latin-America.

8. L. Villaseñor (President of the Mexican Division of Particles and Fields), BUAP, México

Mexican participation in experimental high energy and cosmic ray physics

In the mid 1980s a group of theoretical physicists from Cinvestav started a joint program with Fermilab to send graduate students and young post-docs to spend a summer working closely with experimentalists on some of the experiments taking place at Fermilab. This project of sending students to HEP laboratories for a summer was undertaken by the Mexican Division of Particles and Fields (DPyC) in the 1990s, and it was extended to include CERN, DESY, and other laboratories around the world. At present, an ad-hoc committee from DPC, formed mainly by experimental physicists, performs the selection of senior and first-year graduate students by means of interviews and exams in different host Universities each December. This program has been very successful in motivating young students to pursue graduate studies in experimental high energy or cosmic ray physics as members of some of the most important experiments in these fields.

The present involvement of Mexico in experimental high energy and cosmic ray physics consists of around 25 experimental physicists with a Ph.D. degree working in a dozen universities in Mexico. They participate on some of the most important projects including the Pierre Auger Observatory, ALICE-LHC, CMS-LHC, AMS, D0, CKM, FOCUS, and H1, among others. Most of them have benefited in the past from the above-mentioned program. The universities with PhD professors working in experimental high energy and/or cosmic ray physics include BUAP in Puebla, Cinvestav-DF, Cinvestav-Merida, Instituto Politecnico Nacional in Mexico City, UASLP in San Luis Potosi, Universidad de Guanajuato in Leon, Universidad Michoacana de San Nicolas de Hidalgo in Morelia and UNAM in Mexico City, among others.

In the late 1980s and early 1990s, the participation of Mexican groups in experiments on HEP and in cosmic ray physics was mainly in activities related to data analysis and/or construction of sub-detectors in association with other groups from other countries. However, in the late 1990s and early 2000s their participation has changed to gradually include full responsibilities in the construction of sub-detectors.

9. N. Martinic, U. Mayor de San Andrés, Bolivia

Bolivia has for more than four decades the Chacaltaya Observatory in La Paz with international collaborations, specially with Brasil and Japan.

The big problem we have is the lack of scholarships for postgraduate (basically Master's) students. Without this it will be very difficult to develop the High Energy Physics in our country.

10. C. Javier Solano S., UNI, Perú

Peru has a fast growing High Energy Physics community with participation of Peruvian physicists in many HEP laboratories (AUGER, FERMILAB, CERN, SUPER-KAMIOKANDE) and, for the theoretical part, working in many institutions from Brazil (IFT, CBPF, USP, UFMT), Chile (PUC), USA (U. Houston, Texas), etc

Our great problem is that most of this Peruvian community is not working in Peru, but our great advantage is that most of them are trying to help Peru in establishing collaborations, especially with UNI and PUC of Lima, and UNT of Trujillo.

Peru has similar problems to Bolivia in the lack of scholarships for Master's and PhD students, and similar problems to Argentina in the repatriation of our best students. Nevertheless, we have very hard working physicists in Peru, and we expect to reinforce our collaborations with the American HEP community. This Symposium will be, for sure, a very good help.

11. W. Ponce, U. de Antioquia, Colombia

SILAFAE was created ten years ago, and the first meeting was held in Merida, Mexico in 1996. Meetings have been held every two years since then, in Puerto Rico, Colombia, and Brazil. SILAFAE VI will take place in the Fall of 2006 in Puerta Vallarta, Mexico. This meeting will be following by symposia in Argentina and Chile in the coming years.

One way to help promote collaboration in the field would be for people to cooperate and agree to work together. There are instances where some Latin American physicists do not even cite other groups working in Latin America. A more congenial approach would help the field to develop in all our countries.

SILAFAE Organization

LECTURES

AN INTRODUCTION TO STRINGS
AND SOME OF ITS PHENOMENOLOGICAL ASPECTS*

G. ALDAZABAL

Centro Atómico, 8400, Bariloche, Argentina
aldazaba@cab.cnea.gov.ar

In these notes we present a brief introduction to string theory. As a motivation we start by exposing some of the many striking achievements of the Standard model of fundamental interactions but also some of its limitations. We introduce string theory through a discussion of perturbative closed bosonic string. Since no knowledge of supersymmetry is assumed the different, consistent, superstrings are just qualitatively motivated. D-branes and open string are then discussed and some rough idea about dualities and non perturbative effects is advanced. We end by illustrating the idea of *brane worlds* as a possible framework to embed the Standard model in a string theory context.

1. The Standard Model of Fundamental Interactions and Beyond

The present understanding of the fundamental components of matter and their interactions is based on two very different theoretical descriptions, the Standard model and General Relativity (GR). Both descriptions have produced impressive results. However, at some point, they appear as mutually inconsistent.

The fruitful interplay of theory and experiment, over several decades, lead to what is known today as the Standard Model of fundamental interactions. Namely, a theory which is able to describe, in a very successful way, the components of matter and their strong and electroweak interactions.

There are several reasons of different degree of "necessity" that lead to the belief that the Standard Model of particle physics (plus classical

---

*This work is supported by ANPCyT grant 03-11064.

general relativity) is an effective QFT that must be extended to a more fundamental theory [8,6]. We distinguish:

(1) **Theoretical reasons**: A quantum theory of gravity (QG) must exist. We know that QFT is not enough and that new ingredients are required. We will deal, in what follows, with String theory, the strongest candidate for Quantum Gravity.

(2) **Experimental reasons**: Although the SM is extremely successful in explaining most experimental facts, there are several observations that point to new physics. Some experiments are actually compelling in this sense.

Clearly, neutrino masses and mixings must be included, as established from atmospheric, solar and reactor experiments [5].

Compelling evidence for the existence of Dark matter (with a non baryonic component) and Dark energy is accumulating from cosmology experiments.

Cosmic baryon asymmetry cannot be accounted from SM.

Certainly this does not mean that radically new ideas (like supersymmetry, grand unification, extradimensions, extended objects) must be used to solve these empirical problems. If no attention is paid to aspects like naturalness, hierarchy problem or other "ideological-aesthetical" issues, as mentioned in third point, minimal extensions of SM, by adding some few degrees of freedom and interactions, would be enough to account for these experimental facts (see for instance [9]).

(3) **"Ideological" reasons**: It is a matter of fact that the Standard Model describes nature, at least up to the M_Z scale (and that QG must manifest at Planck scale) but, is there any underlying logic for its structure? Again, the $why?'s$ about the structure of the SM have different degrees of "compellness" and they are somehow biased by the knowledge of possible answers to some of them. Let us briefly mention some of the main questions

- Hierarchy problem: It basically refers to the instability of the mass of Higgs particle, and therefore of the electroweak scale under radiative corrections. Loop corrections suggest that Higgs mass should be of the order of the biggest scale in the theory, M_P, for instance. Extreme fine tuning is needed to keep it of the order of M_W. *Supersymmetry* solves the hierarchy problem by introducing higgsino, the fermionic partner of the Higgs, whose mass is protected by chiral symmetry. Large extra dimensions could avoid such problem by lowering the scale of gravity etc.

- Is there a complete (or partial) unification of all interactions at some energy scale. Extrapolation of coupling constants (combined with Susy) point to such a unification scale, at least for SM gauge interactions.

- *Naturalness problems:*

  There are several quantities in the Standard Model that we would expect to be of order one, that, however, must have very small values, in order to account for experimental results. In this sense the SM is not natural. Hierarchy problem above can be included in this category.

  The strong CP problem is another example is Whereas a CP violating term $\theta_{QCD} G_{\mu\nu} \tilde{G}^{\mu\nu}$ can be included in QCD Lagrangian ($G^{\mu\nu}$ is the $SU(3)_c$ gluon field strength and $\tilde{G}$ its dual) with an arbitrary QCD theta parameter, experiments (neutron electric dipole moment) constrains $\theta_{QCD} \leq 10^{-10}$.

  The cosmological constant problem:

  Once gravity enters the play then we must deal with one of the toughest puzzles in fundamental physics, the cosmological constant problem [10]. The cosmological constant Λ is essentially the vacuum energy. It has dimensions of mass to the fourth power. Thus, if the larger scale in nature is the Planck scale then $\Lambda \propto M_{Planck}^4$. However, experimental bounds require $\Lambda \leq 10^{-120} M_{Planck}^4$ and, moreover, there are recent claims that a value of the order

$$\Lambda \simeq 10^{-47} GeV^4 \tag{1}$$

  is required observationally, which is many orders of magnitude smaller than the expected value.

- Why are there 3 generations?
- Is there any rationality relating the 20 free parameters of the SM?
- Is there an explanation for mass hierarchies of generations?

and many other questions. Such kind of questions underly behind the many proposals for extensions of the Standard Model. Supersymmetry, Grand Unified Theories, Extra dimensions, supergravity, . . . , string theory. Each of them, give plausible explanations to some of the items above. Generically, we should say, new questions are opened and some old ones are rephrased.

In these notes we will deal with string theory. Certainly an extension of the SM but different from other approaches in the sense that it is a theory that addresses the most fundamental issue. Namely, *string theory* is thought to provide a consistent quantum mechanical theory of gauge and gravitational interactions.

String theory provides a consistent, order by order (in perturbation theory) finite, ultraviolet completion of general relativity. Einstein theory is obtained as a low energy limit (below a typical M_s string scale).

Moreover, string theory includes gauge interactions and, in certain scenarios to be discussed below, it provides four dimensional gauge interactions and chiral fermions. It imposes constraints to model building. Many ingredients of other proposed extensions of SM are incorporated, now in a well defined consistent theory, which can lead at low energy, to models very close to the Standard Model.

String theory provides a framework to address some fundamental issues like: the black hole information paradox, the origin of chirality, the number of fermion generations, etc.

Needless to say that string theory is not a complete closed theory. Several aspects of it, like the non perturbative behaviour, are only partially known. Only some corners of the full string/M theory are known.

Also, other new questions are now open.

Even if models close to the SM can be obtained at low energies, other, completely different possibilities very far from the real world seem also possible, therefore, rising doubts about its predictivity power, etc. Some questions like the cosmological constant have no answer, yet, in the context of string theory.

The big question of the big bang origin of the universe is still open, etc.

2. A Brief Introduction to Strings

String theory is a theory under construction. There exists no closed description of the theory at the *strong coupling*. However, some non-perturbative relevant information about the structure of the theory is already available. In particular it is known that string theory contains not only strings but other extended objects. The full theory is referred today as *M/string* theory. The so called, perturbative string theories, are understood as different, corner, manifestations of this complete theory.

Still, this perturbative sector is very well known and its study can shed light on the structure of the full theory.

Let us discuss some very qualitative aspects of this perturbative sector (see for instance [11,12,14,15,13,16]).

In QFT particles are identified with point like fundamental objects. String theory proposes the existence of one-dimensional objects, strings. The typical size of such objects is

$$L_S = 1/M_S \qquad (2)$$

such that, at energy scales well below the string scale $E << M_S$, such one dimensional objects are, *effectively*, seen as point-like particles. Namely, the low energy limit (compared to M_S) of string theory should reduce to ordinary QFT.

In some string theory scenarios the string scale is linked to four dimensional Planck scale and $M_S \simeq 10^{18} GeV$, very much bigger that the $1 TeV$ of elementary particles experiments. In terms of lengths (recalling that $\hbar c \simeq 197 MeV \times fermi$) we see that a resolution of $L_S = \frac{\hbar c}{M_S} \propto 10^{-32} cm$ would be needed to see the string structure.

Just like a particle evolving in time describes a world line trajectory, a string sweeps out a two dimensional surface Σ, *the world sheet*. A point on the *world sheet* is parameterized by two coordinates, a "time" t, like in the particle case, and a a "spatial" coordinate σ which parameterizes the extended dimension at fixed t. Thus, a classical configuration of a string in a d-dimensional, Minkowski, space time M_d will be given by the functions

$$X^\mu(t,\sigma) \qquad \mu = 0, \ldots, D-1 \qquad (3)$$

which are the string coordinates in M_d of a (t,σ) point in Σ.

The classical action for string configuration in M_d, with metric $\eta_{\mu\nu}$, can be written as

$$S_{Polyakov} = -\frac{T}{2} \int_\Sigma d\sigma d\tau \sqrt{-g} g^{\alpha\beta} \partial_\alpha X^\mu \partial_\beta X^\nu \eta_{\mu\nu} \qquad (4)$$

where a metric $g_{\alpha\beta}$ on the two dimensional surface Σ has been introduced. At the classical level such action describes the area A of the surface Σ spanned by the string.

It is not the aim of this brief introduction to go to a detailed analysis of this action. Let us stress, nevertheless, that it corresponds to a two dimensional field theory coupled to 2-d gravity. The Lorentz content appears here as an internal symmetry. The constant T is the string tension and fixes the string scale,

$$\pi T = M_S^2 = \frac{1}{2\alpha'} \qquad (5)$$

where the string constant $2\alpha' = L_S^2$ is usually introduced.

The action possesses several global and local symmetries that we will not discuss here. A fundamental one is conformal invariance, namely the invariance of the action under phase redefinitions of 2-d metric ($g_{\alpha\beta} \rightarrow e^{\phi(t,\sigma)}g^{\alpha\beta}$). In particular, the 2-d metric can be locally gauged away such that the action now reads

$$S = -\frac{T}{2}\int_{\Sigma}\partial_{\alpha}X^{\mu}\partial_{\beta}X^{\nu}\eta_{\mu\nu} \tag{6}$$

i.e. a two dimensional free field theory. The equations of motion are

$$\nabla^2 X^{\mu}(t,\sigma) = (\partial_t^2 - \partial_\sigma^2)X^{\mu}(t,\sigma) = 0 \tag{7}$$

which describe a relativistic string (notice the relative coefficient between time and spatial coordinates is $v/c = 1$). Solutions to such equations are just a superposition of non interacting harmonic oscillators, the zero modes describing the centre of mass motion of the string.

The kind of superposition will depend on the boundary conditions. Closed or open string boundary conditions are possible.

3. Closed Strings

If spatial parameter $\sigma \in [0, 2\pi]$ then strings are closed if

$$X^{\mu}(\sigma + 2\pi) = X(^{\mu}\sigma) \tag{8}$$

and the generic solution can be written as a sum of right (R) and left(L) moving contributions

$$X^{\mu}(t,\sigma) = X_L^{\mu}(t+\sigma) + X_R^{\mu}(t-\sigma) \tag{9}$$

$$X_R(t-\sigma)^{\mu} = x_0^{\mu} + L_s^2 p_R^{\mu}(t-\sigma) + \frac{i}{2}L_s\sum_{m\neq 0}\frac{a_m^{\mu}}{\sqrt{m}}e^{-2im(t-\sigma)},$$

$$X_L(t+\sigma)^{\mu} = \tilde{x}_0^{\mu} + L_s^2 p_L^{\mu}(t+\sigma) + \frac{i}{2}L_s\sum_{m\neq 0}\frac{\tilde{a}_m^{\mu}}{\sqrt{m}}e^{-2im(t+\sigma)}, \tag{10}$$

Namely, $X^{\mu}(t,\sigma)$ describes a string with center of mass position $x^{\mu} = x_0^{\mu} + \tilde{x}_0^{\mu}$ and momentum $p^{\mu} = p_R^{\mu} + p_L^{\mu}$, respectively.

By imposing the familiar quantization relations for harmonic oscillators, which read

$$[a_m^{\mu}, a_n^{\nu}] = \eta^{\mu\nu}\delta_{m,-n} \tag{11}$$

(and similarly for left movers) the string can be rather straightforwardly quantized.

Actually constraints are present, associated to 2-d gauge invariance.

We mentioned that, classically, the theory is conformal invariant. However, when it is quantized, a conformal anomaly is generated, i.e. such symmetry is not preserved, generically, at the quantum level.

This is a familiar situation in QFT where, for example, chiral symmetry is broken when the theory is quantized. In such a case we know that chiral symmetry is recovered for very definite content of fields. This is the case of the Standard Model where, for instance, the addition of a hypercharge charged Weyl fermion would make the theory inconsistent.

In a similar way, cancellation of conformal anomaly is also possible if a definite number, 26, of bosonic fields $X^\mu(t, \sigma)$ is considered.

Interestingly enough, the number of fields is here the dimension of space time where the string moves. We see a first example of how a consistency requirement of the world sheet 2-d theory *constrains space time physics*. Extra (more than 4) dimensions are required for the theory to be well defined. At the spectrum level, conformal invariance ensures that norm of states is definite positive.

String excitations just correspond to harmonic oscillators, *left* and *right*, which are obtained by the action of creation operators applied to the vacuum (plus some constraints). The *physical particles*, in d dimensional space, are associated such "harmonic" excitations of the string.

The mass of such states ($p^2 = -M^2$) can be calculated to be

$$M^2 = \frac{2}{\alpha'}[N + \tilde{N} - 2\frac{2-d}{24}] = \frac{4}{\alpha'}[N - \frac{2-d}{24}] \tag{12}$$

where $N = \tilde{N}$, is the Right (Left) oscillator occupation number. We stress that Left and right oscillators are independent besides the "level matching" relation $N = \tilde{N}$ and that

$$M^2 = \frac{4}{\alpha'}(N - 1) \tag{13}$$

for $d = 26$.

For instance, apart from the vacuum we will have the first excited state

$$\tilde{a}_1^{\mu\dagger} a_1^{\nu\dagger} |0> \tag{14}$$

with mass $M^2 = \frac{26-d}{6\alpha'}$

Would such state be massive then it should span a representation of the Lorentz little group $SO(d-1)$. However, it corresponds to a 2-tensor representation (transforming as a product of two vectors) of the Lorentz little group $SO(d-2)$ (like a gauge boson vector spans a vector representations of $SO(d-2)$!). In order for such state to have a correct interpretation we should require it to be massless and therefore $d = 26$. Again, we see that this conformal invariance requirement, ensures a correct interpretation of excitations organized in Lorenz group representations.

The massless 2-tensor can be decomposed into Lorentz irreducible representations. Actually, only 24 transverse degrees of freedom are physical (time and longitudinal degrees of freedom are unphysical, like in the vector boson case). Therefore the decomposition reads

$$\mathbf{24_v} \otimes \mathbf{24_v} = \phi \oplus B_{\mu\nu} \oplus G_{\mu\nu} = \mathbf{1} \oplus \mathbf{276} \oplus \mathbf{299}. \tag{15}$$

where $G_{\mu\nu}$ is the symmetric tracelees part, $B_{\mu\nu}$ an antisymmetric tensor (2-form) and Φ is the trace.

For instance

$$G_{\mu\nu} \propto [\frac{1}{2}(\tilde{a}_1^{\mu\dagger} a_1^{\nu\dagger} + \tilde{a}_1^{\mu\dagger} a_1^{\nu\dagger}) - \tilde{a}_1^{\mu\dagger} a_1^{\nu\dagger}]|0> \tag{16}$$

this is nothing but the graviton excitation. In the effective, low energy, theory it will generate the graviton field $G_{\mu\nu}(X)$ of $d = 26$ Einstein gravity. We thus see an indication that closed strings can contain a graviton like field!

Besides gravitons, we also notice that the massless spectrum of the theory contains a massless scalar, *the dilaton*, and a two form (antisymmetric) field.

When more, higher order, oscillators are considered, a full infinite tower of massive states is generated. They organize into massive representations of the 26 dimensional Lorentz group. The masses are proportional to M_S and, thus, we will not see such particles at energies much below M_S.

A last comment is in order with respect to the first, non oscillating state, $|0>$. This state is a (Lorentz scalar) tachyon with mass $M^2 = -\frac{4}{\alpha'}$. It signals an instability of the bosonic string (X coordinates are bosonic here). It is not present in superstring theories.

3.1.   *The closed superstring*

Superstrings are constructed by adding fermionic degrees of freedom besides the bosonic coordinates X^μ.

Consistency imposes $d = 10$ in such cases. Spinors in $d = 10$ can have two different chiralities, we name them as $\mathbf{8_s}$ and $\mathbf{8_c}$.

A useful way ($SO(8)$ Fock space representation) in which we can visualize them is as state vectors of the form

$$| \pm \frac{1}{2}, \pm \frac{1}{2}, \pm \frac{1}{2}, \pm \frac{1}{2} > \tag{17}$$

where one chirality spinor, lets say $\mathbf{8_s}$ is obtained by choosing an odd number of minus signs while an even number corresponds to $\mathbf{8_c}$.[a] In this representation a vector, which in $d = 10$ dimensions has 8 (transverse) degrees of freedom reads

$$\mathbf{8_v} \equiv |\underline{\pm 1, 0, 0, 0} > \tag{18}$$

where underlining means permutations of ± 1 entry.

Different closed consistent superstring theories can be constructed, namely, Type IIB, Type IIA and heterotic $E_8 \times E_8$ and $SO(32)$ superstrings. The different cases arise, essentially, because of the freedom to work with left and right movers independently. We will not discuss the details of such constructions but, just to have a feeling of how they work, let us present the massless content in these cases.

Type IIB closed string

Besides the bosonic degrees of freedom, so called the Neveu-Schwarz (NS) sector, leading to $\mathbf{8_v}$ vector representation in $d = 10$ fermionic spinor partners (so called Ramond sector) $\mathbf{8_s}$ are introduced. This is done in both left and right movers sectors. Thus, the massless spectrum can be organized into :

NS-NS sector

$$\mathbf{8_v} \otimes \mathbf{8_v} = \phi \oplus B_{\mu\nu} \oplus G_{\mu\nu} = \mathbf{1} \oplus \mathbf{28} \oplus \mathbf{35}, \tag{19}$$

which is the same kind of structure we discussed in Eq. (15) but now in $d = 10$ dimensions.

R-R sector

$$\mathbf{8_s} \otimes \mathbf{8_s} = \phi' \oplus B'_{\mu\nu} \oplus D_{\mu\nu\rho\sigma} = \mathbf{1} \oplus \mathbf{28} \oplus \mathbf{35}_+, \tag{20}$$

which correspond to bosonic (fermion times fermion) $[n]$-forms, $n = 0, 2, 4$. (where the 4-form is a self dual, $D =^* D$, 8-dimensional ϵ-tensor).

The NS-NS and R-R spectra together, form the bosonic components of $D = 10$ IIB (chiral) supergravity.

---

[a] In $d = 4$ dimensions we would have $| \pm \frac{1}{2} >$ for left and right chiral fermions of $SO(2)$.

In the NS-R and R-NS sectors we have the products

$$\mathbf{8_v} \otimes \mathbf{8_s} = \mathbf{8_c} \oplus \mathbf{56_s}. \tag{21}$$

There are two $\mathbf{56_s}$ identifying the two Type IIB $N = 2$ (16+16 supersymmetry charges) gravitini, with one vector and one spinor index.

Notice that invariance under the exchange of left and right sectors shows up in the closed Type IIB massless spectrum.

Type IIA closed string

Formally, the same steps as in the Type IIB case are followed. However, opposite chirality spinors are considered in left and right sectors. Type IIA theory is non chiral. The masslees spectrum is organized as

$$[\mathbf{8_v} \oplus \mathbf{8_s}] \otimes [\mathbf{8_v} \oplus \mathbf{8_c}] = \mathbf{1} \oplus \mathbf{28} \oplus \mathbf{35} \oplus \mathbf{8_s} \oplus \mathbf{8_c} \oplus \mathbf{56_s} \oplus \mathbf{56_s}. \tag{22}$$

Heterotic string

The heterotic string is constructed in a rather peculiar way. It is built by combining the left sector of the bosonic string with the right sector of the Type II superstring!. We learned that conformal anomaly cancellation in superstring requires $d = 10$ why 26 bosonic fields X are required in the bosonic string. Such apparent inconsistency is solved by taking 10 X^μ bosonic fields with space-time $\mu = 0, \ldots, 9$ indices and 16 X^I bosonic fields where $I = 1, \ldots, 16$ are internal (no space-time indices). Roughly speaking, like states carrying space time indices μ arrange into representations of Lorentz Group $SO(1,9)$ internal indices now span representations of a gauge group G of rank 16. Again, world sheet consistency conditions limit such groups to be $SO(32)$ or $E_8 \times E_8$.

The massles field content of consistent string theories in $d = 10$ is presented in Table 1 (including Type I string to be discussed later below)

It is worth noticing that the closed bosonic sector of all these theories contain the graviton and dilaton fields. Except Type I theory, all other string theories also contain a two $NS - NS$ antisymmetric form $B_{\mu\nu}$.

4. String Interactions

We have identified strings as fundamental objects and its excitations as particles. Still interactions among different strings must be considered. In fact, a well defined prescription for computing scattering among string states

Table 1. Consistent string theories in $D = 10$.

	N	bosonic spectrum	
IIA	2	NS-NS	$g_{\mu\nu}, b_{\mu\nu}, \phi$
		R-R	$A_\mu, C_{\mu\nu\rho}$
IIB	2	NS-NS	$g_{\mu\nu}, b_{\mu\nu}, \phi$
		R-R	$c^*_{\mu\nu\rho\sigma}, b'_{\mu\nu}, \phi'$
heterotic $E_8 \times E_8$	1	$g_{\mu\nu}, b_{\mu\nu}, \phi$ A^a_μ in adjoint of $E_8 \times E_8$	
heterotic $SO(32)$	1	$g_{\mu\nu}, b_{\mu\nu}, \phi$ A^a_μ in adjoint of $SO(32)$	
type I $SO(32)$	1	NS-NS	$g_{\mu\nu}, \phi$
		open string	A^a_μ in adjoint of $SO(32)$
		R-R	$B'_{\mu\nu}$

can be given. Qualitatively two closed strings evolving in time can join into one closed string and then split again and so o and so forth.

The basic "interaction vertex" is the *short pants*, a Φ^3 like vertex where two strings join into one (or viceversa). Interestingly enough, the interaction

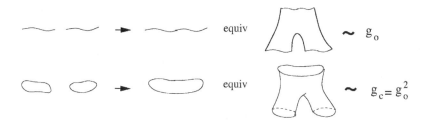

Fig. 1. Basic interaction vertices for open and closed *(pants-like)* strings.

is delocalized in space-time in a region of size of the order of L_S. This is the origin of the finiteness of string amplitudes in contrast with QFT where interactions are pointlike.

Of course, a precise mathematical description of string interactions can be give.

Like in QFT a coupling constant g_c is expected to appear associated to each vertex. In, perturbative, string theory the order of a given diagram is related to the topology of of the surface spanned by the string evolving in time. For instance, in closed string theory it is given by the number of handles h.

Actually, when we wrote down string action in 4 we could have included an extra, topological, term compatible with the symmetries of the theory. Namely, the term

$$S_{euler} = \frac{1}{4\pi\alpha'} \int_\Sigma d\sigma d\tau \sqrt{-g}[\alpha' R^{(2)} \Phi_0] \tag{23}$$

where $R^{(2)}$ is the 2d curvature. This term is a total derivative and it is not zero if the surface Σ is not trivial. Actually

$$\chi = \frac{1}{4\pi} \int_\Sigma \sqrt{-g}[\alpha' R^{(2)}] \tag{24}$$

is the Euler characteristic of the surface. For a closed surface, it counts the number of handles[b] $\chi = 2(1 - h)$. We thus see, that a factor $e^{-\phi_0 \chi}$ will appear associated to a given topology of Euler characteristic χ. If $g_c = e^{\phi_0}$ is small, the perturbative expansion, given in terms of handles, makes sense. The most relevant contribution will be a sphere (g_c^{-2}), then a torus (g_c^0) , a two handle surface (g_c^2), etc.

It is possible to see that ϕ_0 is not a new parameter of the theory but rather can be interpreted as the vacuum expectation value of the dilaton field.

In the limit when the energies under consideration are $E \ll M_S$ these interactions become pointlike and can be described by an effective quantum field theory in $d = 10$ dimensions. This is the point particle limit of string theory.

For instance, for the heterotic string, the (bosonic) part of the effective field theory action (at tree-level (sphere) and up to two-derivative terms) reads

$$S^{het} = \frac{4}{\alpha'^4} \int d^{10}x \sqrt{G} e^{-2\Phi} \left[R + (\nabla\Phi)^2 - \frac{1}{12}\hat{H}^2 - \frac{\alpha'}{4}F^2 \right]. \tag{25}$$

which is nothing but Einstein general relativity in $d = 10$ dimensions, with metric $G_{\mu\nu}$ coupled to the scalar dilaton field Φ and the antisymmetric form $B_{\mu\nu}$ through the three form $H = dB + ...$ and $F_{\mu\nu}$ is the field strength

---

[b]If b boundaries and c crosscaps are present, as it is the case for open strings then $\chi = 2(1 - h) - b - c$.

of the $SO(32)$ or $E_8 \times E_8$ gauge theory. Actually, by a field redefinition

$$\tilde{G}_{\mu\nu} = e^{2\omega} G_{\mu\nu} \tag{26}$$

$$\omega = \frac{1}{4}(\Phi - \Phi_0) \tag{27}$$

the action can be rewritten in a more familiar form in the so called Einstein frame as

$$S_E^{\text{het}} = \frac{1}{2k^2} \int d^{10}x \sqrt{g} \left[R - \frac{1}{8}(\nabla \Phi)^2 + \dots \right] \tag{28}$$

where $2k^2 = 8\pi G_N = \frac{\alpha'^4}{4} e^{2\Phi_0}$.

Thus, Planck mass and string scales are related as

$$M_P^{(d=10)} = M_S^8 e^{-2\Phi_0} = \frac{M_S^8}{g_c^2} \tag{29}$$

In particular we see that the gravitational coupling constant k depends on the value of ϕ_0. Actually, the shift of the field ϕ by ϕ_0 as above is the correct redefinition to perform if Φ acquires the vacuum expectation value Φ_0. We thus see that this coupling constant value is not an additional, external parameter of the theory but, a value dynamically fixed by the theory itself.

It is worth noticing that a similar observation is valid for other "constants" appearing in string theory. Such constants are actually fixed by the expectation values of the fields of the theory.

5. Compactification

All the consistent string theories we have presented are well defined in $d = 10$ dimensional space time. However, string theories in lower, in particular $D = 4$, dimensions can be consistently considered. Actually, when we defined the string world sheet action in Eq4 we assumed strings propagating in flat Minkowski space. Namely we chosed a space time metric $G_{\mu\nu} = \eta_{\mu\nu}$. However, a more generic situation of strings propagating in curved backgrounds can be studied. Indeed, a general metric tensor $G_{\mu\nu}(X(t,\sigma))$ can be considered.

Consistency conditions, as conformal invariance on the world sheet, will constrain the possible space time backgrounds. Such allowed backgrounds solutions define the so called *string vacua*. In $d = 10$ dimensions the 4 closed superstring theories plus the open string theory are the only possibilities. In particular, in order to make contact with the physics we know, it is important to look at string theories that, in the low energy limit, lead to effective theories in $d = 4$ space time dimensions.

The key idea is to look at strings propagating in ten dimensional curved X_{10} space-times backgrounds that look like Minkowski four dimensional spacetime times an internal compact 6-dimensional manifold, a so called compactification. Namely

$$X_{10} = M_4 \times X_6 \tag{30}$$

Notice that, whenever such no flat situations are allowed, the world sheet action does no longer describe free oscillators but rather becomes an interacting theory. This world sheet action describes a complicated non-linear sigma model that, generically, we are only able to study as an expansion in α'. Thus, in many cases we only know how to deal with the point particle limit.

However, it is possible to consider exact solutions. The simplest cases correspond to toroidal like compactifications. In such cases the internal space is essentially flat Euclidean six dimensional space but where coordinates close up to a torus lattice translation. Namely $X_6 = T^6 \equiv R^6/\Lambda$ where Λ is a six dimensional lattice. As an example consider the simplest case of compactification to $D = 9$ dimensions on a one dimensional torus, namely a circle S_1 of radius R. Then ten dimensional space should look as $X_{10} = M_9 \times S_1$. Thus, let us say, the ninth string coordinate satisfies

$$X^9(\sigma + 2\pi) = X(^9\sigma) + 2w\pi R \tag{31}$$

with w an integer, the *winding number* , instead of 8. We see that the solutions to the string theory wave equations are as in the above, flat case, given in Eq. (10). The difference is in the boundary condition for the ninth coordinate. This condition tells us that this string coordinate can wind w times around the circle of radius R before closing on itself.

Once this condition is imposed we can compute the mass of string excitations $p^\mu p_\mu$ but now with $\mu = 0, 8$. We obtain

$$\frac{\alpha'}{2} M^2 = N + \tilde{N} - 2 + \alpha' \frac{n^2}{4R^2} + \frac{1}{\alpha'} w^2 R^2, \qquad N_R - N_L = wn \tag{32}$$

This expression contains very interesting features that also manifest in other more involved compactifications.

The last term encodes the fact that for increasing w, the mass of the string excitations must increase since the string is stretched up around the circle. This is a *stringy* effect, not present in field theory.

The second term is just (p^9) momentum quantization as expected from quantization on a box $(e^{ip^9 X^9} = e^{ip^9(X^9 + 2\pi R)})$. By varying n we obtain an infinite tower of massive states with masses $\sim 1/R$; these are the standard 'momentum states' of Kaluza-Klein compactifications in field theory.

In particular, the massless states with $n = m = 0$ and one oscillator in the compact direction

$$a^\dagger_\mu \tilde{a}^\dagger_9 |0> \qquad (33)$$

(and the same thing if we exchange left and right) are massless vector fields in the extra dimensions which give rise to a $U(1)_L \otimes U(1)_R$ Kaluza-Klein gauge symmetry. Notice that left and right oscillators, both in compact dimension, give rise to a $9d$ scalar field.

Interestingly enough, for special values of $w \neq 0$ and n extra massless states may appear. In particular for $w = n = \pm 1$ we can see that at the special radius $R^2 = \frac{1}{2}\alpha'$, massless less states with a single oscillator $N = 1, \tilde{N} = 0$ are obtained. Together with the above KK vectors, they contribute to enhance the gauge group to $SU(2)_R \times SU(2)_L$. Again, the radius here is not a new parameter of the theory but rather the expectation value of a dynamical field. If, somehow, the special value $R^2 = \frac{1}{2}\alpha'$ is selected, we refer to it as a point of enhanced symmetry in "moduli space". Since $w \neq 0$, this enhancing is a *stringy* effect.

Finally, let us point out another interesting feature that manifest in the mass expression. Namely, the spectrum is invariant under the exchange

$$R \leftrightarrow \frac{\alpha'}{2R} \qquad w \leftrightarrow n. \qquad (34)$$

such a transformation is known as T-duality transformation and it is also a stringy property. It exchanges small with large distances and momentum (Kaluza-Klein) states with winding states at the same time. This symmetry can be shown to hold not only for the spectrum but also for the interactions and therefore it is an exact symmetry of string perturbation theory.

The simple case of the circle we have considered can be extended to compactified six extra dimensions. Strings will be able to wind around the different cycles of the six torus.

In order to make contact with the effective field theory, notice that for low energies and when the volume of the compactification manifold is large compared to string scale,(for instance $\alpha'/R^2 \propto L_S/R \ll 1$)such that no windings or other stringy effects can manifest, we expect the effective (point particle) field theory to be valid and that essentially only the massless modes will be relevant.

As a very simple example consider a $d = 10$ dimensional scalar field $\Phi(x^0, \ldots, x^9)$ on a flat background $X_{10} = M_9 \times S_1$,(this could be for instance the dilaton piece in the heterotic effective action 28)

$$S_{10} = \int_{M_9 \times S_1} d^{10} (\nabla \Phi)^2 \tag{35}$$

The coordinate $x^9 \in [0, 2\pi R]$ parameterizes the circle here and, therefore, we can expand Φ into its Fourier components

$$\Phi(x^0, \ldots, x^9) = \sum_{nZ} e^{inx^9/R} \Phi^{(n)}(x^\mu) \tag{36}$$

with $\mu = 0, \ldots, 8$. The field equation just reads

$$[P_{9d}^2 + P_9^2] \Phi^{(n)}(x^\mu) = 0 \tag{37}$$

with $P_9 = n/R$

Thus, from the $d = 9$ dimensional point of view, we have an infinite set of 9d fields $\Phi^{(n)}(x^\mu)$ labelled by the compact 9d momentum n with 9d mass given by

$$M_{9d}^2 = (\frac{n}{R})^2 \tag{38}$$

These are the Kaluza Klein modes that we identified in Eq. (32). For energies $E << M_C \propto 1/R$, much lower than the compactification scale, massive modes are not reachable and only the zero mode $\Phi^{(0)}(x^\mu)$ is observable. In such case, the ten dimensional action reduces to an effective $d = 9$ dimensional action

$$S_9^{eff} = \int_{M_9} d^9 (2\pi R) \Phi^{(0)} \partial_\mu \partial^\mu \Phi^{(0)} \tag{39}$$

Notice that, since the x^9 dependence dropped out, the volume of the compact manifold, $V_C = 2\pi R$, appears.

We can proceed similarly when fields with non trivial Lorentz transformation are present. The procedure to follow is to decompose the original Lorentz group into the lower dimensional one and then perform the Kaluza Klein reduction. For instance, if we deal with $d = 10$ dimensional metric tensor, $G_{MN}(x^0, \ldots, x^9)$ we obtain

$$G_{\mu\nu}(x^0, \ldots, x^9) \rightarrow G_{\mu\nu}^{(0)}(x^\mu) \tag{40}$$

$$G_{\mu 9}(x^0, \ldots, x^9) \rightarrow G_{\mu 9}^{(0)}(x^\mu) \tag{41}$$

$$G_{99}(x^0, \ldots, x^9) \rightarrow G_{99}^{(0)}(x^\mu) \tag{42}$$

these massless modes correspond to the $D = 9$ dimensional metric, the $U(1)$ vector field and the scalar we found in Eq. (32).

5.1. *Scales*

By properly choosing the compact manifold X_6 sensible semirealistic theories, close to the Standard Model in the low energy limit can be obtained, for instance, by starting with the heterotic string. We briefly refer to them in last chapter. Here we just want to comment about mass scales.

In fact, it is interesting to notice, by following the same steps as above, that the Planck and string scales must be close to each other in such kind of compactifications. Namely, by compactifying the heterotic string action on X_6, with volume V_6, from (28) we obtain obtain

$$S_{d=4}^{eff.} \propto \int d^4x \frac{M_S^8 V_6}{g_s^2} [R_4 + \ldots] + \frac{M_S^6 V_6}{g_s^2} F^2 . \tag{43}$$

Thus, we can express the Planck mass and gauge coupling constant in $d = 4$ dimensions in terms of the string scale and the coupling constant

$$M_P = \frac{M_S^8 V_6}{g_s^2} \simeq 10^{19} GeV \tag{44}$$

$$g_{YM} = \frac{M_S^6 V_6}{g_s} \simeq 0.1 \tag{45}$$

therefore

$$M_S = M_P g_{YM} \simeq 10^{18} GeV \tag{46}$$

6. Open Strings and D-Branes

The world sheet action in (4) admits, besides the closed string boundary condition (8), open string boundary conditions. In fact, variation of Polyakov action leads to

$$\delta S_{Polyakov} = -\frac{T}{2} \int_{-\infty}^{\infty} dt (g^{\alpha\beta} \delta X^\mu \partial_\beta X_\mu)|_{\sigma=0}^{\sigma=l} + \frac{T}{2} \int_{\Sigma} d^2 \zeta \delta X^\mu \partial_\alpha (g^{\alpha\beta} \partial_\beta X_\mu) \tag{47}$$

where the second term leads to the string equations which have a mode expansion solution as in Eq. (9). In order for the first term to vanish, boundary conditions must be satisfied.

In particular we notice that there are two types of boundary conditions that lead to open strings [11,12,14,15,17]. Namely, Neumann boundary conditions

$$\partial_\sigma X = 0 \tag{48}$$

at $\sigma = 0, \pi$ or Dirichlet boundary conditions

$$\delta X = 0 \tag{49}$$

at $\sigma = 0, \pi$.

Let us assume that the coordinates X^μ ($\mu = 0, \ldots, p$) satisfy Neumann conditions on both ends, so called NN (Neumann-Neumann) boundary conditions, whereas, coordinates X^I ($I = p+1, \ldots, 9$) satisfy DD boundary conditions. Imposing such conditions in mode expansions we obtain

$$X^\mu(t,\sigma) = x' + 2\alpha'pt + i\sqrt{2\alpha'} \sum_{m\neq0} \frac{a_m^\mu}{\sqrt{m}} cos(m\sigma)e^{-i\pi t} \qquad (50)$$

and

$$X^I(t,\sigma) = x' + \frac{\delta X'}{\pi}\sigma + \sqrt{2\alpha'} \sum_{m\neq0} \frac{a_m^\mu}{\sqrt{m}} sin(m\sigma)e^{-i\pi t} \qquad (51)$$

We see that the end points of X^I string coordinates are fixed. Namely

$$X^I(\sigma = 0) = x_a^I \qquad (52)$$
$$X^I(\sigma = \pi) = x_a^I + \delta x_I = x_b^I \qquad (53)$$
$$(54)$$

while the string is free to move in the other directions.

Thus, the picture that comes out is that of $p+1$ dimensional hyperplanes, so called Dp-branes (D for Dirichlet) where string end points are constrained to live (see figure 2). One Dp brane sits at x_a^I while the other at x_b^I. X^I are

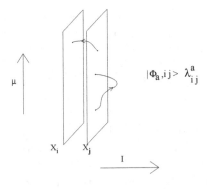

Fig. 2.   D-branes and open strings.

the coordinates transverse to the Dp-brane.[c]

---

[c]A more general case is possible with endpoints on Dd and Dq-branes respectively.

We see that no momentum is allowed to propagate in transverse dimensions and, therefore, the corresponding particles, associated to string excitations, can only propagate along the $p+1$ dimensional "world-volume" of the brane. Notice that for $p = 25$ ($p = 9$ in the susy case) all boundary conditions are NN and the endpoints of the string are free to move in $d = 26$ ($d = 9$) spacetime. However, it is still worth talking about a D25(9)-brane in this case.

The quantization proceeds similarly to the closed string case but now just one kind of oscillators do appear (in fact, boundary conditions require here $a = \tilde{a}$). Also, a string state will be characterized, not only by its world sheet degrees of freedom, i.e. oscillator number, but also by two extra indices a, b, known as Chan-Paton indices, denoting the brane in which the end points of the string are. Namely, an open string state will be characterized by a state vector of the form

$$|\Psi, ab\rangle \tag{55}$$

In principle we could have an arbitrary number N of Dp-branes and, thus, $a, b = 1, \ldots, N$.

The mass operator of an open string state can be computed to be

$$M^2 = \frac{1}{\alpha'}(N - 1) + \sum_{I=p+1}^{25} \left(\frac{x_b^I - x_a^I}{2\pi\alpha'}\right)^2 \tag{56}$$

The last term tells us that stretching the string between two branes requires an energy proportional to brane separation, as expected.

Massless states thus correspond to $N = 1$ oscillators and $x_b^I = x_a^I$ for all (or some of the) possible N values of a, b.

Assume that all Dp-branes coincide on top of each other, then, massless states are

$$\tilde{a}_1^{\mu\dagger}|0, ab\rangle \qquad \mu = 0, \ldots, p \tag{57}$$
$$\tilde{a}_1^{I\dagger}|0, ab\rangle \qquad I = p+1, \ldots, 25 \tag{58}$$

where the first state describes a gauge vector boson in $p + 1$ dimensional space-time. Moreover, since there are N^2 such bosons they will give rise to a unitary $U(N)$ gauge theory in space-time.

Actually a given quantum state vector should be expressable as a combination of the base states $|\psi, ab\rangle$ above. Namely, $|\psi\rangle = \Lambda_{ab}|\psi, ab\rangle$ with Λ_{ab} a hermitian $N \times N$ matrix. Such Chan-Paton matrix spans a representation of $U(N)$.

The other massless states, associated to the transverse coordinates, are scalars transforming in the adjoint representation of the gauge group. The presence of transverse oscillations is an indication that the Dp-brane is not just a fixed hyperplane but a dynamical object.

Such $25 - p$ scalars can be interpreted as Goldstone bosons associated to the breaking of translational invariance by the Dp-brane.

Suppose that we separate some of the D-branes such that M of the locations of the D-branes in transverse space coincide, then, the associated $|ab\rangle$ states are massless while the rest become massive. The gauge group associated to the stack of coincident branes thus brakes from $U(N)$ down to $U(M)$. In particular notice that, even if all D-branes are separated apart, the $|aa\rangle$ states are massless and the gauge group is thus $U(1)^M$.

We see that moving continuously the locations of the D-branes away from each other leads to the breaking of $U(M) \to U(1)^M$. Such a breaking corresponds to a Higgs breaking generated by diagonal components of scalar field on the D-branes acquiring vev's (recall that such scalars are associated to transverse coordinates).

As we found in the closed bosonic string a tachyon scalar $N = 0$ appears here. The tachyon is not present when supersymmetry is included.[d]

We have introduced the basic ideas about open strings and D branes through the bosonic formulation. Similar steps can be followed in order to introduce a fermionic sector in a *supersymmetric* way (a consistent GSO projection must be performed). Indeed, supersymmetry plays a crucial role in the discussion of D-branes which appear as stable BPS states. The gauge theory defined on the world volume is a supersymmetric gauge theory. In particular, a D3-brane, with four dimensional world volume, defines a $d = 4$, $N = 4$ supersymmetric gauge theory.

We mentioned that presence of open strings necessarily implies the existence of closed ones. In particular notice that an open string, with end points at p-branes, when propagating on a loop, describes a cylinder extended in between the branes. Interestingly enough, such a cylinder can be reinterpreted as a closed string propagating between the branes.

These two ways of interpreting these amplitude are a manifestation of what is known as a *open-closed duality*.

From the closed string point of view, closed string states, that can couple to the Dp-brane, do propagate along the cylinder. In fact, such states

---

[d]It is worth mentioning that open tachyons have a clearer interpretation, as brane instabilities, than closed ones [18].

can be isolated by stretching the cylinder. Which is the closed string theory from which such closed string states arise? It can be shown that, if supersymmetry is required and the closed-string duality is ensured, such closed string states correspond to the *Type IIB superstring theory* whenever p is *odd*, whereas *Type IIA superstring theory* states are found for even p.

This observation has deep roots. Indeed, it can be shown (we briefly discuss it below) that p-brane extended objects, with p odd (even), do appear as classical non trivial solitonic solutions of the effective field theory associated to type IIB (A) closed superstring theory (TypeIIB(A) sugra). D-branes do provide a microscopic description of solitonic solutions of such sugra theories!.

From the space time point of view we expect a p extended object, the Dp-brane, to be charged with respect to a p-form (antisymmetric p-tensor fields) of the theory. Namely, terms of the type

$$Q_p \int_{W_{p+1}} C_{p+1} \tag{59}$$

should appear with Q_p the Dp-brane charge with respect to the C_{p+1} form of the closed sector. This is nothing but a generalization of the coupling of the electromagnetic vector field, a 1-form (A^μ) to the world line of a charged particle, i.e. the electron, which can be thought as a $p = 0$ (0-brane) extended object. Another example is the $B_{\mu\nu}$ 2-form coupling to the $p = 1$ string. C_{p+1} forms are present in Type IIB(A) theory, with p even (odd), they arise from RR sectors (we saw some of them above) that is why we refer to them as RR forms.

In particular, a C_{10} is present in Type IIB which couples to D9-branes as,

$$N_9 \int_{W_{9+1}} C_{10} \tag{60}$$

where N_9 is the number of D9-branes (a D9-brane charge is normalized to 1).

This term needs some explanation since we are talking about a [10]-form potential which we did not find when we studied the massless spectrum.

The [10]-form potential is rather peculiar. In fact, notice that it is not possible to construct a field strength $F = dA$ from this form since it would be an [11]-form in $D = 10$. Therefore a kinetic $\int^* F.F$ term in the action is not allowed and thus C_{10} does not represent propagating particles, even if the above couplings are possible. On the other hand, the ten dimensional coupling above is acceptable from Lorentz invariance.

Interestingly enough, the variation of the action (60) would lead to inconsistent equations of motion unless $N_9 = 0$ and, therefore, no branes should be present. Such an inconsistency can be directly observed from the computation of the closed string amplitude. It can be interpreted as a tadpole divergence.

The conclusion is that it is not possible to couple Type IIB closed string to an open string sector in a consistent, 10d, Poincare invariant way.

There are ways to overcome this difficulty.

If $d = 10$ Poincare invariance is not required then consistent models can be constructed. Tadpole cancellation equations, associated to lower dimensional [p]-form RR charge cancellation, will still impose restrictions on brane configurations.

Anti-Dp branes, which are similar to Dp-branes but carrying opposite RR-charge can also be included. Notice that charge cancellation would require now $N_9 = \bar{N}_9$ if $\bar{N}_9$ is the number of antibranes. However, unless some obstruction appears, this is an unstable situation and branes and antibranes do annihilate.

A way out to this situation is to consider unoriented strings. They give rise to the consistent Type I open string theory (which also contains a closed sector). Its massless spectrum is summarized in Table 1. We present a very brief description below.

7. Type I Open Theory

Let us introduce a formal operator Ω that, when acting on open strings, exchanges string orientation. In terms of world sheet parameters it corresponds to $\sigma \to -\sigma$.

The open string end points transform as $(a, b) \to (b, a)$. An unoriented string theory is generated if strings related by Ω are identified.

Such unoriented open strings do not couple to Type IIB but to a truncation of it obtained by implementing an Ω projection on the closed sector. Such projection is referred as *orientifolding* the theory.

Actually, Ω orientifold action on closed sector exchanges Left and Right moving sectors. In terms of world sheet parameters it corresponds to $t+\sigma \to t - \sigma$. Namely, two states of the form

$$|\alpha >_L |\beta >_R \qquad |\beta >_L |\alpha >_R \qquad (61)$$

present in Type IIB theory (recall that Type IIB theory as well as closed bosonic theory is invariant under $L \leftrightarrow R$) are equivalent in the projected theory.

For bosonic string oscillators, for instance,

$$\Omega: \quad \alpha_m^\mu \leftrightarrow \tilde{\alpha}_m^\mu. \tag{62}$$

and therefore, projecting onto invariant states (achieved by introducing the projector $\frac{1}{2}(1 + \Omega)$) we see, for instance that from the original massless spectrum in 15, Graviton and dilaton field are kept but the antisymmetric tensor is projected out. The same considerations are valid for Type IIB. However, in the RR sectors, exchanging Left and Right introduces an extra minus sign since R states are fermionic. Thus, for instance, the antisymmetric form is kept in the RR sector (see Table 1).

The computation of closed string amplitudes over L-R invariant states can be achieved by just introducing the projector $\frac{1}{2}(1 + \Omega)$ in the Type IIB amplitudes. In particular, for the torus vacuum amplitude, while the first term produces just the original torus amplitude (times a half factor) the second term introduces a completely different topology, the Klein bottle. It describes the evolution of an initial left-right state that glues back to itself up to the action of Ω. It can be also interpreted as a closed string propagating between planes, orientifold O_9 planes, which are left invariant under the orientifold action.

The Klein bottle closed string amplitude is ill defined. It contains tadpole like divergences. It is just these divergences that can cancel out against D9-brane tadpole ones to render the full theory (closed plus open sectors) consistent. In fact such tadpoles can be associated to RR charges of the orientifold O_9 planes. An explicit calculation gives $Q_{O_9} = -32$ and therefore tadpoles are absent if

$$N_9 = 32 \tag{63}$$

D9 branes are introduced in the theory!

A stack of N_9 D9 branes would give rise to a $U(N_9)$ unitary gauge theory in the case of oriented strings. However, since open strings with reversed end points must be identified, a subgroup, orthogonal $SO(32)$ subgroup of the original unitary group is obtained.

If we represent the action of the group operator Ω as a unitary matrix γ_Ω, then, the original open string states encoded in Chan-Paton Λ factors ($N_9 \times N_9$ hermitian matrices) which are left invariant by the orientifold projection must satisfy

$$\Lambda = -\gamma_\Omega \Lambda^T \gamma_\Omega^{-1} \tag{64}$$

Since $\Lambda = 1$ can be chosen (from tadpole cancellation conditions), then

$$\Lambda = -\Lambda^T \tag{65}$$

which tells us that the gauge group is $SO(32)$.

8. Beyond Perturbation Theory: p-Branes and Duality

We have found that Dp-branes are extended objects whose fluctuations are described by open string excitations. Such extended objects interact with closed string states. In particular a Dp-brane couples to the graviton (such coupling defining the brane tension) and it carries charge with respect to RR forms.

Interestingly enough such objects do appear as non-trivial, solitonic solutions, of closed string theory effective actions. Namely, by considering the effective low energy actions corresponding to different closed string theories, it is possible to show that finite energy classical solutions to the equations of motion exist. Such solutions correspond to objects which look like lumps of localized energy in some, lets say $d - p - 1$ dimensional space, while they extend in the other p spatial dimensions. They are called p-branes. Their energy per unit volume (tension) is proportional to the inverse of the coupling constant and, therefore, these branes are intrinsically non-perturbative (some references for these subjects are [11,14,19,20]). As an example, a 3-brane solution in Type IIB theory reads

$$ds^2 = f(r)^{-1/2}[(dx_0)^2 + \ldots (dx_3)^2] + f(r)^{1/2}[(dx_4)^2 + \ldots (dx_9)^2] \quad (66)$$

$$f(r) = 1 + 4\pi g_s \alpha'^2 N \frac{1}{r^2} \quad (67)$$

with $r^2 = (x_4)^2 + \cdots + (x_9)^2$ (an F_5 form solution is also present) These solutions have many relevant and striking properties.

On the one hand for each effective closed string theory action there exist a p-brane solution for values of p that correspond to the $p + 1$ forms present in the (perturbative) closed spectrum of the string theory.

Moreover, p-branes carry electric charge under $p + 1$ forms and they are magnetically charged under the dual $(7 - p)$ forms.

The 3-brane above is an example since we know that a 4-form field is present in Type IIB. The 3-brane is identified in this case with the D3-brane that we have found above, that couples to Type IIB closed states.

For instance, for heterotic string we have a 1-brane (F1) and a 5-brane (NS5) that couple to a 2 form B_2 and a 6-form $\tilde{B}_6$, respectively. As mentioned, the energy per unit volume of these solitonic objects is of the order of $\frac{M_s}{g_s}$ or $\frac{M_s}{g_s^2}$. For weak coupling $g_s \ll 1$ they are non perturbative and extremely massive.

A crucial result, that we can just sketch here, is that these solutions are 1/2 BPS states. Roughly speaking that means that these solutions are kept invariant by half of the total supersymmetry generators of the theory. Like in any group theory representation, representations of supersymmetry are constructed by keeping together all states that mix up under the symmetry group. Since in this case only half of the susy generators are effective, the corresponding multiplets are shorter than a generic susy multiplet, in fact, they contain half of the number of states.

Why are 1/2 BPS states so important? The fact is that we have studied solutions of effective theories which are actually found in the low energy $\alpha' \to 0$ limit. The action is written in terms of the light modes of the corresponding string theory. Therefore it appears controversial to interpret solutions of this action, describing a particular regime, as solutions of the full string theory. Interestingly enough, if we continuously increase the value of α' we cannot expect a discrete jump from a BPS multiplet to a state which contains twice the degrees of freedom. Namely, such states remain BPS even if α' is turned on. They are stable.[e] The idea is that if a BPS solution of the effective supergravity action is found, with its corresponding mass, charge, etc. there will exist a corresponding BPS state with the same properties in the full string theory. It is said that BPS states are protected by supersymmetry.

9. p-Brane Democracy

Several results point towards the conclusion that different p-branes should be considered on equal footing. Namely, even if some of such branes could appear as more fundamental in some regimes and others as solitonic objects their role could be inverted. For instance, 1-branes, "strings" appear as solitonic solutions. Moreover, in some cases they couple (electrically) to a NS-NS 2-form with the same charge the usual string couples to it. For instance, the solitonic 1-brane of the heterotic effective action must be interpreted as the *fundamental* heterotic string. However, the fact that we see it as fundamental, in the sense that we can write a world sheet action and quantize it etc., is interpreted as due to the fact that we are in a regime where the coupling constant is small and thus allows us to perform such an expansion in string oscillation modes. If we move to a different regime, lets say of strong coupling, such a perturbative expansion will not be possible

---

[e]Moreover, it can be shown that the mass (tension) of the BPS state coincides with the charge of the central extension of susy with $M = Q$.

and such string will not look as fundamental any more. Other objects could look as more fundamental in this other regime.

Thus, the idea that emerges is that there is a unique underlying theory that contains different kinds of BPS p-brane extended objects. Such theory is named M-theory (mother, magic, mysterious?). The different perturbative string theories are just descriptions of such underlying theory in some special regime. This is what is pictorially shown in the, by now even popular, drawing in 3.

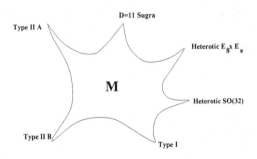

Fig. 3.   M-theory and its perturbative corners.

If one moves from one of these weak couplings regimes (let's say described by theory A) to a strong coupling regime, then the massive solitonic p-brane degrees of freedom can become weakly coupled and thus look as fundamental in this new regime (described by weakly interacting theory B).

This equal footing is referred to as p-brane democracy. The fact that two perturbative corners, namely two apparently different string theories, of M-theory could be connected by continuously changing the string coupling constant is a manifestation of what is called *string duality*.

Since there exist no closed description of the full M-theory the above picture can not be fully proved. Nevertheless, there are several non-trivial indications that support the whole idea. In particular, the identification of BPS states is crucial since they are stable, regardless the value of coupling constant.

Just as an example of the present description let us look at the heterotic SO(32)-Type I corners. We have seen in Table 1 that both theories have a similar massless spectrum. Moreover, the low energy effective action for the heterotic string given in(28) reads, in the Einstein frame

$$S_E^{\text{het}} = \int d^{10}x\sqrt{g}\left[R - \frac{1}{8}(\nabla\Phi)^2 - \frac{1}{4}e^{-\Phi/4}F^2 - \frac{1}{12}e^{-\Phi/2}\hat{H}^2\right] \qquad (68)$$

while the Type I effective action in such frame reads,

$$S_E^I = \int d^{10}x\sqrt{g}\left[R - \frac{1}{8}(\nabla\Phi)^2 - \frac{1}{4}e^{\Phi/4}F^2 - \frac{1}{12}e^{\Phi/2}\hat{H}^2\right]. \qquad (69)$$

Thus, the two actions are related by $\Phi \to -\Phi$ while keeping the other fields invariant. Since e^Φ is the string coupling constant such relation suggests that the weak coupling of heterotic string is the strong coupling Type I and vice versa.[f]

Recall that the two theories have perturbative expansions that are very different.

Interestingly enough, the D1 string of Type I has the quantum numbers of the $SO(32)$ heterotic fundamental string. The D5 brane of Type I maps into the so called NS5 brane of heterotic.

Similar considerations are valid for other corners of the diagram.

It is important to stress that duality relations span an intricate web when compactifications to lower dimensions are considered.

10. Brane Worlds

In the mid-eighties a new way to look at Particle Physics phenomenology, from a *string theory* point of view, emerged. The main reason for the development of a plausible *String Phenomenology* was the fact that all the ingredients required to embed the observed standard model (SM) physics inside a fully unified theory with gravity were, in principle, present in string theory. Before the so called "duality revolution" the standard framework considered compactifications of the heterotic string. By starting, for instance, with the $E_8 \times E_8$ in $D = 10$ dimensions, compactifications allowed a reduction of the number of dimensions, supersymmetries and the gauge group leading to a massless spectrum as similar as possible to the SM. The guide lines of this approach were presented in Candelas et al.[21] where compactifications on a particular type of manifolds, the Calabi-Yau manifolds,

---

[f]The fact that the two actions are related by a field redefinition is not a surprise. It is known that $N = 1$ ten-dimensional supergravity is completely fixed once the gauge group is chosen. It is interesting though, that the field redefinition here is just an inversion of the ten-dimensional coupling.

were considered. Other constructions using compact orbifolds or fermionic string models followed essentially the same strategy [22].

We have briefly discussed the compactification idea. Strings propagating in ten dimensional curved X_{10} space-times backgrounds that look like a Minkowski four dimensional spacetime times an internal compact 6-dimensional manifold

$$X_{10} = M_4 \times X_6 \qquad (70)$$

must be considered. X_6 must be adequately chosen in order to produce a spectrum close to the Standard Model or some extension of it. In particular, the number of fermionic generations is related to the topology of this manifold.[g] For instance, if compactification on a six torus is considered, the states of the ten dimensional theory must be reexpressed in terms of $D = 4$ Poincare representations. However, since the torus is a trivial flat manifold, it is easy to see that no fermion is projected out. The same number of left and right fermions (in the same gauge representation) appear in $D = 4$ and, thus, the theory is non chiral. Calabi-Yau manifolds are such type of non trivial manifold that, moreover, ensure that one supersymmetry is preserved in the compactification process. One important consequence of perturbative heterotic string compactifications is that the string scale is close to the Planck scale, as we discussed in section (3.3).

We will not follow this road here but rather present new scenarios in the context of p-branes.

By now it must be clear that the different classes of p-branes (e.g. D-branes) play a fundamental role in the structure of the full theory of strings. In particular, a fundamental fact is that branes localize gauge interactions on their worldvolume without any need for compactification at this level.

We know that, for example, Type IIB D3-branes have gauge theories with matter fields living in their worldvolume. These fields are localized in the four-dimensional world-volume, and their nature and behaviour depends only on the local structure of the string configuration in the vicinity of that four-dimensional subspace.

Thus, as far as gauge interactions are concerned, a sensible approach should be to first look for *localized* D-brane configurations with world volume field theories resembling as much as possible the SM field theory, even before any compactification of the six transverse dimensions . Our world (SM) would thus be described by a set of branes sitting at some particular

---

[g]Zero modes of the Dirac equation on a non trivial X_6 "box".

point (if Dp-branes, with $p > 3$ are considered the extra dimensions must be compactified) we call it *brane world*.

However the problem of chirality arises also here. Namely, when branes sit a smooth point the spectrum is non-chiral. For instance, open string massless fermionic (Ramond) states $|a\rangle\lambda^a$ on a D3-brane are represented by

$$|s_{st}, s_1, s_2, s_3\rangle\lambda^a \qquad (71)$$

where $s_{st}, s_i = \pm\frac{1}{2}$ (an odd number of minus signs). $s_{st} = -\frac{1}{2}$ is a negative chirality spinor while $s_{st} = \frac{1}{2}$ corresponds to a positive chirality one. Thus, we have four left and four right chiral fermions. The theory is non-chiral. Actually, an $N = 4$ supersymmetric $U(N)$ gauge theory is obtained if N D3-branes sit on the top of each other.

This result can be easily interpreted. If branes sit on a smooth point we have no obstruction in separate them continuously apart. States arising from branes stretching between two different stacks of branes will become massive. Something that would be forbidden if the theory were chiral.

Essentially two approaches have been followed in order to obtain chiral theories:

(1) **Branes at singularities**
(2) **intersecting branes**

11. Branes at Singularities

When a stack of D3-branes sits at a singular point some of the possible fermion states are projected out (see i.e. [23] and references therein). For instance, if a C_3/Z_N Z_N orbifold like singularity is considered, states do transform under a Z_N orbifold rotation. If θ is the generator associated to such rotations, its action on a spinor state reads

$$\theta|s_{st}, s_1, s_2, s_3\rangle = e^{2i\pi s_i \frac{a_i}{N}}|s_{st}, s_1, s_2, s_3\rangle \qquad (72)$$

where a_i are integers (which are, generically, further restricted). Thus, if such a state is required to be invariant under the Z_M action then

$$s_i\frac{a_i}{N} = 0 \quad mod \quad integer \qquad (73)$$

and some fermions are projected out. This is the basic mechanism that leads to chirality.

Actually, when open string states are considered Z_M action manifests, not only, on world sheet degrees of freedom but also on Chan-Paton factors

λ^a. The twist action is represented by a unitary $N \times N$ matrices γ_θ. Thus, a θ twist on an open string state leads to

$$\theta(\Psi_a)\lambda^a)) = (\theta\Psi_a)\rangle\gamma_\theta\lambda^a)\gamma_\theta^{-1} \tag{74}$$

For a gauge boson, with space-time indices only, world sheet rotation is trivial and, therefore, invariance of the open string state under orbifold action leads to the restriction

$$\lambda^a) = \gamma_\theta\lambda^a)\gamma_\theta^{-1} \tag{75}$$

Namely, the original gauge group $U(N)$, encoded in hermitian Chan-Paton matrices λ^a is broken. A generic consistent γ_θ generically leads to the breaking

$$U(N) \rightarrow U(n_0) \times \cdots \times U(n_{M-1}) \tag{76}$$

with $n_0 + \cdots + n_{M-1} = N$.

By consistently choosing the eigenvalues $a_i(i = 1, 2, 3)$ an $N = 1$ supersymmetric, chiral theory can be obtained.

As an interesting example consider $(a_1, a_2, a_3) = (1, 1, -2)$ and a Z_3 action. The following, chiral, spectrum is obtained

$$U(n_0) \times U(n_1) \times U(n_2) \tag{77}$$
$$3[(n_0, \bar{n}_1, 1) + (1, n_1, \bar{n}_2) + (\bar{n}_0, 1, n_2)] \tag{78}$$

with 3 generations of chiral fields. The 3 is associated to Z_3. Recall that $n_0 = 3, n_1 = 2, n_2 = 1$ would look quite close to Standard Model.

However, it is worth noticing that we have considered just part of the whole construction and different consistent conditions, associated to tadpole cancellation must be imposed. In model above such conditions imply $n_0 = n_1 = n_2$. Interestingly enough these are the conditions that ensure that non chiral anomalies are present. The model is not phenomenologically relevant.

Nevertheless, more sophisticated models are available, for instance, when D7-branes containing the orbifold singularity are present, that lead to much interesting models.

A pictorial representation of a Standard Model like construction is given in 4.

12. Intersecting Branes

There exists another kind of brane configurations that may lead to chiral families. This is the case when branes intersect at angles as schematically

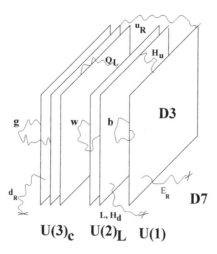

Fig. 4. D-brane configuration of a SM $\mathbb{Z}_3$ orbifold model. Six D3-branes (with worldvolume spanning Minkowski space) are located on a $\mathbb{Z}_3$ singularity and the symmetry is broken to $U(3) \times U(2) \times U(1)$. For the sake of visualization the D3-branes are depicted at different locations, even though they are in fact on top of each other. Open strings starting and ending on the same sets of D3-branes give rise to gauge bosons; those starting in one set and ending on different sets give rise to the left-handed quarks, right-handed U-quarks and one set of Higgs fields. Leptons, and right-handed D-quarks correspond to open strings starting on some D3-branes and ending on the D7-branes (with world-volume filling the whole figure).

shown in figure 5 see [24] and references therein.). More general cases will involve several stacks of branes intersecting at different angles. In fact, the geometry of the configuration is encoded in such angles. Open strings with both end-points at the same stack of N branes will lead to $U(N)$ gauge bosons. Interestingly enough, strings stretching between two different stacks can lead to chiral fermions (see fig.5). Computation of string modes and quantization proceeds as in previous cases. The difference arises in the boundary conditions.

For, lets say k stacks of branes the gauge group and fermionic matter (left chiral here)

$$\prod_a^k U(n_a) \tag{79}$$

$$\sum_{a<b} I_{ab}(n_a, \bar{n}_b) \tag{80}$$

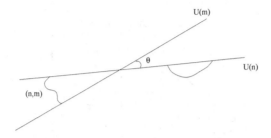

Fig. 5.   Chiral fermions appear at intersections.

where n_a is the number of branes in stack a whereas I_{ab} is the intersection number which counts the number of times that branes in stack a intersect branes in stack b. This number is not necessarily one (or zero) if branes wrap around cycles on compact dimensions.

Interestingly enough, in this picture, the number of families is associated to the number of times that cycles intersect in the compact manifold, as shown in figure 6.4. Let us present an explicit interesting example of a

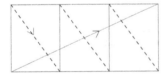

Fig. 6.   Two stacks of branes intersect on a T^2 torus. First stack wraps once on each torus cycle (one with negative orientation), denoted by $(1, -1)$ while the second one wraps once on e_1, vertical direction and three times on horizontal direction e_2, denoted by $(1, 3)$. Intersection number is thus $I_{12} = 4$.

Standard like model. Consider a toroidal compact manifold $X_6 = T^2 \times T^2 \times T^2$ defined as factorized two dimensional tori.

Assume, for simplicity that each torus T_i^2 with $i = 1, 2, 3$ is described by two unit vectors (e_1^i, e_2^i). We indicate with (n_a^i, m_a^i) the number of times that stack a wraps around each cycle. Take six stacks of D6 branes with $N_1 = 3$, leading to QCD group, $N_2 = 2$ and $N_3 = N_4 = N_5 = N_6 = 1$. Wrapping numbers are given in Table 2 and lead to the following, non-zero, intersection number. They lead to the following non-zero intersection numbers $I_{12} = 3 = I_{56} = I_{25} = -I_{13} = -I_{16} = -I_{35}$ and $I_{24} = I_{46} = 6 = -I_{34}$.

Table 2. Wrapping numbers.

$N_1 = 3$	(1,0)	(1,-1)	(1,-1)
$N_2 = 2$	(2,1)	(1,2)	(1,0)
$N_3 = 1$	(2,1)	(-1,-2)	(1,0)
$N_4 = 1$	(1,0)	(1,-1)	(-2,2)
$N_5 = 1$	(1,0)	(1,-1)	(-1,1)
$N_6 = 1$	(2,1)	(-1,-2)	(1,0)

By recalling Eq. (80) we find the gauge group

$$SU(3) \times SU(2) \times U(1)_Y (\times U(1)'s) \tag{81}$$

with the chiral fermion spectrum

$$3(3,2,1/6)_{(1,-1,0,0,0,0)} + 3(\bar{3},1,-2/3)_{(-1,0,1,0,0,0)} + 3(\bar{3},1,1/3)_{(-1,0,0,0,0,1)} +$$
$$3(1,2,1/2)_{(0,1,0,0,-1,0)} + 3(1,2,-1/2)_{(0,1,0,-1,0,0,0)} +$$
$$3(1,1,1)_{(0,0,0,1,0,0,-1)} + 3(1,1,-1)_{(0,0,-1,0,1,0)} +$$
$$3(1,1,0)_{(0,0,0,0,1,-1)} + 3(1,1,0)_{(0,0,-1,1,0,0,0)}$$

where underlining means permutation (notice multiwrapping on third torus for N_4) and hypercharge is defined as a linear combination of $U(1)$ generators

$$Y = -(\frac{Q_1}{3} + \frac{Q_2}{2} + Q_5 + Q_6) \tag{82}$$

where Q_a is the $U(1)$ generator in $U(N_a)$.

13. Scales

In compactifications where D-branes are present the propagation of gauge particles is confined to the low dimensional world volumes of the branes while gravitational interactions propagate in the bulk, ten dimensional space.

This separation of bulk and world volume interactions has very important consequences on the string scale.

Consider a $D = 10$ dimensional space is compactified to $D = 4$ dimensional space $M_4 \times X_6$. Let us assume that the gauge sector propagates on the world volume of a Dp-brane while gravity propagates in 10d bulk. Three spatial dimensions of the brane will fill M_4 while the rest will be wrapped on a $p - 3$ internal cycle (manifold). Before compactification to four dimensions the structure of the effective action reads

$$S_{d=10}^{eff} \propto \int d^{10}x \frac{M_S^8}{g_s^2}[R_{10} + \ldots] + \int d^{p+1}x \frac{M_S^{p+1}}{g_s}F^2 . \quad (83)$$

thus, when compactification on X_6 is considered, following the steps as above, each integral acquires a different volume factor. Namely,

$$S_{d=4}^{eff} \propto \int d^4x \frac{M_S^8 V_6}{g_s^2}[R_4 + \ldots] + \int d^{p+1}x \frac{M_S^{p+1}V_{||}}{g_s}F^2 . \quad (84)$$

Thus, we can express the Planck mass and gauge coupling constant in $D = 4$ dimensions in terms of string scale and coupling constants

$$M_P = \frac{M_S^8 V_6}{g_s^2} \simeq 10^{19}GeV \quad (85)$$

$$g_{YM} = \frac{M_S^{p+1}V_{||}}{g_s} \simeq 0.1 \quad (86)$$

Thus, if $V_6 = V_{||}V_\perp$ then

$$M_P g_{YM} = \frac{M_S^{11-p}V_\perp}{g_s} \quad (87)$$

We see that a large Planck mass, with a consistent Yang-Mills coupling constant is obtainable even for low string scales provided the compactification volume is sufficiently large.

In this picture large extra dimensions could be detected as deviations from Newton's inverse square law for the gravitational force, which is valid only if $d = 4$. Actually, due to the small value of Newton's coupling constant, it is very difficult to measure gravitational force at low distances for small masses. From experimental results it is known that inverse square law is valid down to $1/10mm$. A violation at shorter distances would be consistent with experimental bounds.

Interestingly enough scales down to the electroweak scale, of the order of $1TeV$, can be considered. Recall that in this framework no hierarchy problem would arise. However, the problem is transmuted into a geometrical one of explaining why compactification lengths should stabilize at such large values.

These interesting results have given new impetus to experiments aiming to detect deviations from Newton law. Most of them are based on modern technique reformulations of the Cavendish null experiment. Also, if compactification radii are large massive Kaluza Klein replicas of the graviton, for instance, could be detected at accelerators.

Nevertheless, it is important to keep in mind that such low TeV scale is not compulsory but just a consistent possibility. Much larger energy scales are also possible and therefore with a lower chance of direct detection of strings in the near future.

Acknowledgments

I am grateful to SILAFAE organizers for this stimulating meeting. My participation was partially supported by ANPCyT grant 03-11064.

References

1. The references here are very general and incomplete. More detailed ones should be found in the cited papers, books and reviews
2. M. Carena, talk at this SILAFAE school.
3. C. Wagner, talk at this SILAFAE school.
4. E. Ma, talk at this SILAFAE school.
5. E. Roulet, talk at this SILAFAE school.
6. E. Roulet, *Beyond the standard model*, hep-ph/0112348
7. Particle Data Group, http://pdg.lbl.gov/
8. Juan M. Maldacena, *Gravity, particle physics and their unification*, Int. J. Mod. Phys. A15S1: 840–852, 2000.
   F. Quevedo, *Lectures on superstring phenomenology*, hep-th/9603074
   L. Ibañez, *Recent development in physics beyond the Standard Model*, hep-ph/9901292.
9. H. Davoudiasl, R. Kitano, T. Li, H. Murayama, *The New Minimal Standard Model*, hep-ph/0405097
10. M. Dine, Tasi Lectures on M theory Phenomenology, hep-th/0003175
11. J. Polchinski *String Theory* Vol. I y II;Cambridge Univ. Press, 1998
12. M. Green, J. Schwarz and E. Witten, *Superstring Theory* Vol.I y II;Cambridge Univ. Press, 1987)
13. B. Zwiebach, A first course in String Theory, Cambridge, UK: Univ. Pr. (2004).
14. A. Uranga, *Notes on String Theory* http://gesalerico.ft.uam.es /paginaspersonales/angeluranga/index.html
15. E. Kiritsis, *Introduction to Superstring Theory*, hep-th/9709062.
16. S. Forste, J. Louis, Duality in String Theory, Nucl. Phys. Proc. Suppl. 61A: 3-22, 1998, hep-th/9612192.
17. G. Aldazabal, *D = 4 orientifold models*. Lecture notes of Latinamerican School on Strings, Mexico (2000).
   Atish Dabholkar, *Lecture notes on orientifolds and duality*. hep-th/9804208.
18. A. Sen, *Tachyon Dynamics in String Theory*, hep-th/0410103.
19. C. M. Hull, P. K. Townsend, *Nucl. Phys.* **B438** (1995) 109, hep-th/9410167.
20. E. Witten, *Nucl. Phys.* **B443** (1995) 65, hep-th/9503124.
21. P. Candelas, G. Horowitz, A. Strominger, E. Witten, *Nucl. Phys.* **B258** 85.

22. Some possible reviews on string phenomenology with reference to the original literature are:

F. Quevedo, hep-ph/9707434; hep-th/9603074;

K. Dienes, hep-ph/0004129; hep-th/9602045;

J. D. Lykken, hep-ph/9903026; hep-th/9607144;

M. Dine, hep-th/0003175;

G. Aldazabal, hep-th/9507162;

L. E. Ibáñez, hep-ph/9911499;hep-ph/9804238;hep-th/9505098;

Z. Kakushadze and S.-H. H. Tye, hep-th/9512155;

I. Antoniadis, hep-th/0102202;

E. Dudas, hep-ph/0006190;

D. Bailin, G. Kraniotis, A. Love, hep-th/0108127.

23. G. Aldazabal, L. E. Ibáñez, F. Quevedo, A. M. Uranga, *JHEP* **0008**(2000) 002, hep-th/0005067.

24. M. Berkooz, M. R. Douglas, R. G. Leigh, *Nucl.Phys.* **B480**: 265–278, 1996; hep-th/9606139.

R. Blumenhagen, L. Goerlich, B. Kors, D. Lust, *JHEP* **0010**:006, 2000; hep-th/0007024.

R. Blumenhagen, B. Kors, D. Lust, *JHEP* **0102**:030, 2001; hep-th/0012156.

G. Aldazabal, S. Franco, Luis E. Ibanez, R. Rabadan, A. M. Uranga, *JHEP* **0102**:047, 2001; hep-ph/0011132.

NEUTRINO PHENOMENOLOGY*

ESTEBAN ROULET

CONICET, Centro Atómico Bariloche
Av. Bustillo 9500, 8400, Bariloche, Argentina
roulet@cab.cnea.gov.ar

A general overview of neutrino physics is given, starting with a historical account of the development of our understanding of neutrinos and how they helped to unravel the structure of the Standard Model. We discuss why it is so important to establish if neutrinos are massive and the indications in favor of non-zero neutrino masses are discussed, including the recent results on atmospheric and solar neutrinos and their confirmation with artificial neutrino sources.

1. The Neutrino Story

1.1. *The hypothetical particle*

One may trace back the appearance of neutrinos in physics to the discovery of radioactivity by Becquerel one century ago. When the energy of the electrons (beta rays) emitted in a radioactive decay was measured by Chadwick in 1914, it turned out to his surprise to be continuously distributed. This was not to be expected if the underlying process in the beta decay was the transmutation of an element X into another one X' with the emission of an electron, i.e. $X \to X' + e$, since in that case the electron should be monochromatic. The situation was so puzzling that Bohr even suggested that the conservation of energy may not hold in the weak decays. Another serious problem with the 'nuclear models' of the time was the belief that nuclei consisted of protons and electrons, the only known particles by then. To explain the mass and the charge of a nucleus it was then necessary that it had A protons and $A - Z$ electrons in it. For instance, a ^{4}He nucleus

---

*Work partially supported by Fundación Antorchas.

would have 4 protons and 2 electrons. Notice that this total of six fermions would make the ^{4}He nucleus to be a boson, which is correct. However, the problem arouse when this theory was applied for instance to ^{14}N, since consisting of 14 protons and 7 electrons would make it a fermion, but the measured angular momentum of the nitrogen nucleus was $I = 1$.

The solution to these two puzzles was suggested by Pauli only in 1930, in a famous letter to the 'Radioactive Ladies and Gentlemen' gathered in a meeting in Tubingen, where he wrote: 'I have hit upon a desperate remedy to save the exchange theorem of statistics and the law of conservation of energy. Namely, the possibility that there could exist in nuclei electrically neutral particles, that I wish to call neutrons, which have spin 1/2 ...'. These had to be not heavier than electrons and interacting not more strongly than gamma rays.

With this new paradigm, the nitrogen nucleus became ^{14}N$= 14p + 7e +$ 7'n', which is a boson, and a beta decay now involved the emission of two particles $X \rightarrow X' + e + $'$n$', and hence the electron spectrum was continuous. Notice that no particles were created in a weak decay, both the electron and Pauli's neutron 'n' were already present in the nucleus of the element X, and they just came out in the decay. However, in 1932 Chadwick discovered the real 'neutron', with a mass similar to that of the proton and being the missing building block of the nuclei, so that a nitrogen nucleus finally became just ^{14}N$= 7p + 7n$, which also had the correct bosonic statistics.

In order to account now for the beta spectrum of weak decays, Fermi called Pauli's hypotetised particle the neutrino (small neutron), ν, and furthermore suggested that the fundamental process underlying beta decay was $n \rightarrow p + e + \nu$. He wrote [1] the basic interaction by analogy with the interaction known at the time, the QED, i.e. as a vector×vector current interaction:

$$H_F = G_F \int \mathrm{d}^3 x [\bar{\Psi}_p \gamma_\mu \Psi_n][\bar{\Psi}_e \gamma^\mu \Psi_\nu] + h.c..$$

This interaction accounted for the continuous beta spectrum, and from the measured shape at the endpoint Fermi concluded that m_ν was consistent with zero and had to be small. The Fermi coupling G_F was estimated from the observed lifetimes of radioactive elements, and armed with this Hamiltonian Bethe and Peierls [2] decided to compute the cross section for the inverse beta process, i.e. for $\bar{\nu} + p \rightarrow n + e^+$, which was the relevant reaction to attempt the direct detection of a neutrino. The result, $\sigma = 4(G_F^2/\pi)p_e E_e \simeq 2.3 \times 10^{-44} \mathrm{cm}^2 (p_e E_e/m_e^2)$ was so tiny that they concluded

'... This meant that one obviously would never be able to see a neutrino.'. For instance, if one computes the mean free path in water (with density $n \simeq 10^{23}/\text{cm}^3$) of a neutrino with energy $E_\nu = 2.5$ MeV, typical of a weak decay, the result is $\lambda \equiv 1/n\sigma \simeq 2.5 \times 10^{20}$ cm, which is 10^7 AU, i.e. comparable to the thickness of the Galactic disk.

It was only in 1956 that Reines and Cowan were able to prove that Bethe and Peierls had been too pessimistic, when they measured for the first time the interaction of a neutrino through the inverse beta process[3]. Their strategy was essentially that, if one needs 10^{20} cm of water to stop a neutrino, having 10^{20} neutrinos a cm would be enough to stop one neutrino. Since after the second war powerful reactors started to become available, and taking into account that in every fission of an uranium nucleus the neutron rich fragments beta decay producing typically 6 $\bar{\nu}$ and liberating ~ 200 MeV, it is easy to show that the (isotropic) neutrino flux at a reactor is

$$\frac{d\Phi_\nu}{d\Omega} \simeq \frac{2 \times 10^{20}}{4\pi} \left(\frac{\text{Power}}{\text{GWatt}}\right) \frac{\bar{\nu}}{strad}.$$

Hence, placing a few hundred liters of water (with some Cadmium in it) near a reactor they were able to see the production of positrons (through the two 511 keV γ produced in their annihilation with electrons) and neutrons (through the delayed γ from the neutron capture in Cd), with a rate consistent with the expectations from the weak interactions of the neutrinos.

1.2. *The vampire*

Going back in time again to follow the evolution of the theory of weak interactions of neutrinos, in 1936 Gamow and Teller [4] noticed that the $V \times V$ Hamiltonian of Fermi was probably too restrictive, and they suggested the generalization

$$H_{GT} = \sum_i G_i[\bar{p}O_i n][\bar{e}O^i \nu] + h.c.,$$

involving the operators $O_i = 1$, γ_μ, $\gamma_\mu \gamma_5$, γ_5, $\sigma_{\mu\nu}$, corresponding to scalar (S), vector (V), axial vector (A), pseudoscalar (P) and tensor (T) currents. However, since A and P only appeared here as $A \times A$ or $P \times P$, the interaction was parity conserving. The situation became unpleasant, since now there were five different coupling constants G_i to fit with experiments, but however this step was required since some observed nuclear transitions which were forbidden for the Fermi interaction became now allowed with its generalization (GT transitions).

The story became more involved when in 1956 Lee and Yang suggested that parity could be violated in weak interactions[5]. This could explain why the particles theta and tau had exactly the same mass and charge and only differed in that the first one was decaying to two pions while the second to three pions (e.g. to states with different parity). The explanation to the puzzle was that the Θ and τ were just the same particle, now known as the charged kaon, but the (weak) interaction leading to its decays violated parity.

Parity violation was confirmed the same year by Wu [6], studying the direction of emission of the electrons emitted in the beta decay of polarised ^{60}Co. The decay rate is proportional to $1 + \alpha \vec{P} \cdot \hat{p}_e$. Since the Co polarization vector $\vec{P}$ is an axial vector, while the unit vector along the electron momentum $\hat{p}_e$ is a vector, their scalar product is a pseudoscalar and hence a non–vanishing coefficient α would imply parity violation. The result was that electrons preferred to be emitted in the direction opposite to $\vec{P}$, and the measured value $\alpha \simeq -0.7$ had then profound implications for the physics of weak interactions.

The generalization by Lee and Yang of the Gamow Teller Hamiltonian was

$$H_{LY} = \sum_i [\bar{p}O_i n][\bar{e}O^i(G_i + G'_i \gamma_5)\nu] + h.c..$$

Now the presence of terms such as $V \times A$ or $P \times S$ allows for parity violation, but clearly the situation became even more unpleasant since there are now 10 couplings (G_i and G'_i) to determine, so that some order was really called for.

Soon the bright people in the field realized that there could be a simple explanation of why parity was violated in weak interactions, the only one involving neutrinos, and this had just to do with the nature of the neutrinos. Lee and Yang, Landau and Salam [7] realized that, if the neutrino was massless, there was no need to have both neutrino chirality states in the theory, and hence the handedness of the neutrino could be the origin for the parity violation. To see this, consider the chiral projections of a fermion

$$\Psi_{L,R} \equiv \frac{1 \mp \gamma_5}{2} \Psi.$$

We note that in the relativistic limit these two projections describe left and right handed helicity states (where the helicity, i.e. the spin projection in the direction of motion, is a constant of motion for a free particle), but in general an helicity eigenstate is a mixture of the two chiralities. For a massive particle, which has to move with a velocity smaller than the speed of light, it is always possible to make a boost to a system where the helicity is reversed, and hence the helicity is clearly not a Lorentz invariant while the chirality is (and hence has the desireable properties of a charge to which a gauge boson can be coupled). If we look now to the equation of motion for a Dirac particle as the one we are used to for the description of a charged massive particle such as an electron $((i\not\partial - m)\Psi = 0)$, in terms of the chiral projections this equation becomes

$$i\not\partial \Psi_L = m\Psi_R$$

$$i\not\partial \Psi_R = m\Psi_L$$

and hence clearly a mass term will mix the two chiralities. However, from these equations we see that for $m = 0$, as could be the case for the neutrinos, the two equations are decoupled, and one could write a consistent theory using only one of the two chiralities (which in this case would coincide with the helicity). If the Lee Yang Hamiltonian were just to depend on a single neutrino chirality, one would have then $G_i = \pm G'_i$ and parity violation would indeed be maximal. This situation has been described by saying that neutrinos are like vampires in Dracula's stories: if they were to look to themselves into a mirror they would be unable to see their reflected images.

The actual helicity of the neutrino was measured by Goldhaber et al. [8]. The experiment consisted in observing the K-electron capture in ^{152}Eu $(J = 0)$ which produced ^{152}Sm* $(J = 1)$ plus a neutrino. This excited nucleus then decayed into ^{152}Sm $(J = 0) + \gamma$. Hence the measurement of the polarization of the photon gave the required information on the helicity of the neutrino emitted initially. The conclusion was that '...Our results seem compatible with ... 100% negative helicity for the neutrinos', i.e. that the neutrinos are left handed particles.

This paved the road for the $V - A$ theory of weak interactions advanced by Feynman and Gell Mann, and Marshak and Soudarshan [9], which stated that weak interactions only involved vector and axial vector currents, in the combination $V - A$ which only allows the coupling to left handed fields, i.e.

$$J_\mu = \bar{e}_L \gamma_\mu \nu_L + \bar{n}_L \gamma_\mu p_L$$

with $H = (G_F/\sqrt{2})J_\mu^\dagger J^\mu$. This interaction also predicted the existence of purely leptonic weak charged currents, e.g. $\nu + e \to \nu + e$, to be experimentally observed much later[a].

The current involving nucleons is actually not exactly $\propto \gamma_\mu (1 - \gamma_5)$ (only the interaction at the quark level has this form), but is instead $\propto \gamma_\mu (g_V - g_A \gamma_5)$. The vector coupling remains however unrenormalised ($g_V = 1$) due to the so called conserved vector current hypothesis (CVC), which states that the vector part of the weak hadronic charged currents ($J_\mu^\pm \propto \bar{\Psi} \gamma_\mu \tau^\pm \Psi$, with $\tau^\pm$ the raising and lowering operators in the isospin space $\Psi^T = (p, n)$) together with the isovector part of the electromagnetic current (i.e. the term proportional to τ_3 in the decomposition $J_\mu^{em} \propto \bar{\Psi} \gamma_\mu (1 + \tau_3) \Psi$) form an isospin triplet of conserved currents. On the other hand, the axial vector hadronic current is not protected from strong interaction renormalization effects and hence g_A does not remain equal to unity. The measured value, using for instance the lifetime of the neutron, is $g_A = 1.27$, so that at the nucleonic level the charged current weak interactions are actually "$V - 1.27A$".

With the present understanding of weak interactions, we know that the clever idea to explain parity violation as due to the non-existence of one of the neutrino chiralities (the right handed one) was completely wrong, although it lead to major advances in the theory and ultimately to the correct interaction. Today we understand that the parity violation is a property of the gauge boson (the W) responsible for the gauge interaction, which couples only to the left handed fields, and not due to the absence of right handed fields. For instance, in the quark sector both left and right chiralities exists, but parity is violated because the right handed fields are singlets for the weak charged currents.

1.3. *The trilogy*

In 1947 the muon was discovered in cosmic rays by Anderson and Neddermeyer. This particle was just a heavier copy of the electron, and as was suggested by Pontecorvo, it also had weak interactions $\mu + p \to n + \nu$ with the same universal strength G_F. Hincks, Pontecorvo and Steinberger showed that the muon was decaying to three particles, $\mu \to e\nu\nu$, and the

---

[a]A curious fact was that the new theory predicted a cross section for the inverse beta decay a factor of two larger than the Bethe and Peierls original result, which had already been confirmed in 1956 to the 5% accuracy by Reines and Cowan. However, in an improved experiment in 1959 Reines and Cowan found a value consistent with the new prediction, what shows that many times when the experiment agrees with the theory accepted at the moment the errors tend to be underestimated.

question arose whether the two emitted neutrinos were similar or not. It was then shown by Feinberg [10] that, assuming the two particles were of the same kind, weak interactions couldn't be mediated by gauge bosons (an hypothesis suggested in 1938 by Klein). The reasoning was that if the two neutrinos were equal, it would be possible to join the two neutrino lines and attach a photon to the virtual charged gauge boson (W) or to the external legs[b], so as to generate a diagram for the radiative decay $\mu \to e\gamma$. The resulting branching ratio would be larger than 10^{-5} and was hence already excluded at that time. This was probably the first use of 'rare decays' to constrain the properties of new particles.

The correct explanation for the absence of the radiative decay was put forward by Lee and Yang, who suggested that the two neutrinos emitted in the muon decay had different flavour, i.e. $\mu \to e + \nu_e + \nu_\mu$, and hence it was not possible to join the two neutrino lines to draw the radiative decay diagram. This was confirmed at Brookhaven in the first accelerator neutrino experiment[11]. They used an almost pure $\bar{\nu}_\mu$ beam, something which can be obtained from charged pion decays, since the $V - A$ theory implies that $\Gamma(\pi \to \ell + \bar{\nu}_\ell) \propto m_\ell^2$, i.e. this process requires a chirality flip in the final lepton line which strongly suppresses the decays $\pi \to e + \bar{\nu}_e$. Putting a detector in front of this beam they were able to observe the process $\bar{\nu} + p \to n + \mu^+$, but no production of positrons, what proved that the neutrinos produced in a weak decay in association with a muon were not the same as those produced in a beta decay (in association with an electron). Notice that although the neutrino fluxes are much smaller at accelerators than at reactors, their higher energies make their detection feasible due to the larger cross sections ($\sigma \propto E^2$ for $E \ll m_p$, and $\sigma \propto E$ for $E \gtrsim m_p$).

In 1975 the τhird charged lepton was discovered by Perl at SLAC, and being just a heavier copy of the electron and the muon, it was concluded that a third neutrino flavour had also to exist. The direct detection of the τ neutrino has been achieved by the DONUT experiment at Fermilab, looking at the short τ tracks produced by the interaction of a ν_τ emitted in the decay of a heavy meson (containing a b quark) produced in a beam dump. Furthermore, we know today that the number of light weakly interacting neutrinos is precisely three (see below), so that the proliferation of neutrino species seems to be now under control.

---

[b]This reasoning would have actually also excluded a purely leptonic generalisation of a Fermi's theory to describe the muon decay.

1.4. *The gauge theory*

As was just mentioned, Klein had suggested that the short range charged current weak interaction could be due to the exchange of a heavy charged vector boson, the $W^\pm$. This boson exchange would look at small momentum transfers ($Q^2 \ll M_W^2$) as the non renormalisable four fermion interactions discussed before. If the gauge interaction is described by the Lagrangian $\mathcal{L} = -(g/\sqrt{2})J_\mu W^\mu + h.c.$, from the low energy limit one can identify the Fermi coupling as $G_F = \sqrt{2}g^2/(8M_W^2)$. In the sixties, Glashow, Salam and Weinberg showed that it was possible to write down a unified description of electromagnetic and weak interactions with a gauge theory based in the group $SU(2)_L \times U(1)_Y$ (weak isospin × hypercharge, with the electric charge being $Q = T_3 + Y$), which was spontaneously broken at the weak scale down to the electromagnetic $U(1)_{em}$. This (nowadays standard) model involves the three gauge bosons in the adjoint of $SU(2)$, V_i^μ (with $i = 1, 2, 3$), and the hypercharge gauge field B^μ, so that the starting Lagrangian is

$$\mathcal{L} = -g\sum_{i=1}^{3} J_\mu^i V_i^\mu - g' J_\mu^Y B^\mu + h.c.,$$

with $J_\mu^i \equiv \sum_a \bar{\Psi}_{aL}\gamma_\mu(\tau_i/2)\Psi_{aL}$. The left handed leptonic and quark isospin doublets are $\Psi^T = (\nu_{eL}, e_L)$ and (u_L, d_L) for the first generation (and similar ones for the other two heavier generations) and the right handed fields are SU(2) singlets. The hypercharge current is obtained by summing over both left and right handed fermion chiralities and is $J_\mu^Y \equiv \sum_a Y_a \bar{\Psi}_a \gamma_\mu \Psi_a$.

After the electroweak breaking one can identify the weak charged currents with $J^\pm = J^1 \pm iJ^2$, which couple to the W boson $W^\pm = (V^1 \mp iV^2)/\sqrt{2}$, and the two neutral vector bosons V^3 and B will now combine through a rotation by the weak mixing angle θ_W (with $\mathrm{tg}\theta_W = g'/g$), to give

$$\begin{pmatrix} A_\mu \\ Z_\mu \end{pmatrix} = \begin{pmatrix} c\theta_W & s\theta_W \\ -s\theta_W & c\theta_W \end{pmatrix} \begin{pmatrix} B_\mu \\ V_\mu^3 \end{pmatrix} \tag{1}$$

We see that the broken theory has now, besides the massless photon field A_μ, an additional neutral vector boson, the heavy Z_μ, whose mass turns out to be related to the W boson mass through $s^2\theta_W = 1 - (M_W^2/M_Z^2)$. The electromagnetic and neutral weak currents are given by

$$J_\mu^{em} = J_\mu^Y + J_\mu^3$$

$$J_\mu^0 = J_\mu^3 - s^2\theta_W J_\mu^{em},$$

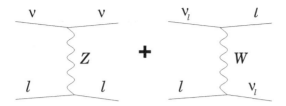

Fig. 1.   Neutral and charged current contributions to neutrino lepton scattering.

with the electromagnetic coupling being $e = g \, s\theta_W$.

The great success of this model came in 1973 with the experimental observation of the weak neutral currents using muon neutrino beams at CERN (Gargamelle) and Fermilab, using the elastic process $\nu_\mu e \to \nu_\mu e$. The semileptonic processes $\nu N \to \nu X$ were also studied and the comparison of neutral and charged current rates provided a measure of the weak mixing angle. From the theoretical side t'Hooft proved the renormalisability of the theory, so that the computation of radiative corrections became also meaningful.

The Hamiltonian for the leptonic weak interactions $\nu_\ell + \ell' \to \nu_\ell + \ell'$ can be obtained, using the Standard Model just presented, from the two diagrams in figure 1. In the low energy limit ($Q^2 \ll M_W^2$, M_Z^2), it is just given by

$$H_{\nu_\ell \ell'} = \sqrt{2} G_F [\bar{\nu}_\ell \gamma_\mu (1 - \gamma_5) \nu_\ell][\bar{\ell'} \gamma^\mu (c_L P_L + c_R P_R) \ell']$$

where the left and right couplings are $c_L = \delta_{\ell\ell'} + s^2\theta_W - 0.5$ and $c_R = s^2\theta_W$. The $\delta_{\ell\ell'}$ term in c_L is due to the charged current diagram, which clearly only appears when $\ell = \ell'$. On the other hand, one sees that due to the B component in the Z boson, the weak neutral currents also couple to the charged lepton right handed chiralities (i.e. $c_R \neq 0$). This interaction leads to the cross section (for $E_\nu \gg m_{\ell'}$)

$$\sigma(\nu + \ell \to \nu + \ell) = \frac{2G_F^2}{\pi} m_\ell E_\nu \left[c_L^2 + \frac{c_R^2}{3} \right],$$

and a similar expression with $c_L \leftrightarrow c_R$ for antineutrinos. Hence, we have the following relations for the neutrino elastic scatterings off electrons

$$\sigma(\nu_e e) \simeq 9 \times 10^{-44} \text{cm}^2 \left(\frac{E_\nu}{10 \text{ MeV}} \right) \simeq 2.5\sigma(\bar{\nu}_e e) \simeq 6\sigma(\nu_{\mu,\tau} e) \simeq 7\sigma(\bar{\nu}_{\mu,\tau} e).$$

Regarding the angular distribution of the electron momentum with respect to the incident neutrino direction, in the center of mass system of the process $d\sigma(\nu_e e)/d\cos\theta \propto 1 + 0.1[(1 + \cos\theta)/2]^2$, and it is hence almost isotropic.

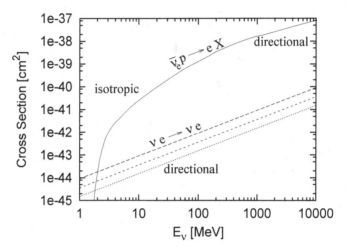

Fig. 2.   Neutrino nucleon and neutrino lepton cross sections (the three lines correspond, from top to bottom, to the ν_e, $\bar{\nu}_e$ and $\nu_{\mu,\tau}$ cross sections with electrons).

However, due to the boost to the laboratory system, there will be a significant correlation between the neutrino and electron momenta for $E_\nu \gg$ MeV, and this actually allows to do astronomy with neutrinos. For instance, water cherenkov detectors such as Superkamiokande detect solar neutrinos using this process, and have been able to reconstruct a picture of the Sun with neutrinos. It will turn also to be relevant for the study of neutrino oscillations that these kind of detectors are six times more sensitive to electron type neutrinos than to the other two neutrino flavours.

Considering now the neutrino nucleon interactions, one has at low energies (1 MeV$< E_\nu <$ 50 MeV) the cross section for the quasi-elastic process[c]

$$\sigma(\nu_e n \to pe) \simeq \sigma(\bar{\nu}_e p \to ne^+) \simeq \frac{G_F^2}{\pi} c^2 \theta_C (g_V^2 + 3g_A^2) E_\nu^2,$$

where we have now introduced the Cabibbo mixing angle θ_C which relates, if we ignore the third family, the quark flavour eigenstates q^0 to the mass eigenstates q, i.e. $d^0 = c\theta_C d + s\theta_C s$ and $s^0 = -s\theta_C d + c\theta_C s$ (choosing a flavour basis so that the up type quark flavour and mass eigenstates coincide).

At $E_\nu \gtrsim$ 50 MeV, the nucleon no longer looks like a point-like object for

---

[c]Actually for the $\bar{\nu}_e p$ CC interaction, the threshold energy is $E_\nu^{th} \simeq m_n - m_p + m_e \simeq$ 1.8 MeV.

the neutrinos, and hence the vector (v_μ) and axial (a_μ) hadronic currents involve now momentum dependent form factors, i.e.

$$\langle N(p')|v_\mu|N(p)\rangle = \bar{u}(p')[\gamma_\mu F_V + \frac{i}{2m_N}\sigma_{\mu\nu}q^\nu F_W]u(p)$$

$$\langle N(p')|a_\mu|N(p)\rangle = \bar{u}(p')[\gamma_\mu\gamma_5 F_A + \frac{\gamma_5}{m_N}q_\mu F_P]u(p),$$

where $F_V(q^2)$ can be measured using electromagnetic processes and the CVC relation $F_V = F_V^{em,p} - F_V^{em,n}$ (i.e. as the difference between the proton and neutron electromagnetic vector form factors). Clearly $F_V(0) = 1$ and $F_A(0) = 1.27$, while F_W is related to the anomalous magnetic moments of the nucleons. The q^2 dependence has the effect of significantly flattening the cross section. In the deep inelastic regime, $E_\nu \gtrsim$ GeV, the neutrinos interact directly with the quark constituents. The cross section in this regime grows linearly with energy, and this provided an important test of the parton model. The main characteristics of the neutrino cross section just discussed are depicted in figure 2. For even larger energies, the gauge boson propagators enter into the play (e.g. $1/M_W^2 \to 1/q^2$) and the growth of the cross section is less pronounced above 10 TeV ($\sigma \propto E^{0.36}$).

The most important test of the standard model came with the direct production of the $W^\pm$ and Z gauge bosons at CERN in 1984, and with the precision measurements achieved with the Z factories LEP and SLC after 1989. These e^+e^- colliders working at and around the Z resonance ($s = M_Z^2 = (91 \text{ GeV})^2$) turned out to be also crucial for neutrino physics, since studying the shape of the $e^+e^- \to f\bar{f}$ cross section near the resonance, which has the Breit–Wigner form

$$\sigma \simeq \frac{12\pi\Gamma_e\Gamma_f}{M_Z^2}\frac{s}{(s-M_Z^2)^2 + M_Z^2\Gamma_Z^2},$$

it becomes possible to determine the total Z width Γ_Z. This width is just the sum of all possible partial widths, i.e.

$$\Gamma_Z = \sum_f \Gamma_{Z\to f\bar{f}} = \Gamma_{vis} + \Gamma_{inv}.$$

The visible (i.e. involving charged leptons and quarks) width Γ_{vis} can be measured directly, and hence one can infer a value for the invisible width Γ_{inv}. Since in the standard model this last arises from the decays $Z \to \nu_i\bar{\nu}_i$, whose expected rate for decays into a given neutrino flavour is $\Gamma_{Z\to\nu\bar{\nu}}^{th} = 167$ MeV, one can finally obtain the number of neutrinos coupling to the Z as $N_\nu = \Gamma_{inv}/\Gamma_{Z\to\nu\bar{\nu}}^{th}$. The present best value for this quantity is $N_\nu =$

2.994±0.012, giving then a strong support to the three generation standard model.

Going through the history of the neutrinos we have seen that they have been extremely useful to understand the standard model. On the contrary, the standard model is of little help to understand the neutrinos. Since in the standard model there is no need for ν_R, neutrinos are massless in this theory. There is however no deep principle behind this (unlike the masslessness of the photon which is protected by the electromagnetic gauge symmetry), and indeed in many extensions of the standard model neutrinos turn out to be massive. This makes the search for non-zero neutrino masses a very important issue, since it provides a window to look for physics beyond the standard model. Indeed, solid evidence has accumulated in the last years indicating that neutrinos are massive, and this makes the field of neutrino physics even more exciting now than in the long historic period that we have just reviewed.

2. Neutrino Masses

2.1. *Dirac or Majorana?*

In the standard model, charged leptons (and also quarks) get their masses through their Yukawa couplings to the Higgs doublet field $\phi^T = (\phi_+, \phi_0)$

$$-\mathcal{L}_Y = \lambda \bar{L} \phi \ell_R + h.c. ,$$

where $L^T = (\nu, \ell)_L$ is a lepton doublet and ℓ_R an SU(2) singlet field. When the electroweak symmetry gets broken by the vacuum expectation value of the neutral component of the Higgs field $\langle \phi_0 \rangle = v/\sqrt{2}$ (with $v = 246$ GeV), the following 'Dirac' mass term results

$$-\mathcal{L}_m = m_\ell(\bar{\ell}_L \ell_R + \bar{\ell}_R \ell_L) = m_\ell \bar{\ell}\ell,$$

where $m_\ell = \lambda v/\sqrt{2}$ and $\ell = \ell_L + \ell_R$ is the Dirac spinor field. This mass term is clearly invariant under the $U(1)$ transformation $\ell \to \exp(i\alpha)\ell$, which corresponds to the lepton number (and actually in this case also to the electromagnetic gauge invariance). From the observed fermion masses, one concludes that the Yukawa couplings range from $\lambda_t \simeq 1$ for the top quark up to $\lambda_e \simeq 10^{-5}$ for the electron.

Notice that the mass terms always couple fields with opposite chiralities, i.e. requires a $L \leftrightarrow R$ transition. Since in the standard model the right handed neutrinos are not introduced, it is not possible to write a Dirac mass term, and hence the neutrino results massless. Clearly the simplest way to give the neutrino a mass would be to introduce the right handed fields just for this purpose (having no gauge interactions, these sterile states would be essentially undetectable and unproduceable). Although this is a logical possibility, it has the ugly feature that in order to get reasonable neutrino masses, below the eV, would require unnaturally small Yukawa couplings ($\lambda_\nu < 10^{-11}$). Fortunately it turns out that neutrinos are also very special particles in that, being neutral, there are other ways to provide them a mass. Furthermore, in some scenarios it becomes also possible to get a natural understanding of why neutrino masses are so much smaller than the charged fermion masses.

The new idea is that the left handed neutrino field actually involves two degrees of freedom, the left handed neutrino associated with the positive beta decay (i.e. emitted in association with a positron) and the other one being the right handed 'anti'-neutrino emitted in the negative beta decays (i.e. emitted in association with an electron). It may then be possible to write down a mass term using just these two degrees of freedom and involving the required $L \leftrightarrow R$ transition. This possibility was first suggested by Majorana in 1937, in a paper named 'Symmetric theory of the electron and positron', and devoted mainly to the problem of getting rid of the negative energy sea of the Dirac equation[12]. As a side product, he found that for neutral particles there was 'no more any reason to presume the existence of antiparticles', and that 'it was possible to modify the theory of beta emission, both positive and negative, so that it came always associated with the emission of a neutrino'. The spinor field associated to this formalism was then named in his honor a Majorana spinor.

To see how this works it is necessary to introduce the so called antiparticle field, $\psi^c \equiv C\bar{\psi}^T = C\gamma_0^T\psi^*$. The charge conjugation matrix C has to satisfy $C\gamma_\mu C^{-1} = -\gamma_\mu^T$, so that for instance the Dirac equation for a charged fermion in the presence of an electromagnetic field, $(i\slashed{\partial} - e\slashed{A} - m)\psi = 0$ implies that $(i\slashed{\partial} + e\slashed{A} - m)\psi^c = 0$, i.e. that the antiparticle field has the same mass but charges opposite to those of the particle field. Since for a chiral projection one can show that $(\psi_L)^c = (P_L\psi)^c = P_R\psi^c = (\psi^c)_R$, i.e. this conjugation changes the chirality of the field, one has that ψ^c is related to the CP conjugate of ψ. Notice that $(\psi_L)^c$ describes exactly the same two degrees of freedom described by ψ_L, but somehow using a CP reflected formalism. For instance for the neutrinos, the ν_L operator annihilates the left handed neutrino and creates the right handed antineutrino, while the $(\nu_L)^c$ operator annihilates the right handed antineutrino and creates the left handed neutrino.

We can then now write the advertised Majorana mass term, as

$$-\mathcal{L}_M = \frac{1}{2}m[\overline{\nu_L}(\nu_L)^c + \overline{(\nu_L)^c}\nu_L].$$

This mass term has the required Lorentz structure (i.e. the $L \leftrightarrow R$ transition) but one can see that it does not preserve any $U(1)$ phase symmetry, i.e. it violates the so called lepton number by two units. If we introduce the Majorana field $\nu \equiv \nu_L + (\nu_L)^c$, which under conjugation transforms into itself ($\nu^c = \nu$), the mass term becomes just $\mathcal{L}_M = -m\bar{\nu}\nu/2$.

Up to now we have introduced the Majorana mass by hand, contrary to the case of the charged fermions where it arose from a Yukawa coupling in a spontaneously broken theory. To follow the same procedure with the neutrinos presents however a difficulty, because the standard model neutrinos belong to $SU(2)$ doublets, and hence to write an electroweak singlet Yukawa coupling it is necessary to introduce an $SU(2)$ triplet Higgs field $\vec{\Delta}$ (something which is not particularly attractive). The coupling $\mathcal{L} \propto \overline{L^c}\vec{\sigma}L \cdot \vec{\Delta}$ would then lead to the Majorana mass term after the neutral component of the scalar gets a VEV. Alternatively, the Majorana mass term could be a loop effect in models where the neutrinos have lepton number violating couplings to new scalars, as in the so-called Zee models or in the supersymmetric models with R parity violation. These models have as interesting features that the masses are naturally suppressed by the loop, and they are attractive also if one looks for scenarios where the neutrinos have relatively large dipole moments, since a photon can be attached to the charged particles in the loop.

However, by far the nicest possibility to give neutrinos a mass is the so-called see-saw model[13]. In this scenario, which naturally occurs in grand unified models such as $SO(10)$, one introduces the $SU(2)$ singlet right handed neutrinos. One has now not only the ordinary Dirac mass term, but also a Majorana mass for the singlets which is generated by the VEV of an $SU(2)$ singlet Higgs, whose natural scale is the scale of breaking of the grand unified group, i.e. in the range 10^{12}–10^{16} GeV. Hence the Lagrangian will contain

$$\mathcal{L}_M = \frac{1}{2}\overline{(\nu_L, (N_R)^c)}\begin{pmatrix} 0 & m_D \\ m_D & M \end{pmatrix}\begin{pmatrix} (\nu_L)^c \\ N_R \end{pmatrix} + h.c..$$

The mass eigenstates are two Majorana fields with masses $m_{light} \simeq m_D^2/M$ and $m_{heavy} \simeq M$. Since $m_D/M \ll 1$, we see that $m_{light} \ll m_D$, and hence the lightness of the known neutrinos is here related to the heaviness of the sterile states N_R.

If we actually introduce one singlet neutrino per family, the mass terms in eq. (2.1) are 3×3 matrices. Notice that if m_D is similar to the up-type quark masses, as happens in $SO(10)$, one would have $m_{\nu_3} \sim m_t^2/M \simeq 0.04$ eV$(10^{15}$ GeV$/M)$. It is clear then that in these scenarios the observation of neutrino masses near 0.1 eV would point out to new physics at about the GUT scale, while for $m_D \sim$ GeV this would correspond to singlet neutrino masses at an intermediate scale $M \sim 10^{10}$–10^{12} GeV, also of great theoretical interest.

2.2. *Neutrino mixing and oscillations*

If neutrinos are massive, there is no reason to expect that the mass eigenstates (ν_k, with $k = 1, 2, 3$) would coincide with the flavour (gauge) eigenstates (ν_α, with $\alpha = e, \mu, \tau$, where we are adopting the flavor basis such that for the charged leptons the flavor eigenstates coincide with the mass eigenstates), and hence, in the same way that quark states are mixed through the Cabibbo, Kobayashi and Maskawa matrix, neutrinos would be related through the Maki, Nakagawa and Sakita [14] (and Pontecorvo) mixing matrix, i.e. $\nu_\alpha = V_{\alpha k}^* \nu_k$. This matrix can be parametrized as ($c_{12} \equiv \cos\theta_{12}$, etc.)

$$V = \begin{pmatrix} c_{12}c_{13} & c_{13}s_{12} & s_{13}e^{-i\delta} \\ -c_{23}s_{12} - c_{12}s_{13}s_{23}e^{i\delta} & c_{12}c_{23} - s_{12}s_{13}s_{23}e^{i\delta} & c_{13}s_{23} \\ s_{23}s_{12} - c_{12}c_{23}s_{13}e^{i\delta} & -c_{12}s_{23} - c_{23}s_{12}s_{13}e^{i\delta} & c_{13}c_{23} \end{pmatrix} \begin{pmatrix} e^{i\frac{\alpha_1}{2}} & 0 & 0 \\ 0 & c^{i\frac{\alpha_2}{2}} & 0 \\ 0 & 0 & 1 \end{pmatrix}.$$

The phases $\alpha_{1,2}$ here cannot be removed by a rephasing of the fields (as is done for quarks) if the neutrinos are Majorana particles, since such rotations would then introduce a complex phase in the neutrino masses.

The possibility that neutrino flavour eigenstates be a superposition of mass eigenstates allows for the phenomenon of neutrino oscillations. This is a quantum mechanical interference effect (and as such it is sensitive to quite small masses) and arises because different mass eigenstates propagate differently, and hence the flavor composition of a state can change with time.

To see this consider a flavour eigenstate neutrino ν_α with momentum p produced at time $t = 0$ (e.g. a ν_μ produced in the decay $\pi^+ \to \mu^+ + \nu_\mu$). The initial state is then

$$|\nu_\alpha\rangle = \sum_k V_{\alpha k}^* |\nu_k\rangle.$$

We know that the mass eigenstates evolve with time according to $|\nu_k(t, x)\rangle = \exp[i(px - E_k t)]|\nu_k\rangle$. In the relativistic limit relevant for neutrinos, one has that $E_k = \sqrt{p^2 + m_k^2} \simeq p + m_k^2/2E$, and thus the different mass eigenstates will acquire different phases as they propagate. Hence, the probability of observing a flavour ν_β at time t is just

$$P(\nu_\alpha \to \nu_\beta) = |\langle \nu_\beta | \nu(t) \rangle|^2 = |\sum_k V_{\alpha k}^* e^{-i\frac{m_k^2}{2E}t} V_{\beta k}|^2.$$

Taking into account the explicit expression for V, it is easy to convince oneself that the Majorana phases $\alpha_{1,2}$ do not enter into the oscillation probability, and hence oscillation phenomena cannot tell whether neutrinos are Dirac or Majorana particles.

In the case of two generations, V can be taken just as a rotation R_θ with mixing angle θ, so that one has

$$P(\nu_\alpha \to \nu_\beta) = \sin^2 2\theta \, \sin^2 \left(\frac{\Delta m^2 x}{4E}\right),$$

which depends on the squared mass difference $\Delta m^2 = m_2^2 - m_1^2$, since this is what gives the phase difference in the propagation of the mass eigenstates. Hence, the amplitude of the flavour oscillations is given by $\sin^2 2\theta$ and the oscillation length of the modulation is $L_{osc} \equiv 4\pi E/\Delta m^2 \simeq 2.5$ m $E[\text{MeV}]/\Delta m^2[\text{eV}^2]$. We see then that neutrinos will typically oscillate with a macroscopic wavelength. For instance, putting a detector at ~ 1 km from a reactor (such as in the CHOOZ experiment) allows to test oscillations of ν_e's to another flavour (or into a singlet neutrino) down to $\Delta m^2 \sim 10^{-3}$ eV^2 if $\sin^2 2\theta$ is not too small (≥ 0.1). These kind of experiments look essentially for the disappearance of the reactor ν_e's, i.e. to a reduction in the original ν_e flux. When one uses neutrino beams from accelerators, it becomes possible also to study the disappearance of muon neutrinos into another flavor, and also the appearance of a flavour different from the original one, with the advantage that one becomes sensitive to very small oscillation amplitudes (i.e. small $\sin^2 2\theta$ values), since the observation of only a few events is enough to establish a positive signal. At present there is one experiment (LSND) claiming a positive signal of $\nu_\mu \to \nu_e$ conversion, but this highly suspected result is expected to be clarified unambiguously by the MINI-BOONE experiment at Fermilab during 2005. The appearance of ν_τ's out of a ν_μ beam was searched by CHORUS and NOMAD at CERN without success, but these experiments were only sensitive to squared mass differences larger than $\sim$ eV^2. There are two experiments which have obtained recently solid evidence of neutrino oscillations (K2K and Kamland), but let's however, following the historical evolution, start with the discussion of solar and atmospheric neutrinos and the clues they have given in favor of non-vanishing neutrino masses.

2.3. *Solar neutrinos and oscillations in matter*

The Sun gets its energy from the fusion reactions taking place in its interior, where essentially four protons combine to form a He nucleus. By charge conservation this has to be accompanied by the emission of two positrons and, by lepton number conservation in the weak processes, two ν_e's have to be produced. This fusion liberates 27 MeV of energy, which is eventually emitted mainly (97%) as photons and the rest (3%) as neutrinos. Knowing the energy flux of the solar radiation reaching us ($k_\odot \simeq 1.5$ kW/m^2), it is then simple to estimate that the solar neutrino flux at Earth is $\Phi_\nu \simeq 2k_\odot/27$ MeV $\simeq 6 \times 10^{10}\nu_e$/cm^{2}s, which is a very large number indeed. Since there are many possible paths for the four protons to lead to an He nucleus, the solar neutrino spectrum consists of different components: the so-called *pp* neutrinos are the more abundant, but have very small energies (< 0.4 MeV), the ^8B neutrinos are the more energetic ones (< 14 MeV) but are much less in number, there are also some monochromatic lines (^7Be and *pep* neutrinos), and then the *CNO* and *hep* neutrinos.

Many experiments have looked for these solar neutrinos: the radio-chemical experiments with ^{37}Cl at Homestake and with gallium at SAGE, GALLEX and GNO, and the water Cherenkov real time detectors (Super-) Kamiokande and more recently the heavy water Subdury Neutrino Observatory (SNO)[d]. The result which has puzzled physicists for almost thirty years is that only between 1/2 to 1/3 of the expected fluxes were observed. Remarkably, Pontecorvo [15] noticed even before the first observation of solar neutrinos by Davies that neutrino oscillations could reduce the expected rates. We note that the oscillation length of solar neutrinos ($E \sim 0.1$– 10 MeV) is of the order of 1 AU for $\Delta m^2 \sim 10^{-11}$ eV^2, and hence even those tiny neutrino masses could have had observable effects if the mixing angles were large (this would be the 'just so' solution to the solar neutrino problem). Much more interesting became the possibility of explaining the puzzle by resonantly enhanced oscillations of neutrinos as they propagate outwards through the Sun. Indeed, the solar medium affects ν_e's differ-

---

[d]See the Neutrino 2004 homepage at http://neutrino2004.in2p3.fr for this year's results.

ently than $\nu_{\mu,\tau}$'s (since only the first interact through charged currents with the electrons present), and this modifies the oscillations in a beautiful way through an interplay of neutrino mixings and matter effects, in the so called MSW effect [16].

To see how this effect works it is convenient to write the effective CC interaction (after a Fierz rearangement) as

$$H^{CC} = \sqrt{2} G_F \bar{e} \gamma_\mu (1 - \gamma_5) e \bar{\nu}_{eL} \gamma^\mu \nu_{eL}. \tag{2}$$

Since the electrons in a normal medium (such as the Sun or the interior of the Earth) are non-relativistic, one can see that $\bar{e} \gamma_\mu e \to (N_e, \vec{0})$, where N_e is the electron density, while for the axial vector part one gets $\bar{e} \gamma_\mu \gamma_5 e \to (0, \vec{S}_e)$, which vanishes for an unpolarised medium, as is the case of interest here. This means that the electron neutrinos will feel a potential

$$V_{CC} = \langle e \nu_e | H^{CC} | e \nu_e \rangle \simeq \sqrt{2} G_F N_e \tag{3}$$

(and for the antineutrinos the potential will have a minus sign in front). The evolution of the neutrino states will hence be determined by a Schroedinger like equation of the form (for the case of just two flavor mixing, i.e. $\alpha = \mu$ or τ)

$$i \frac{d}{dt} \begin{pmatrix} \nu_e \\ \nu_\alpha \end{pmatrix} = \left\{ R_\theta \begin{bmatrix} p + \frac{m_1^2}{2E} & 0 \\ 0 & p + \frac{m_2^2}{2E} \end{bmatrix} R_\theta^T + \begin{bmatrix} \sqrt{2} G_F N_e & 0 \\ 0 & 0 \end{bmatrix} + NC \right\} \begin{pmatrix} \nu_e \\ \nu_\alpha \end{pmatrix} \tag{4}$$

where θ is the vacuum mixing angle and

$$R_\theta \equiv \begin{pmatrix} \cos\theta & \sin\theta \\ -\sin\theta & \cos\theta \end{pmatrix}. \tag{5}$$

The terms indicated as NC correspond to the effective potential induced by the neutral current interactions of the neutrinos with the medium, but since these are flavor blind, this term is proportional to the identity matrix and hence does not affect the flavor oscillations, so that we can ignore it in the following (these terms could be relevant e.g. when studying oscillations into sterile neutrinos, which unlike the active ones do not feel NC interactions).

To solve this equation it is convenient to introduce a mixing angle in matter θ_m and define the neutrino matter eigenstates

$$\begin{pmatrix} \nu_1^m \\ \nu_2^m \end{pmatrix} \equiv R_{\theta_m}^T \begin{pmatrix} \nu_e \\ \nu_\alpha \end{pmatrix} \tag{6}$$

such that the evolution equation becomes

$$i \frac{d}{dt} \begin{pmatrix} \nu_e \\ \nu_\alpha \end{pmatrix} = \left\{ \frac{1}{4E} R_{\theta_m} \begin{bmatrix} -\Delta\mu^2 & 0 \\ 0 & \Delta\mu^2 \end{bmatrix} R_{\theta_m}^T + \lambda I \right\} \begin{pmatrix} \nu_e \\ \nu_\alpha \end{pmatrix} \tag{7}$$

where the diagonal term λI is again irrelevant for oscillations. It is simple to show that to have such 'diagonalisation' of the effective Hamiltonian one needs

$$\Delta\mu^2 = \Delta m^2 \sqrt{(a - c2\theta)^2 + s^2 2\theta} \tag{8}$$

$$s^2\theta_m = \frac{s^2 2\theta}{(c2\theta - a)^2 + s^2 2\theta} \tag{9}$$

where $a \equiv 2\sqrt{2}G_F N_e E_\nu / \Delta m^2$ is just the ratio between the effective CC matter potential and the energy splitting between the two vacuum mass eigenstates. It is clear then that there will be a resonant behaviour for $a = c2\theta$, i.e. for

$$\Delta m^2 c2\theta = 2\sqrt{2}G_F N_e E_\nu \simeq \left(\frac{Y_e}{0.5}\right)\left(\frac{E_\nu}{10 \text{ MeV}}\right)\frac{\rho}{100 \text{ g/cm}^3} 10^{-4} \text{ eV}^2, \tag{10}$$

where for the second relation we used that $N_e = Y_e \rho / m_p$, with $Y_e \equiv N_e/(N_n + N_p)$ (for the Sun $Y_e \sim 0.7\text{--}0.8$). One can see that at the resonance the matter mixing angle becomes maximal, i.e. $\theta_m = \pi/4$, while at densities much larger than the resonant one it becomes $\sim \pi/2$ (i.e. one gets for densities much larger than the resonance one that $\nu_e \simeq \nu_2^m$).

In the case of the Sun, one has that the density decreases approximately exponentially

$$\rho \simeq 10^2 \frac{\text{g}}{\text{cm}^3} \exp(-r/h) \tag{11}$$

with the scale height being $h \simeq 0.1 R_\odot$ in terms of the solar radius $R_\odot$. Hence, solar neutrinos, which are produced near the center, will cross a resonace in their way out only if $\Delta m^2 < 10^{-4}$ eV^2, and moreover only if $\Delta m^2 > 0$. For the opposite sign of Δm^2 only antineutrinos, which are however not produced in fusion processes, could meet a resonance in normal matter[e].

One can also associate a width δ_R to the resonance, corresponding to the density for which $|a - c2\theta| \simeq s2\theta$, i.e. $|da/dr|_R \delta_R \simeq s2\theta$. This leads to $\delta_R \simeq h \, \text{tg}2\theta$. This width is useful to characterize the two basic regimes of resonant flavor conversions, which are the adiabatic one, taking place when

---

[e]Alternatively, one can stick to positive values of Δm^2 and consider vacuum mixing angles in the range $0 \le \theta \le \pi/2$, with the range $\pi/4 \le \theta \le \pi/2$ sometimes called the 'dark side' of the parameter space.

the oscillation length in matter is much smaller than the resonance width, i.e. for

$$\frac{4\pi E_\nu}{\Delta\mu^2|_R} = \frac{4\pi E_\nu}{\Delta m^2 s2\theta} < h \ \text{tg}2\theta, \tag{12}$$

and the opposite one, which is called non-adiabatic, for which the resonance is so narrow that the oscillating neutrinos effectively don't see it and hence no special flavor transition occurs at the resonant crossing. The adiabatic condition can be rewritten as

$$\frac{s^2 2\theta}{c2\theta} > \left(\frac{E_\nu}{10 \ \text{MeV}}\right)\left(\frac{6 \times 10^{-8}\text{eV}^2}{\Delta m^2}\right). \tag{13}$$

To better understand the flavor transition during resonance crossing, it proves convenient to write down the evolution equation for the matter mass eigenstates, which is easily obtained as (ignoring terms proportional to the identity)

$$i\frac{d}{dx}\begin{pmatrix}\nu_1^m \\ \nu_2^m\end{pmatrix} = \begin{pmatrix} -\frac{\Delta\mu^2}{4E} & -i\frac{d\theta_m}{dx} \\ i\frac{d\theta_m}{dx} & \frac{\Delta\mu^2}{4E}\end{pmatrix}\begin{pmatrix}\nu_1^m \\ \nu_2^m\end{pmatrix} \tag{14}$$

We see that in the adiabatic case, the off diagonal terms in this equation are negligible and hence during the resonance crossing the matter mass eigenstates remain themselves so that the flavor of the neutrinos changes just following the change on the matter mixing angle with the varying electron density. This adiabatic behaviour is also relevant for the propagation of neutrinos in a medium of constant density (as is sometimes a good approximation for the propagation through the Earth), and in this case the matter effects just change the mixing angle and the frequency of the oscillations among neutrinos. When the propagation is non-adiabatic, the off-diagonal terms in Eq. (14) induce transitions between the different matter mass eigenstates as the resonance is crossed. Indeed the probability of jumping from one eigenstate to the other during resonance crossing for an exponential density profile can be written as

$$P_c(\nu_1^m \to \nu_2^m) = \frac{\exp\left(-\gamma\sin^2\theta\right) - \exp\left(-\gamma\right)}{1 - \exp\left(-\gamma\right)}, \tag{15}$$

where the adiabaticity parameter is $\gamma \equiv \pi h\Delta m^2/E$ (notice that for an electron density varying in a more general way, not just exponentially, it is usually a good approximation to replace h in the above formulas by $|(dN_e/dr)/N_e|_R^{-1})$.

The many observations of the solar neutrino fluxes by the different experiments, which having different energy thresholds are sensitive to the oscillation probabilities in different energy ranges (moreover water Cherenkov detectors can measure the neutrino spectrum directly above their thresholds), and also the non-observation of a possible diurnal modulation induced by the matter effects when neutrinos have to cross the Earth before reaching the detectors during night-time observations, have converged over the years towards the so-called large mixing angle (LMA) solution as the one required to account for the solar neutrino observations. This one corresponds to mixing of ν_e with some combination of ν_μ and ν_τ flavors involving a mass splitting between the mass eigenstates of[f] $\Delta m^2_{sol} = +(7.9^{+0.6}_{-0.5}) \times 10^{-5}$ eV^2 with a mixing angle given by $\tan^2 \theta_{sol} = 0.40^{+0.10}_{-0.07}$. This values imply that the resonance layer is actually at large densities near the center of the Sun, and that it is quite wide, so that matter oscillations are well in the adiabatic regime.

Another crucial result obtained in 2002 was the independent measurement of the CC and NC interactions of the solar neutrinos with the heavy water experiment SNO[17]. The result is that the NC rates, which are sensitive to the three flavors of active neutrinos, are indeed consistent with the solar model expectations for ν_e alone in the absence of oscillations, while the CC rates, which are sensitive to the electron neutrinos alone, show the deficit by a factor ~ 3, indicating that the oscillations have occured and that they convert electron neutrinos into other active neutrino flavors ($\nu_{\mu,\tau}$).

The last remarkable result that has confirmed this picture has been the observation of oscillations of reactor neutrinos (from a large number of japanese reactors) using a huge 1 kton scintillator detector (KAMLAND), measuring oscillations over distances of $\sim 10^2$ km, and the reduction found from expectations just agrees with those resulting from the LMA parameters, and have actually restricted the mass splitting involved to the narrow range just mentioned, as is shown in figure 3 (from the Kamland experiment).

2.4. *Atmospheric neutrinos*

When a cosmic ray (proton or nucleus) hits the atmosphere and knocks a nucleus a few tens of km above ground, an hadronic (and electromagnetic) shower is initiated, in which pions in particular are copiously produced.

---

[f]This number includes also the results from the Kamland experiment[18].

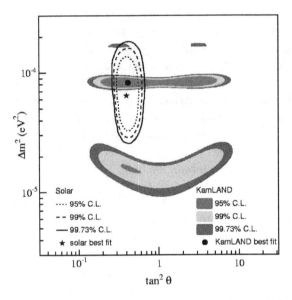

Fig. 3.   Bounds from the KAMLAND experiment. The region favored by solar neutrino observations is that with unfilled contours.

The charged pion decays are the main source of atmospheric neutrinos through the chain $\pi \to \mu\nu_\mu \to e\nu_e\nu_\mu\nu_\mu$. One expects then twice as many ν_μ's than ν_e's (actually at very high energies, $E_\nu \gg$ GeV, the parent muons may reach the ground and hence be stopped before decaying, so that the expected ratio $R \equiv (\nu_\mu + \bar\nu_\mu)/(\nu_e + \bar\nu_e)$ should be even larger than two at high energies). However, the observation of the atmospheric neutrinos by e.g. IMB, Kamioka, Soudan, MACRO and Super-Kamiokande indicates that there is a deficit of muon neutrinos, with $R_{obs}/R_{th} \simeq 2/3$.

More remarkably, the Super-Kamiokande experiment[19] observes a zenith angle dependence indicating that neutrinos coming from above (with pathlengths $d \sim 20$ km) had not enough time to oscillate, especially in the multi-GeV sample for which the neutrino oscillation length is larger, while those from below ($d \sim 13000$ km) have already oscillated (see figure 4). The most plausible explanation for these effects is an oscillation $\nu_\mu \to \nu_\tau$ with maximal mixing $\sin^2 2\theta_{atm} = 1.00 \pm 0.04$ and $\Delta m^2 \simeq (2.5 \pm 0.4) \times 10^{-3}$ eV^2, and as shown in fig. 4 (from the Super-Kamiokande experiment) the fit to the observed angular dependence is in excellent agreement with the oscillation hypothesis. Since the electron flux shape is in good agreement with the

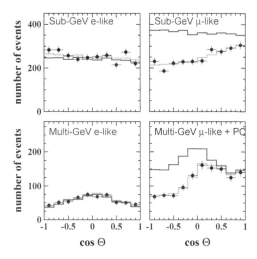

Fig. 4.   Distribution of the contained and partially contained event data versus cosine of the zenith angle ($\cos\theta = -1$ being up-going, while $+1$ being down-going) for two energy ranges, from Super-Kamiokande data. The solid line corresponds to the expectations with no oscillations, while the lighter line is for $\nu_\mu \rightarrow \nu_\tau$ oscillations with maximal mixing and $\Delta m^2 = 0.003$ eV^2.

theoretical predictions[g], this means that the oscillations from $\nu_\mu \rightarrow \nu_e$ can not provide a satisfactory explanation for the anomaly (and furthermore they are also excluded from the negative results of the CHOOZ reactor search for oscillations). On the other hand, oscillations to sterile states would be affected by matter effects (ν_μ and ν_τ are equally affected by neutral current interactions when crossing the Earth, while sterile states are not), and this would modify the angular dependence of the oscillations in a way which is not favored by observations. The oscillations into active states (ν_τ) is also favored by observables which depend on the neutral current interactions, such as the π_0 production in the detector or the 'multi ring' events.

An important experiment which has confirmed the oscillation solution to the atmospheric neutrino anomaly is K2K[20], consisting of a beam of muon neutrinos sent from KEK to the Super-Kamiokande detector (baseline of

---

[g]The theoretical uncertainties in the absolute flux normalisation may amount to $\sim 25\%$, but the predictions for the ratio of muon to electron neutrino flavours and for their angular dependence are much more robust.

250 km). The results indicate that there is a deficit of muon neutrinos at the detector (150.9±10 events expected with only 108 observed), consistent with the expectations from the oscillation solution.

It is remarkable that the mixing angle involved seems to be maximal, and this, together with the large mixing angle required in the solar sector, is giving fundamental information about the new physics underlying the origin of the neutrino masses, which seem to be quite different from what is observed in the quark sector.

2.5. *The direct searches for the neutrino mass*

Already in his original paper on the theory of weak interactions Fermi had noticed that the observed shape of the electron spectrum was suggesting a small mass for the neutrino. The sensitivity to m_{ν_e} in the decay $X \rightarrow X' + e + \bar{\nu}_e$ arises clearly because the larger m_ν, the less available kinetic energy remains for the decay products, and hence the maximum electron energy is reduced. To see this consider the phase space factor of the decay, $d\Gamma \propto d^3p_e d^3p_\nu \propto p_e E_e dE_e p_\nu E_\nu dE_\nu \delta(E_e + E_\nu - E_0)$, with E_0 being the total energy available for the leptons in the decay: $E_0 \simeq M_X - M_{X'}$ (neglecting the nuclear recoil). This leads to a differential electron spectrum proportional to $d\Gamma/dE_e \propto p_e E_e (E_0 - E_e)\sqrt{(E_0 - E_e)^2 - m_\nu^2}$, whose shape near the endpoint ($E_e \simeq E_0 - m_\nu$) depends on m_ν (actually the slope becomes infinite at the endpoint for $m_\nu \neq 0$, while it vanishes for $m_\nu = 0$).

Since the fraction of events in an interval ΔE_e around the endpoint is $\sim (\Delta E_e/Q)^3$ (where $Q \equiv E_0 - m_e$), to enhance the sensitivity to the neutrino mass it is better to use processes with small Q-values, what makes the tritium the most sensitive nucleus ($Q = 18.6$ keV). Experiments at Mainz and Troitsk have allowed to set the bound $m_{\nu_e} \leq 2.2$ eV[21]. It is important to keep in mind that in the presence of flavor mixing, as is indicated by solar and atmospheric neutrino observations, the bound from beta decays actually applies to the quantity $m_\beta \equiv \sqrt{\sum |V_{ei}|^2 m_i^2}$, since the beta spectrum will actually be an incoherent supperposition of spectra (weighted by $|V_{ei}|^2$) with different endpoints, but which are however not resolved by the experimental apparatus. Hence, given the constraints we already have on the mixing angles, and the mass splittings observed, these results already constrain significantly all three neutrino masses.

Regarding the muon neutrino, a direct bound on its mass can be set by looking to its effects on the available energy for the muon in the decay of a pion at rest, $\pi^+ \rightarrow \mu^+ + \nu_\mu$. From the knowledge of the π and μ masses, and measuring the momentum of the monochromatic muon, one can get the neutrino mass through the relation

$$m_{\nu_\mu}^2 = m_\pi^2 + m_\mu^2 - 2m_\pi \sqrt{p_\mu^2 + m_\mu^2}.$$

The best bounds at present are $m_{\nu_\mu} \leq 170$ keV from PSI, and again they are difficult to improve through this process since the neutrino mass comes from the difference of two large quantities. There is however a proposal to use the muon $(g-2)$ experiment at BNL to become sensitive down to $m_{\nu_\mu} \leq 8$ keV.

Finally, the direct bound on the ν_τ mass is $m_{\nu_\tau} \leq 17$ MeV and comes from the effects it has on the available phase space of the pions in the decay $\tau \to 5\pi + \nu_\tau$ measured at LEP.

To look for the electron neutrino mass, besides the endpoint of the ordinary beta decay there is another interesting process, but which is however only sensitive to a Majorana (lepton number violating) mass. This is the so called double beta decay. Some nuclei can undergo transitions in which two beta decays take place simultaneously, with the emission of two electrons and two antineutrinos ($2\beta2\nu$ in fig. 5). These transitions have been observed in a few isotopes (^{82}Se, ^{76}Ge, ^{100}Mo, ^{116}Cd, ^{150}Nd) in which the single beta decay is forbidden, and the associated lifetimes are huge (10^{19}–10^{24} yr). However, if neutrinos were Majorana particles, the virtual antineutrino emitted in one vertex could flip chirality by a mass insertion and be absorbed in the second vertex as a neutrino, as exemplified in fig. 5 ($2\beta0\nu$). In this way only two electrons would be emitted and this could be observed as a monochromatic line in the added spectrum of the two electrons. The non observation of this effect has allowed to set the bound $m_{\nu_e}^{Maj} \equiv |\sum V_{ei}^2 m_i| \leq$ eV (by the Heidelberg–Moscow collaboration at Gran Sasso). A reanalysis of the results of this experiment even suggest a mass

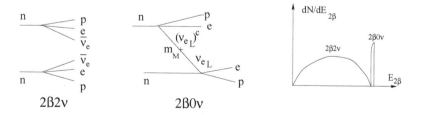

Fig. 5.   Double beta decay with and without neutrino emission, and qualitative shape of the expected added spectrum of the two electrons.

in the range 0.2–0.6 eV, but this controversial claim is expected to be re-explored by the next generation of double beta decay experiments (such as CUORE). There are even projects to improve the sensitivity of $2\beta0\nu$ down to $m_{\nu_e} \sim 10^{-2}$ eV, and we note that this is quite relevant since as we have seen, if neutrinos are indeed massive, it is somehow theoretically favored (e.g. in the see saw models) that they are Majorana particles.

References

1. E. Fermi, Z. Phys. **88** (1934) 161.
2. H. Bethe and R. Peierls, Nature **133** (1934) 532.
3. F. Reines and C. Cowan, Phys. Rev. **113** (1959) 273.
4. G. Gamow and E. Teller, Phys. Rev. **49** (1936) 895.
5. T. D. Lee and C. N. Yang, Phys. Rev. **104** (1956) 254.
6. C. S. Wu et al., Phys. Rev. **105** (1957) 1413.
7. T. D. Lee and C. N. Yang, Phys. Rev. **105** (1957) 1671; L. D. Landau, Nucl. Phys. **3** (1957) 127; A. Salam, Nuovo Cimento **5** (1957) 299.
8. M. Goldhaber, L. Grodzins and A. W. Sunyar, Phys. Rev. **109** (1958) 1015.
9. R. Feynman and M. Gell-Mann, Phys. Rev. **109** (1958) 193; E. Sudarshan and R. Marshak Phys. Rev. **109** (1958) 1860.
10. G. Feinberg, Phys. Rev. **110** (1958) 1482.
11. G. Danby et al., Phys. Rev. Lett. **9** (1962) 36.
12. E. Majorana, Nuovo Cimento **14** (1937) 170.
13. P. Minkowski, Phys. Lett. **B 67** (1977) 421; M. Gell-Mann, P. Ramond and R. Slansky, in *Supergravity*, p. 135, Ed. P. van Nieuwenhuizen and D. Freedman (1979); T. Yanagida, Proc. of the *Workshop on unified theory and baryon number in the universe*, KEK, Japan (1979).
14. Z. Maki, M. Nakagawa and S. Sakata, Prog. Theoret. Phys. **28** (1962) 870.
15. B. Pontecorvo, Sov. Phys. JETP **26** (1968) 984.
16. S. P. Mikheyev and A. Yu. Smirnov, Sov. J. Nucl. Phys. **42** (1985) 913; L. Wolfenstein, Phys. Rev. **D17** (1979) 2369.
17. S. N. Ahmed et al. (SNO Collaboration) Phys. Rev. Lett. **92**: 181301 (2004).
18. T. Araki et al. (Kamland Collaboration) hep-ex/0406035.
19. Y. Ashie et al. (Super-Kamiokande Collaboration) Phys. Rev. Lett **93**: 101801 (2004).
20. M. H. Ahn et al. (K2K Collaboration) Phys. Rev. Lett. **90** 041801 (2003).
21. C. Weinheimer et al., hep-ex/0210050; V. Lobashev et al., Nucl. Phys. Proc. Suppl. **91** (2001) 280.

AN INTRODUCTION TO COSMOLOGY*

D. J. H. CHUNG

Department of Physics
University of Wisconsin
1150 University Ave.
Madison, WI 53706, USA
danielchung@wisc.edu

An introduction to early universe cosmology at an advanced undergraduate level is presented. After a discussion of current problems of cosmology, topics in inflationary cosmology, baryogenesis, and MSSM dark matter are discussed.

1. Introduction

The purpose of these lectures are to present, at a level understandable by advanced undergraduates, the basics, current status, and future prospects of early universe cosmology with regards to its connection to the physics beyond the standard model of particle physics.

The plan of these lectures is as follows. First, we will introduce basics of cosmology at the most elementary level and its current difficulties. Then we will focus on several selected advanced topics: inflationary cosmology, baryogenesis, and dark matter. The context of minimal supersymmetric standard model (MSSM) is emphasized as I believe it will bear fruit in the near future. Throughout these lectures, unless the Planck's constant $M_{pl} = 1.22 \times 10^{19}$ GeV appears explicitly, we will set $M_{pl}/\sqrt{8\pi} = 1$.

2. Basic Cosmology

Cosmology is the study of the origin and large scale structure of the universe. Large scale here corresponds to a scale larger than 10kpc ($\approx 30,000$

---

*This work is supported in part by the US National Science Foundation.

lyr which is our galaxy size) which means that cosmological questions essentially reach their limit when one asks questions that resolve the details of the galactic structure.

Because of the observed expansion of the universe as a function of time, one can infer that the current state of the universe filled with 2.7 °K microwave radiation is a result of cooling and gravitational collapse from a initial condition of hotter and more homogeneous state. Hence, traditional cosmology focuses on reconstructing the gravitational and thermal history of the universe.

By now, cosmology is a well developed branch of physics with many observational foundations. By electromagnetic observations of galaxies, quasars, supernovae, and hot gases, we can measure the redshift as a function of brightness which allows us to infer the Hubble expansion rate. We have very accurately measured the ambient $T_0 \approx 2.73$ °K microwave radiation and its temperature fluctuation spectrum whose amplitude is of order $\frac{\Delta T}{T} \sim 10^{-5}$. We also have data on light element abundance from light emission and absorption spectra. From measurements of electromagnetic radiation from galaxies and gases, we can infer the distribution of visible matter (and thereby baryons), which helps us to understand structure formation. Gravitational lensing observations tell us about the density distribution of clumped invisible matter. We have data on diffuse gamma ray, X-ray, and other electromagnetic spectrum which allows us to infer the astrophysical source of electromagnetic radiation. Finally, we have cosmic rays (including neutrino, proton, positron, antiproton, etc.) which give us a method to probe the astrophysical sources at cosmological distances and to understand the cosmological medium through which the cosmic rays propagate.

The first principles used in constructing the standard model of cosmology are Einstein's equations, Boltzmann equations, and the Standard Model (SM) of particle physics. The fundamental action is

$$S = \frac{-1}{2} \int d^4x \sqrt{g} R + \int d^4x \sqrt{g} L_{SM} \tag{1}$$

where L_{SM} is the standard model Lagrangian. Depending on the kinematic regime, either classical equations or quantum field theory computations implied by this action are used to analyze the dynamics. Because first principle computations of the particle distribution function is impossible, one usually employs the leading hierarchical level truncated Boltzmann equations

$$[p^\alpha \frac{\partial}{\partial x^\alpha} - \Gamma^\alpha_{\beta\gamma} p^\beta p^\gamma \frac{\partial}{\partial p^\alpha}] f_i(x^\alpha, p^\alpha) = C[\{f_j\}] \tag{2}$$

where f_i is the classical phase space distribution function of particles of species i and $C[\{f_i\}]$ is the collision integral describing the scattering of various species of particles interacting. It is in this collision integral where the SM dynamics become important, and it is the initial conditions of f_i that cosmology has historical dependence. SM also enters in the types of particle species composing the set $\{f_i\}$. The Boltzmann equations together with the Einstein equations form a coupled set of dynamical equations defining the cosmological model.

The main idea in constructing the standard model of cosmology is to start from a homogeneous and isotropic background metric and to treat all the inhomogeneities as perturbations. This simplifying assumption is well justified by the cosmic microwave background radiation data. The homogeneous and isotropic metric has the form $ds^2 = dt^2 - a^2(t)[\frac{dr^2}{1-qr^2} + r^2 d\theta^2 + r^2 \sin^2 \theta d\phi^2]$ where q characterizes the spatial curvature through $^{(3)}R = \frac{6q}{a^2}$. Currently, this scalar curvature is bounded to be less than 10^{-44}GeV^2. The corresponding homogeneous and isotropic stress tensor is usually assumed to be of the perfect fluid form

$$T_{\mu\nu} = (\rho(t) + P(t))u_\mu u_\nu + P(t)g_{\mu\nu} \tag{3}$$

$$T_{00} = \rho \qquad T_{11} = a^2 P \tag{4}$$

where ρ is the energy density, P is the pressure, and u_μ is the fluid velocity. It is interesting that we still do not know exactly how to compute $\rho(t)$ and $P(t)$ although the approximation of naive volume average is commonly believed to be a good approximation.

The background Einstein equations is given as

$$H^2 + \frac{q}{a^2} = \frac{\rho}{3} \qquad 2\dot{H} + 3H^2 + \frac{q}{a^2} = -P \tag{5}$$

where $H = \frac{\dot{a}}{a}$ is the Hubble expansion rate and q is a constant parameter.[a] One can combine these equations to write

$$\frac{\ddot{a}}{a} = \frac{-1}{6}(\rho + 3P) \tag{6}$$

which says that since ordinary collection of particles has positive pressure and energy, the universe will be decelerating if our universe was filled with mostly ordinary matter. Note that one of the Einstein equations can be replaced with the stress energy conservation equation $d(\rho a^3) = -Pda^3$. At this point, it is worthwhile reminding the reader that equation of state

---

[a]We are assuming that $M_{pl}/\sqrt{8\pi} = 1$.

of energy density is defined as $w \equiv P/\rho$. For matter dominated universe, $\{w = 0, \rho \propto 1/a^3, a \propto t^{2/3}\}$ and for radiation dominated universe $\{w = 1/3, \rho \propto 1/a^4, a \propto t^{1/2}\}$.

Much of the current data[1] gives the following approximate composition for the energy density of the universe: 73% "dark energy", 22% cold dark matter, 4.4% baryons (protons and neutrons), 0.6% neutrinos, 0.005% photons. The CMB data shows that the spatial curvature is 0 to about 1% accuracy and that the energy density of the universe was homogeneous and isotropic to 1 part in about 10^5, 15 billion years ago. Different cosmological data are consistent with $H = 70$km/s/Mpc.

Typical boundary condition for Boltzmann equations starts with equilibrium distribution functions. The number density, the energy density, and the pressure are defined as

$$n = \frac{g}{(2\pi)^3} \int d^3p f(E) \qquad \rho = \frac{g}{(2\pi)^3} \int d^3p E f(E) \qquad p = \frac{g}{(2\pi)^3} \int d^3p \frac{p^2}{3E} f(E)$$

where $E = \sqrt{p^2 + m^2}$ and for equilibrium distributions, $f(E) = f^{eq}(E) = 1/(e^{(E-\mu)/T} \pm 1)$. The photon temperature is typically defined as the temperature of the universe, and its energy density is given by $\rho_\gamma = \frac{\pi^2}{30}2T^4$. All the relativistic particles are considered "radiation", and its energy density is given by

$$\rho_R = \frac{\pi^2}{30}g_*(T)T^4 \qquad g_*(T) \equiv \sum_{i=bosons} g_i(\frac{T_i}{T})^4 + \frac{7}{8} \sum_{i=fermions} g_i(\frac{T_i}{T})^4$$

with g_i being the number of internal degrees of freedom (e.g. for a photon $g_\gamma = 2$). The temperature of the universe is determined as a function of time or scale factor by assuming that entropy is conserved. Entropy density is defined as $s \equiv \frac{\rho+P}{T} = \frac{2\pi^2}{45}g_{*S}(T)T^3$ where

$$g_{*S}(T) \equiv \sum_{i=bosons} g_i(\frac{T_i}{T})^3 + \frac{7}{8} \sum_{i=fermions} g_i(\frac{T_i}{T})^3$$

counts only the **relativistic** degrees of freedom. Note that $g_{*S} \approx g_*$ when $T > 1$ MeV for the SM particle content.

Equilibrium is reached when the reaction rate Γ satisfies $\Gamma \gg H$. There are two kinds of equilibrium depending on which reaction Γ corresponds to. Kinetic equilibrium of particle species X says $X\{Y\} \rightarrow X\{Z\}$ reaction rate is large where $\{Y\}$ and $\{Z\}$ are in kinetic equilibrium with the photon. X in kinetic equilibrium implies that it maintains the same temperature as the photon and that particle number does not change. The second kind of equilibrium is called chemical equilibrium. This requires Γ to correspond to $X\{Y\} \rightarrow \{Z\}$ reaction where $\{Z\}$ is in chemical equilibrium with the photon. In this case, the X particle maintains the same temperature as the photon and the particle number changes as a function of temperature. When a particle goes out of kinetic equilibrium, it is said to have "decoupled," while when a particle goes out of chemical equilibrium, it is said to have "frozen out."

One more comment regarding temperature is appropriate at this time. Suppose a species X is in equilibrium (with the photons) and then it becomes nonrelativistic at temperature T_{NR} where $T_- < T_{NR} < T_+$ and $T_+(T_-)$ corresponds to the temperature sometime before (after) the particle X becomes nonrelativistic . Conservation of total entropy gives $\frac{T_-}{T_+} = \frac{a(T_+)}{a(T_-)}\left[\frac{g_{*S}(T_+)}{g_{*S}(T_-)}\right]^{1/3}$. However, because species X no longer contributes to the entropy (since nonrelativistic), we find $g_{*S}(T_+) > g_{*S}(T_-)$. Hence, although generically, temperature scales like $T \propto 1/a$, when some particle becomes nonrelativistic, the temperature decreases more slowly as the universe expands. Hence, one can think that there is an effective heating of the photons every time a particle becomes nonrelativistic. For example, suppose electrons become nonrelativistic at temperature T_+ when only the photons and electrons are in kinetic equilibrium while the 3 effectively Majorana neutrinos have already decoupled (but have same temperature as the photons thus far due to neutrino entropy conservation), we have

$$g_{*S}(T_+) = 2 + \frac{7}{8}(3+4) \qquad g_{*S}(T_-) = 2 + \frac{7}{8}(3)(\frac{T_\nu}{T_-})^3 \qquad T_\nu = T_+\frac{a(T_+)}{a(T_-)}$$

where we have assumed that T_- is the temperature of the photons just after the neutrinos have become . Hence, we find

$$\frac{T_-}{T_\nu} = \left(\frac{2 + \frac{7}{8}(3+4)}{2 + \frac{7}{8}(\frac{T_\nu}{T_-})^3}\right)^{1/3}$$

which implies $T_\nu/T_- = (4/11)^{1/3}$.

Although the standard model of cosmology is successful, it presents many puzzles. Some of the most important puzzles can be stated as the following: 1) What is the composition of dark energy? 2) What is the composition of cold dark matter (CDM)? 3) Why are there more baryons than antibaryons? 4) If inflation solves the cosmological initial condition problems, what is the inflaton? 5) How are classical singularities of general relativity resolved? 6) Why is the observed cosmological constant small when SM says it should be big? 7) What is the origin of the ultra-high energy cosmic rays which seem to travel cosmological distances? The rest of these lectures will first give an introduction to these problems, then focus on three selected topics of inflation, baryogenesis, and dark matter, the latter two of which can be easily connected with physics beyond the SM in the context of MSSM.

3. Introduction to the Outstanding Problems of Cosmology

In this section, we discuss the outstanding problems of cosmology.

3.1. *Dark energy*

Recall that the dark energy by definition must cause the universe to accelerate. Because of Eq. (6), acceleration (i.e. $\ddot{a} > 0$) implies $P < \frac{-\rho}{3}$. The reason why this is considered unusual is because cosmologists traditionally thought that the energy density is dominated by gas of particles. Since the pressure for gas of particles is given by

$$\rho(t, \vec{x}) = \frac{1}{3} \sum_n \frac{\vec{p}_n^2}{\sqrt{m_n^2 + \vec{p}_n^2}} \delta^{(3)}(\vec{x} - \vec{x}_n(t))$$

a negative pressure is puzzling. However, from everyday experience, negative pressure is commonly found, for example, in the tension of a tightly stretched rope. Generically, the dark energy is associated with field potential energy. For a homogeneous scalar field, pressure is given by $P = \frac{1}{2}\dot{\phi}^2 + V(\phi)$.

The problem of dark energy are two-fold: 1) Why is the dark energy density nearly coincident with the matter density today? 2) How is the zero point energy contribution to the cosmological constant canceled to an energy density that is much smaller than TeV^4 (which is the natural cutoff in SM)? People have tried to solve question 1 using a dynamical scalar field, but it is difficult to protect the timescale of variation. The second problem still has no acceptable solution other than by saying that there is fine tuning.

3.2. *Cold dark matter*

Currently, cold dark matter is the preferred class of dark matter in standard cosmology from the point of view of structure formation. The "coldness" of dark matter is defined to be the property that the matter is nonrelativistic at the time of matter radiation equality. This is the preferred definition from a structure formation point of view because appreciable gravitational clustering occurs only after matter domination. Cold dark matter cannot all be in cold baryons not emitting light because 1) big bang nucleosynthesis puts an upper bound on the amount of baryons; 2) Microlensing data is consistent with the amount of cold baryons implied by big bang nucleosynthesis. Neutrinos cannot be cold dark matter either because a cold neutrino must be much heavier than a keV which would overclose the universe. Hence, physics beyond the SM is necessary to explain cold dark matter.

3.3. *Baryogenesis*

The next problem on the list is that of baryogenesis: how our observable universe came to be dominated by baryons instead of antibaryons. Current CMB and big bang nucleosynthesis data agree that the photon to baryon ratio is $n_B/n_\gamma \equiv (n_b - n_{\bar{b}})/n_\gamma = 6 \times 10^{-10}$. According to the SM, if this number is in $B + L$, then due to the interaction of sphalerons, the natural number is at most $n_B/n_\gamma \sim 10^{-18}$ and this fluctuation should not prefer only one sign of the baryon number throughout the homogeneous universe.

3.4. *Inflation?*

CMB data currently supports inflationary scenario as there is no observed spatial curvature and the density perturbation spectrum is independent of scale. Just as in dark energy scenario, inflation requires a source of negative pressure. Unlike dark energy, however, the source of negative pressure must be time dependent. At this point, there is no compelling candidate for the inflaton.

3.5. *Singularities of general relativity*

According to the Hawking-Penrose theorem, as long as there is nonzero spacetime curvature somewhere and energy is positive, Einstein's theory will produce a singularity somewhere on the 4-manifold. Although there is no direct evidence for a singularity behind the horizon of a black hole, there is evidence for black hole horizons. Also, classically, the FRW universe today must have evolved from a spacelike singularity: $R = -6[\frac{\ddot{a}}{a} + (\frac{\dot{a}}{a})^2 + \frac{k}{a^2}] \sim \frac{1}{t^2} \to \infty$.

3.6. *Cosmological constant*

Another curvature related problem which comes from our incomplete understanding of quantum interactions with gravity is the cosmological constant problem. According to the SM, there are contributions to the cosmological constant arising from the zero point energy of all of the SM fields. Because the natural scales of cutoff Λ of zero point energy is either the Planck scale (10^{19}GeV), GUT scale (10^{16} GeV), see-saw scale (10^{13} GeV), or the perturbative unitarity scale (10^3 GeV), the predicted cosmological constant of $\rho_{\Lambda(theory)} \sim \Lambda^4$ is much larger than that which is observed: $\rho_\Lambda \sim (10^{-12}\text{GeV})^4$. There is currently no satisfactory explanation of the suppression of zero-point energy contribution to the cosmological constant.

3.7. *Ultra-high energy cosmic rays*

The ultra-high energy cosmic ray puzzle is not purely a cosmological puzzle, but because the propagation distance of the primary is most likely over cosmological distances, we can include this in the cosmological puzzle category. The statement of the puzzle is that we observe cosmic ray primary particles with energy above $10^{19.8}$eV, but we do not see a likely source within 40 Mpc for this high energy cosmic ray. The difficulty with such a scenario is that the photopion production reaction $\gamma p \to \pi p$ threshold is at $10^{19.8}$eV which means that there should be sharp flux cutoff at about that energy referred to as the GZK cutoff energy. Hence, the measured flux of cosmic rays above the GZK cutoff implies that either the primaries are not protons or the sources of these cosmic rays are closer than 40 Mpc. Another question is what the acceleration mechanism for such high energy cosmic rays is since for protons, it is difficult to find astrophysical explanation for energies above 10 eV.

4. Inflation

Now we will switch gears and focus on more details about inflationary cosmology (currently the most favored model of the universe).[2] The primary motivation for inflation is to solve initial condition problems of cosmology. The most notable initial condition problems are 1) Why is the universe flat? 2) Why is universe homogeneous and isotropic on "acausal" scales? 3) Why don't we observe relics from GUT phase transitions?

Among these, the most difficult problem to solve without inflation is problem 1), the flatness problem. Generally, Friedmann equation gives $\frac{q}{H^2 a^2} = \Omega - 1$ where we have defined q to be the spatial curvature parameter. This implies that during BBN

$$\frac{q}{(H_e a_e)^2} = \frac{q}{(H_0 a_0)^2}\left(\frac{a_r}{a_0}\right)\left(\frac{a_e}{a_r}\right)^2 < \frac{q}{(H_0 a_0)^2}(10^{-4})(10^{-6})^2 = \frac{q}{(H_0 a_0)^2}10^{-16}$$

where a_r is the matter-radiation equality time scale factor, a_0 is the scale factor today, and a_e is the scale factor during big bang nucleosynthesis. Since according to WMAP data the spatial curvature today is constrained by $\frac{q}{(H_0 a_0)^2} < 10^{-2}$, we must have $\frac{q}{(H_e a_e)^2}$ smaller than 10^{-18} during BBN. The origin of this number is called the flatness problem. Another way to state this problem is why is the universe old and flat when the only natural scale in the dynamical system is M_{pl}.

Another problem which inflation solves effectively is the horizon problem which can be stated as "Why is the universe homogeneous and isotropic on acausal scales?" To see this problem quantitatively, consider the horizon distance which is given by $d_H = a(t) \int_0^t \frac{dt'}{a(t')} \sim t$. On the other hand, physical density perturbations have wavelengths that scale as

$$a(t)/k \sim t^{2/[3(1+w)]}/k \tag{14}$$

where $w \equiv \frac{p}{\rho}$ is the equation of state and k is the comoving wave vector. Hence, for any collection of particles, we have $w \geq 0$ which implies that sometime in the past, the perturbation wavelength was outside of the horizon. Hence, the question arises why we seem to see approximate homogeneity and isotropy (i.e. strong correlation) on scales that are beyond the horizon distance.

The final commonly quoted initial condition problem for standard cosmology is that of topological defects. Suppose SM is embedded in a theory with a larger gauge group G_1 and as the universe cools from a high temperature, there is a successive breaking of the gauge symmetry down from G_1 to the SM gague group as $G_1 \rightarrow G_2 \rightarrow ... \rightarrow SU(3)_c \times SU(2)_L \times U(1)_Y$. During one of these phase transitions, if $\Pi_2(G_i/G_j) \neq I$ (i.e. nontrivial homotopy of the coset) , then at that symmetry breaking temperature T_c, $n_M/s \sim (T_c^6/M_{pl}^3)/T_c^3 \sim 10^{-15}(\frac{T_c}{10^{14}\text{GeV}})^3$ which yields $\Omega_M \sim 10^{11}$ when the mass of the monopole $m_M \sim 10^{16}\text{GeV}$ and $T_c \sim 10^{14}\text{GeV}$. Here, it was important to assume that there was one monopole produced per Hubble volume $1/H^3$ during the phase transition (this is called the Kibble mechanism).

Inflation solves all these problems by blowing up a small flat patch of the universe into the entire universe. Flat patches being blown up solves the flatness problem. The horizon problem is solved by requiring that $w < -1/3$ for sufficiently long time such that perturbation wavelength that are longer than the horizon size at some time could have been shorter than the horizon size at an earlier time (see Eq. (14)). Finally, even if monopoles are produce with the abundance of one per Hubble volume, since one Hubble patch blows up to become the entire observable universe, there will be only about one monopole in the entire universe.

What makes inflation interesting is not only that it solves three problems through one cosmological era in which $w < -1/3$, it generically predicts a scale invariant density perturbation spectrum. Recent CMB experiments have confirmed the approximate scale invariance of the spectrum. The measured amplitude of the spectrum is $\frac{\delta \rho_k}{\rho} \sim 10^{-5}$.

To achieve $w < -1/3$, people usually construct models based on homogeneous scalar fields. Homogeneous scalar fields are preferred over vectors or fermions because the latter would spontaneously break Lorentz invariance locally. For a single scalar field, $w < -1/3$ can be translated into the constraints on the scalar field effective potential by noting $P = \frac{1}{2}\dot{\phi}^2 - V$ and $\rho = \frac{1}{2}\dot{\phi}^2 + V$. The number of efolds during inflation is read off from $a(t) = a(t_i)\exp[\int_{t_i}^t dt' H(t')]$ where H is determined by the potential V from ϕ equation of motion and Friedmann equation Eq. (5).

Quantitatively, for a single inflaton field ϕ, people usually employ what is called "slow roll approximation" $\ddot{\phi} \ll H\dot{\phi}$ to find analytic solutions to the density perturbation spectrum. This yields the density perturbation spectrum in the long wavelength limit (when $k/a \to 0$) as

$$P^{(\zeta)}(k) \approx \frac{V(\phi_k)}{24\pi^2 \epsilon(\phi_k)} \qquad \epsilon(\phi) \equiv \frac{1}{2}\left(\frac{V'(\phi)}{V(\phi)}\right)^2$$

where ϕ_k corresponds to the value of the inflaton field when $k/a = H$ for a fixed comoving wave vector k of interest.[b] Here, the power spectrum is defined as

$$P^{(\zeta)}(t,k) \equiv \frac{k^3}{2\pi^2}\int d^3 r e^{-i\vec{k}\cdot\vec{r}}\langle\zeta(t,\vec{x})\zeta(t,\vec{y})\rangle \qquad \zeta \equiv -\Phi + \frac{\delta\rho}{3(\rho + P)} \quad (16)$$

where $\vec{r} \equiv \vec{x} - \vec{y}$ and Φ_k corresponds to the metric perturbation in the gauge

$$ds^2 = a^2(\eta)(d\eta^2[1 + 2\Phi(t,\vec{x})] - d\vec{x}^2[1 - 2\Phi(t,\vec{x})]). \qquad (17)$$

If we parameterize

$$P^{(\zeta)}(k) = A_S^2 k^{n-1} \qquad (18)$$

(where the time variable is taken to be when $k/a \ll H$ and A_S^2 is a constant amplitude), one can express the spectral index n as $n - 1 = 2\eta - 6\epsilon$ where $\eta(\phi_k) \equiv V''(\phi_k)/V(\phi_k)$. The beauty in writing the density perturbations in terms of ζ is that this power spectrum is time invariant when $k/a \ll H$.

---

[b]Note here one must remember $M_{pl}/\sqrt{8\pi} = 1$.

Note that sufficient condition for inflation is $\eta, \epsilon \ll 1$ and the number of efolds N of inflation is computed as $N = \int_{\phi(t_i)}^{\phi(t_f)} \frac{d\phi}{\sqrt{2\epsilon}}$ which must satisfy $N > 60.4 + \frac{1}{3}\ln[g_*(t_{RH})g_{*S}(t_{RH})] + \frac{1}{3}\ln(T_{RH}/V_e^{1/4}) + \ln(V_e^{1/4}/10^{16}\text{GeV}) + \ln(H_0/h/k_{phys})$ if inflation should blow up an initial patch of the universe to a size $1/k_{phys}$ today. Here, t_{RH} stands for the time of reheating. Note that for $k_{phys} \sim 1/10^3 \text{Mpc}^{-1}$, current data suggests $P^{(\zeta)}(k) \sim 10^{-9}$.

After inflation ends, the inflaton must decay to reheat the universe to a temperature of at least 10 MeV. If the inflaton interaction to the decay particles is perturbative and the decay rate per particle is given as Γ_{tot}, the differential equations governing the reheating is given approximately as

$$\ddot{\phi} + (3H + \Gamma_{tot})\dot{\phi} + (m^2 + \Gamma_{tot}^2/4)\phi \approx 0$$

$$\rho_\phi \approx \frac{1}{2}(\dot{\phi}^2 + m^2\phi^2) \qquad \dot{\rho}_R + 4H\rho_R = \Gamma_{tot}\rho_\phi.$$

The reheating temperature as a function of time is given through $\rho_R = \frac{\pi^2}{30}g_*(T)T^4$. By the time radiation domination occurs, the reheating temperature is $T_{RH} \approx 0.2(\frac{200}{g_*})^{1/4}\sqrt{\Gamma_{tot}M_{pl}}$.

Note that although the origin of the density perturbations ζ is quantum, the power spectrum can be considered classical because of a peculiar aspect of field theory in curved spacetime. By perturbing the background metric and the inflaton field about the homogeneous solution, one can arrive at an action of the form

$$S = \int d\eta d^3x \frac{1}{2}[v'^2 - (\partial_i v)^2 + \frac{a'/a}{a\phi'}v^2\partial_\eta^2(\frac{a\phi'}{a'/a})]$$

where $v \equiv a(\delta\phi - \frac{\phi'}{a'/a}\Phi)$ and we have defined all of the quantities previously. Here, $\delta\phi$ corresponds to inhomogeneous perturbation of the inflaton field about the homogeneous background field ϕ. Now, when one quantizes in the usual manner as

$$v = \int \frac{d^3k}{(2\pi)^{3/2}}[a_k v_k(\eta) + a^\dagger_{-k}v_k^*(\eta)]e^{i\vec{k}\cdot\vec{x}},$$

we find a differential equation for v_k as

$$v_k'' + [k^2 - \frac{a'/a}{a\phi'}\partial_\eta^2(\frac{a\phi'}{a'/a})]v_k = 0.$$

For long wavelengths, the equation becomes approximately

$$v_k'' - \frac{a'/a}{a\phi'}\partial_\eta^2(\frac{a\phi'}{a'/a})v_k \approx 0$$

which has an exact solution $v_k = A_k \frac{a\phi'}{a'/a}$ + decreasing solution. If we neglect the decreasing solution, we can write

$$-\zeta_k \frac{a\phi'}{a'/a} = a_k v_k(\eta) + a^\dagger_{-k} v^*_k(\eta)$$

where ζ_k is a constant (hence the statement that this is time independent when $k/a \ll H$). Furthermore, one sees that ζ_k is a classical variable since $[\zeta_k, \zeta^\dagger_l] = 0$. Hence, the power spectrum $P^{(\zeta)}(k)$ can be considered classical variables.

Unfortunately, because an inflationary model is not unique, the observation of scale invariant density perturbations does not by itself substantiate the inflationary paradigm. To give nontrivial evidence for the density perturbations to have arisen from inflation, one must also measure tensor perturbations arising from inflation. In contrast to Eq. (17) which defines scalar perturbations, tensor perturbations are defined to be metric perturbations of the form

$$\delta g^{(T)}_{\mu\nu} = \begin{pmatrix} 0 & 0 \\ 0 & h_{ij} \end{pmatrix}$$

where h_{ij} is a symmetric spatial tensor satisfying the constraints $h^i{}_i = 0 = \partial_i h^{ij}$. Intuitively, this corresponds to gravity waves produced during inflation. If the inflationary tensor perturbation power spectrum is defined as $P^{(T)}(k) \equiv A^2_T k^{n_T - 1}$ similarly as Eqs. (16) and (18), single field slow roll inflationary scenario predicts $P^{(T)} = 2\epsilon P^{(\zeta)}$. In particular, this predicts

$$n_T - 1 = -2\epsilon \tag{27}$$

which is known as a consistency relation. Confirming this would give credence to the picture of density perturbations arising from inflation. In more realistic scenarios involving multifield inflation, the consistency relation reduces only to an inequality

$$1 - n_T \geq \frac{P^{(T)}}{P^{(\zeta)}} \tag{28}$$

which still may be used to rule out inflation.

Until now, all of these density perturbations that we considered are called adiabatic or curvature perturbations because the entropy is assumed to be homogeneous and the spatial curvature of comoving hypersurfaces is nonzero. Although we have no time to discuss in depth, there is another class of perturbations called the isocurvature perturbations which are spatial inhomogeneities in the entropy. Recently, people have been interested in cosmological scenarios in which there is a significant transfer of power from isocurvature perturbations to curvature perturbations. Although this type of isocurvature scenario helps to relax constraints on inflationary models, there is a loss of predictivity in such inflationary scenarios.

Other recent developments in inflationary cosmology include models of inflation constructed from string theory and uncertainties in the inflationary vacuum boundary conditions due to trans-Planckian (quantum gravitational) physics. Models of inflation constructed from string theory unfortunately gives no uniqueness nor improvements in fine tuning. The modified vacuum boundary conditions due to trans-Planckian physics is ill motivated and hence does not seem to be likely in playing a role in our universe.

Physics results to anticipate in the future includes experimental determination of the running of the spectral index $n(k)$ and of the polarization power spectrum completely (the B mode polarization, which is the parity even polarization, has not been measured yet). Although we have no time to explain here, the B mode polarization information will allow us to understand the tensor perturbation contribution to the CMB.[3] The tensor contributions are important for determining whether these density perturbations are really arising from inflation or not as explained around Eqs. (27) and (28). Many theoretical problems still remain unsolved such as 1) What is the inflaton? 2) Is de Sitter space stable against back reaction from quantum particle production? 3) What observables give more evidence to the inflationary picture?

5. Baryogenesis

According to both big bang nucleosynthesis and CMB data, there is a predominance of baryons over antibaryons in our universe with a magnitude

$$\frac{n_B}{n_\gamma} \equiv \frac{n_b - n_{\bar{b}}}{n_\gamma} \sim 6 \times 10^{-10}.$$

This number can be simply considered some type of accident had it not been for the fact that in the SM, there is a $B + L$ violating operator[c] which would erase any n_B/n_γ for temperatures above 100 GeV. Furthermore, any generic extensions of the SM contain operators violating $B - L$. Hence, something prevented the universe from reaching high temperatures during its early history, the expected baryon asymmetry today is expected to be

---

[c]B stands for baryon number while L stands for lepton number.

much smaller than 10^{-10}.[d]

Three necessary conditions for creation of baryon asymmetry are called Sakharov criteria. Suppose there is a particle X carrying 0 baryon number which can decay into particle "a" carrying baryon number b_a and particle "b" carrying baryon number b_b. The branching ratios for the decays can be written as

$$r = \frac{\Gamma(X \to a)}{\Gamma_X} \qquad \bar{r} = \frac{\Gamma(\bar{X} \to \bar{a})}{\Gamma_X} \qquad 1-r = \frac{\Gamma(X \to b)}{\Gamma_X} \qquad 1-\bar{r} = \frac{\Gamma(\bar{X} \to \bar{b})}{\Gamma_X}$$

For any decay process of X and $\bar{X}$, we have

$$\Delta B_X = rb_a + (1-r)b_b \qquad \Delta B_{\bar{X}} = -\bar{r}b_a - (1-\bar{r})b_b$$

$$\Delta B = \Delta B_X + \Delta B_{\bar{X}} = (b_a - b_b)(r - \bar{r}). \tag{32}$$

Note that the $(r - \bar{r})$ factor shows that CP violation is required for baryon number generation and $(b_a - b_b)$ demonstrates that B-violation is required. Finally, unless there is out of equilibrium, the inverse decay process will produce $(b_a - b_b)(\bar{r} - r)$ resulting in zero net baryon asymmetry production.

One of the first realistic baryogenesis scenarios that people have thought about was in the context of of grand unified theories where the baryon number violation is provided by the new gauge bosons of the embedding group. Typical computation of the baryon asymmetry in such scenarios involves solving the Boltzmann equations Eq. (2) where we can write

$$C[f] = -\frac{1}{2} \int d\Pi_a d\Pi_b d\Pi_c (2\pi)^4 \delta^{(4)}(p_X + p_a - p_b - p_c) \times$$

$$[|\mathcal{M}|^2_{Xa \to bc} f_X f_a (1 \pm f_b)(1 \pm f_c) - |\mathcal{M}|^2_{bc \to Xa} f_b f_c (1 \pm f_X)(1 \pm f_a)]$$

$$d\Pi_i \equiv \frac{g_i}{(2\pi)^3} \frac{d^3 p_i}{2E_i}$$

where the matrix elements $\mathcal{M}$ contains the short distance physics information for the CP and baryon number violating scattering. These Boltzmann equations can be simply solved if the final states $\{b, c\}$ are in chemical and kinetic equilibrium:

$$\text{chemical: } f_b = f_b^{eq} \qquad\qquad f_c = f_c^{eq}$$

$$\text{kinetic: } f_X = F(t)f_X^{eq} \qquad\qquad f_a = A(t)f_a^{eq}$$

---

[d]If we consider freezeout of $p\bar{p}$ annihilation, the expected proton to photon density ratio is of order 10^{-18}. We would still have to account for how to separate the baryons and the antibaryons such that we observe mostly baryons today.

where $\{A, F\}$ are real functions of time and $f^{eq} = (e^{(E-\mu)/T} \pm 1)^{-1}$. Once this simplification can be made, the number density of particles can be simply solved for through the equation

$$\partial_t \int d^3p f + 3H \int d^3p f = \int \frac{d^3p}{E} C[f].$$

To see how CP violation in the Lagrangian enters the baryon number generation, consider the toy model interaction Lagrangian:

$$\begin{aligned} L_I = &|m|^2(|\phi_1|^2 + |\phi_2|^2) - |m|(e^{i\rho}\psi_1\psi_2 + e^{-i\rho}\bar{\psi}_2\bar{\psi}_1) \\ &+ |M_3|(e^{i\theta_b}\lambda^a\lambda^a + e^{-i\theta_b}\bar{\lambda}^a\bar{\lambda}^a) + |m_{LR}|^2(e^{i\phi}\phi_2\phi_1 + e^{-i\phi}\phi_1^*\phi_2^*). \end{aligned}$$

In this Lagrangian, there is only one physical phase (phase that cannot be removed by field redefinition):

$$\delta_{phys} = \phi - \theta_\lambda - \rho. \tag{37}$$

CP violation in physical processes manifest in interference of transition amplitudes:

$$|\mathcal{M}|^2 = |\mathcal{M}_1 + \mathcal{M}_2 e^{i\delta_{phys}}|^2 = |\mathcal{M}_1|^2 + |\mathcal{M}_2|^2 + 2\text{Re}(\mathcal{M}_1\mathcal{M}_2 e^{-i\delta_{phys}})$$

$$|\mathcal{M}^{CP}|^2 = |\mathcal{M}_1 + \mathcal{M}_2 e^{-i\delta_{phys}}|^2 = |\mathcal{M}_1|^2 + |\mathcal{M}_2|^2 + 2\text{Re}(\mathcal{M}_1\mathcal{M}_2 e^{i\delta_{phys}})$$

Hence, the transition probability for $Xa \to bc$ and $\bar{X}\bar{a} \to \bar{b}\bar{c}$ are different by an interference term involving the physical phase δ_{phys} . (Note that $\bar{X}\bar{a} \to \bar{b}\bar{c}$ can be written as $bc \to Xa$ if CPT is valid).

For example, in Eq. (32), the quantity $r - \bar{r}$ is proportional to an integral involving

$$|\mathcal{M}|^2 - |\bar{\mathcal{M}}|^2 = \text{Re}(\mathcal{M}_1\mathcal{M}_2 e^{i\delta_{phys}} - \mathcal{M}_1\mathcal{M}_2 e^{-i\delta_{phys}})$$

which vanishes unless the non-CP violating part $(\mathcal{M}_1\mathcal{M}_2)$ develops an imaginary part due to virtual states going on shell. Diagrammatically, one can represent this as a "cut" which represents the operation of taking the imaginary part of the diagram. The representation using cuts are theoretically useful as they allow one to estimate the CP violating part of the amplitude.

Several difficulties exist for baryogenesis scenarios involving grand unified theories. Two of the most prominent are 1) any $B + L$ that is generated at temperatures near the GUT scale can be erased by the sphaleron 2) the temperature at which GUT gauge boson mediated baryon number violating interactions are active is generically incompatible with inflation. A similar scenario which does not suffer from these difficulties is called thermal leptogenesis. It has the added attractive feature that light neutrino masses and cosmology favor the existence of lepton number violation. The only uncomfortable aspect of this scenario is that in gravity mediated SUSY breaking scenarios, there is a bound on the maximum temperature in the universe (so-called the "gravitino problem") which strongly constrains the leptogenesis scenario which requires a temperature at least as large as 10^8 GeV in most cases. We will not discuss the thermal leptogenesis further since it is well covered by the lectures of Esteban Roulet.

6. Electroweak Baryogenesis

The thermal (GUT baryogenesis)/leptogenesis that we briefly discussed previously is theoretically attractive but suffers from one important malady common in modern high energy physics: untestability. That is because the resulting baryon asymmetry in leptogenesis can be estimated as

$$\frac{n_B}{n_\gamma} \approx \frac{\delta_{CP} m_\nu M}{(246\text{GeV})^2 g_*} \sqrt{\frac{M}{T_c}} e^{-M/T_c}$$

where δ_{CP} is the CP violating phase akin to Eq. (37) and M is the mass of the heavy particle (e.g. right-handed Majorna mass scale in the thermal leptogenesis). However, because the origin of δ_{CP} involves physics at both the electroweak scale (approximately 246 GeV) and the "intermediate" scale (approximately $M \approx 10^{10}$ GeV), there is little prospect for measuring this phase in a laboratory in the near future.

There is however a baryogenesis scenario which is mostly likely to be testable in the near future: electroweak baryogenesis. The key to testability of this scenario is that all of the microphysics which is responsible for the generation of baryon asymmetry is at an energy scale below 10^3GeV, well within the current technological reach. Indeed, it is very important to note that the electroweak baryogenesis scenario will be well probed by the upcoming LHC experiment in 2007.

The phenomenon of symmetry restoration at high temperatures is generic in the Higgs sector of the SM. What this means is that finite temperature corrections of the form DT^2H^2 (where D is a dimensionless constant and T is the temperature) to the SM Higgs potential makes the global minimum of the potential to be at $\langle H \rangle = 0$ for temperatures far above the critical temperature of $T_c \sim 100$ GeV. Hence, in the SM and its "minimal" extensions, there is undoubtedly a phase transition which can provide the out of equilibrium environment necessary for baryogenesis.

The phase transition is at least weakly first order in character such that the phase transition proceeds through bubble nucleation where the bubble wall separates the broken phase ($\langle H \rangle = 246$GeV) and the unbroken phase ($\langle H \rangle = 0$). Furthermore, although CP violation in the SM is too small to explain the baryon asymmetry (the dimensionless number characterizing the CP violation in minimal SM is around 10^{-22} given by

$$\delta_{CP}^{(SM)} = (\frac{g_W}{2m_W})^{12}(m_t^2 - m_u^2)(m_t^2 - m_c^2)(m_c^2 - m_u^2)(m_b^2 - m_d^2)(m_b^2 - m_s^2)(m_s^2 - m_d^2)j$$

$$j \equiv \Im(V_{cs}V_{us}^*V_{ud}V_{cd}^*) \,,$$

there are plenty of new parameters in physics beyond the SM that can provide sufficient CP violation. Lastly, the baryon number violation ingredient is provided by the $SU(2)_L$ sphalerons, which are unstable Higgs-gauge field configurations which interpolate between two different vacua characterized by different Chern-Simons number. The difference of Chern-Simons number can be associated with change in baryon number of the system.

To get a feeling for how baryon asymmetry is generated, consider a one generation model with just $\{u, d\}$ quarks. During the first order phase transition, the quarks scatter off the EW phase transition bubble walls and pick up an asymmetry in the left handed quarks due to CP violation in the *true vacuum* (broken phase): $n_b^L - n_{\bar{b}}^L \neq 0$. This does not mean that baryon asymmetry is generated in the true vacuum since in the true vacuum, the baryon number violation is exponentially suppressed with the exponent proportional to the height of the potential barrier. Indeed, in the broken phase, the baryon asymmetry in the left handed sector is canceled by the opposite baryon asymmetry in the right handed sector. Hence, at this point, for example, there can be net quark content of $u_R\bar{u}_L$ giving a net baryon number of 0. The quarks then diffuses out in front of the bubble wall into the false vacuum (unbroken phase) and is processed by the unsuppressed sphaleron to be converted into net total baryon number. For example, our original configuration of $u_R\bar{u}_L$ diffuses to the unbroken phase and the sphalerons induce $\bar{u}_L \to d_L d_L \nu_e$ giving a final net configuration of $u_R d_L d_L \nu_e$ which is baryon number 1. The net total baryon number will however relax to zero if the sphaleron processes are allowed to equilibrate. For example, we can have $d_L \to \bar{u}_L \bar{d}_L \bar{\nu}_e$ which results in $u_R d_L \bar{u}_L \bar{d}_L \bar{\nu}_e \nu_e$ giving a net baryon number of 0. Hence, the broken phase bubble must sweep over the net baryon number and protect it.

One well motivated model beyond the SM is Minimal Supersymmetric Standard Model (MSSM).[4] The additional CP violating phase is provided by, for example, "extra" charged fermion sector of winos $\tilde{W}$ and Higgsinos $\tilde{h}$ (superpartners of W and Higgs) through

$$\Delta L = \frac{1}{2} M_2 \tilde{W}_R^\dagger \tilde{W}_L + \mu \tilde{h}_R^\dagger \tilde{h}_L + h.c.$$

which provides a rephasing invariant phase of the form $\Im(M_2\mu)$. Furthermore, MSSM provides the added benefit of making the electroweak phase transition sufficiently strongly first order to produce sufficient amount of baryon asymmetry. This enhancement of the phase transition arises from the fact that the finite temperature effective potential for the lightest Higgs particle has the $\Delta L = ETH^3$ term (where T is the temperature, H is the Higgs field, and E is a dimensionless number) enhanced due to extra light fields which give loop contributions to the coefficient E. A larger cubic term enhances the baryon asymmetry since this term determines the height of the potential barrier protecting the baryon number from erasure once it is formed.

The computation for the baryon asymmetry is performed schematically as follows.[5] First, write down the (s)quarks and Higgs(inos) diffusion equations. Second, assume that the Yukawa interactions and strong sphaleron interactions are much faster than all other processes. Third, solve the (s)quarks and Higgs(inos) diffusion equations for the $SU(2)_L$ charged left handed fermions assuming that the $SU(2)_L$ sphaleron rate is slow enough to neglect it compared to the remaining diffusion processes. Finally, integrate the sphaleron transition sourced by above. The resulting answer for the baryon asymmetry is given as approximately

$$n_B \sim \frac{c_1 \Gamma_{EW}}{v_w} \int_{-\infty}^{0} dz\, n_L(z) e^{c_2 z \Gamma_{EW}/v_w}$$

$$n_L(z) \sim \frac{e^{f_1 z}}{f_3} \int_{0}^{\infty} dx\, S_H(x) e^{-f_2 x}$$

where $c_{1,2}$ are dimensionless numbers and $f_{1,2}$ are constant quantities characterizing the true vacuum bubble wall profile. The reaction rate Γ_{EW} refers to the unsuppressed electroweak sphaleron transition rate:

$$\Gamma_{EW} = 6(k\alpha_w)\alpha_w^4 T$$

where $k\alpha_W \sim O(1)$. The variables z and x refer to the same one dimensional spatial direction perpendicular to the bubble walls where the bubble wall is located at $z = 0$ (or $x = 0$). The quantity S_H is a Higgs (Higgsino) source term which has the functional form

$$S_H \sim D_h \frac{\Im(M_2\mu)}{M_2^2 + \mu^2} \partial_z^2 [(v_1 \partial_z v_2 - v_2 \partial_z v_1)(F_1(z)\partial_z m_1 + F_2(z)\partial_z m_2)]$$

where v_1 and v_2 are the vacuum expectation values of the two Higgs doublets; m_1 and m_2 are chargino mass eigenvalues; F_1 and F_2 are space dependent functions associated with diagonalizing the chargino mass matrix. It is simple to estimate the final answer as

$$\frac{n_B}{n_\gamma} \sim \frac{(k\alpha_w)\alpha_w^4 \delta_{CP}}{g_*} f$$

where δ_{CP} is the rephasing invariant dimensionless quantity characterizing CP violation, g_* is the number of relativistic degrees of freedom during electroweak phase transition, and f is a factor that characterizes the variation of the Higgs expectation value in a moving bubble wall. Typically, since $k\alpha_w^5 \sim 10^{-6}$ and $g_* \sim 10^2$, to achieve successful baryogenesis one must arrange for $\delta_{CP} f \sim 10^{-2}$.

The resulting allowed parameter space for MSSM is as follows:

(1) Strong enough phase transition: $m_{\tilde{t}_R} < m_t$, $A_t \leq 0.4 m_Q$, $m_h < 115$GeV.
(2) Avoid charge and color breaking minimum: 120 GeV $< m_{\tilde{t}_R}$.
(3) $m_h > 114$ GeV as LEP data suggests: $0.2 m_Q \leq A_t$, $\tan\beta > 4$, $m_Q > 1$ TeV.
(4) Sufficient diffusion: $\mu, M_{1,2} < m_Q$
(5) Sufficient CP violation: $\Im(\mu M_{1,2})/T_c^2 > 0.05$
(6) Sufficient density processed by sphaleron: $\mu, M_{1,2} < 2T_c$.
(7) Electric dipole moment bounds of e, n, and Hg: $\text{Arg}(M_2\mu) < 0.05$

Note that conditions 1, 4, 5, and 6 are bounds required by sufficient baryogenesis while the rest of the conditions are bounds obtained from laboratory experiments. Particularly because of the conflict between conditions 5 and 7, electroweak baryogenesis in the MSSM has a very small parameter space that is still viable. However, minimal extensions to MSSM allow viable electroweak baryogenesis scenarios.[6]

7. Future Prospects and Dark Matter

Both high energy physics and standard big bang cosmology suffer from unresolved puzzles they present. In the SM, the unresolved questions are as follows: 1) Why is the Higgs field light? 2) What is the origin of electroweak symmetry breaking? 3) Is it simply an accident that the gauge couplings seem to meet? 4) How is gravity incorporated into the SM? 5) Why is the CP violation from QCD small? It has been the hope of many physicists like me that the search for answers to these questions can be complemented by the search for answers to cosmological questions presented in section 3. Although there has been progress in connecting cosmology and high energy physics, much remains to be done. Here, we will briefly describe one particular speculation that is believed to be very promising by many researchers.

To make any progress in connecting the two sets of problems, a guess must be made about the physics beyond the SM. A well motivated guess about physics beyond the SM is MSSM which solves two of the SM problems: Why the Higgs is light and whether or not the meeting of the gauge coupling constants is an accident. Furthermore, with judicious boundary conditions, it explains the origin of electroweak symmetry breaking.

MSSM makes two "generic" cosmological predictions for cosmology: 1) Enhancement of conditions necessary for electroweak baryogenesis 2) Existence of new cold dark matter candidates. Although both of these predictions do not apply to all viable MSSM models, both are natural enough to be considered generic. We have already discussed the MSSM contribution to the electroweak baryogenesis. Let us now discuss its dark matter implications.

Supersymmetrizing the minimal SM introduces dimension 4 and 5 operators which break baryon number and lead to unacceptable proton decay rate. Fortunately, in the context of supergravity, there is a natural discrete gauge symmetry called R-parity which can eliminate these operators. If one accepts this proton decay solution, then R-parity implies that the lightest supersymmetric particle (LSP) odd under R-parity can be a dark matter candidate. An interesting coincidence is that any particle with mass of order 100 GeV with electroweak scale interactions will give $\Omega_0 \sim O(1)$ [7] and that MSSM provides precisely such LSPs. Because of the current experimental bounds on $U(1)_{EM}$ charged dark matter, the LSP dark matter must be a neutral particle: either a neutralino or a sneutrino. Neutralino is a linear combination of neutral wino, bino, and two neutral Higgsinos.

Currently much work is in progress in trying to "detect" the LSP dark matter.[4] In addition to looking for the dark matter particle production in colliders, one can set up direct elastic scattering recoil detectors (direct detection method) and detectors sensitive to secondary particles produced far away from the detector environment (indirect detection method).

Direct detection methods[8,9] utilize elastic scattering of ambient dark matter particles with large atomic number atoms (such as Ge). The main difficulty in the direct detection method is the smallness of event rate compared to background noise (e.g. scattering of neutrons from ambient radioactive materials). One method to enhance the signal/background ratio is to utilize the fact that sun has a velocity with respect to our galactic center and that when the earth's orbital motion is parallel to the sun's motion the effective flux velocity will be different from from when the earth's orbital motion is antiparallel to the sun's motion. Locking in on the time variation can significantly increase the signal to noise ratio. Other types of time variation include diurnal modulation arising from the Earth's rotation. Theoretical uncertainties of the dark matter matter direct detection is about a factor of 10: 1) a factor of 4-5 arises from unknown local density 2) a factor of 2-3 arises from uncertainties in the nuclear physics of the detector.

The most promising indirect detection method is through the detection of secondary neutrinos produced by the annihilation products of the dark matter in the sun. The dark matter collects in the sun due to elastic scattering off the nucleons of the sun. The collected dark matter can annihilate in the sun and the cascade of final products include neutrinos (e.g. $\chi\chi \to W^+W^- \to \mu^-\bar{\nu}_\mu\mu^+\nu_\mu$). The neutrinos can easily escape the sun's environment while the charged final state particles are trapped in the sun due to charged interactions. Interestingly enough the rate of neutrino production from the sun due to annihilations may not be proportional to the annihilation rate if the capture rate is large: the neutrino production event rate can be written as $\Gamma_A = \frac{C}{2}\tanh^2(t\sqrt{CC_A})$ where C is the capture rate and C_A is the annihilation rate. The theoretical uncertainty in the indirect detection is similar to the direct detection (about a factor of 10) due to the uncertainties in the local density of dark matter.

Other indirect detection probes include gamma rays, radio waves, and charged antimatter. For example, currently, the HEAT experiment reports an "excess" of positrons at an energy of around 10 GeV which may be explained by LSP annihilation if the dark matter density profile is modified in a nonstandard way and if the MSSM parameter space is tuned sufficiently. These positron signals have a greater uncertainty in theoretical modeling (as much as 10^3) because it depends sensitively on the dark matter density at the center of our galaxy.

It is interesting to note that in some region of the parameter space (e.g.

large Higgsino component and LSP is heavy), the only detection method for dark matter is to look for neutralino annihilation line signal: $\chi\chi \to \gamma\gamma$ and $\chi\chi \to \gamma Z$. It is also interesting to note that even if LSP is not the dominant CDM (say 1% of the dark matter is composed of the LSPs), the direct detectors and neutrino telescopes may detect the LSP.

8. Conclusions

In these two short lectures, we have discussed the basic problems of early universe cosmology and touched upon some topics of current interest, including inflation, baryogenesis, and supersymmetric dark matter. There are many speculations for physics beyond the SM, but because supersymmetry is agreed to be the best motivated by many physicists, we have focused on the MSSM and its implications for the electroweak baryogenesis and dark matter. There is much to be done in connecting cosmology and high energy physics, and we are at an exciting time in scientific history when the next generation of collider experiments will allow us to gain new fundamental understanding of cosmology and the next generation of cosmological experiments will allow us to gain new fundamental understanding of high energy physics.

References

1. D. N. Spergel *et al.* [WMAP Collaboration], Astrophys. J. Suppl. **148**, 175 (2003) [arXiv:astro-ph/0302209].
2. D. H. Lyth and A. Riotto, Phys. Rept. **314**, 1 (1999) [arXiv:hep-ph/9807278].
3. A. Kosowsky, Annals Phys. **246**, 49 (1996) [arXiv:astro-ph/9501045].
4. D. J. H. Chung, L. L. Everett, G. L. Kane, S. F. King, J. Lykken and L. T. Wang arXiv:hep-ph/0312378.
5. M. Carena, M. Quiros, M. Seco and C. E. M. Wagner, Nucl. Phys. B **650**, 24 (2003) [arXiv:hep-ph/0208043].
6. S. J. Huber and M. G. Schmidt, Nucl. Phys. B **606**, 183 (2001) [arXiv:hep-ph/0003122].
7. E. W. Kolb and M. S. Turner, *The Early Universe*, Redwood City, USA: Addison-Wesley (1990) (Frontiers in physics, 69).
8. G. Bertone, D. Hooper and J. Silk, Phys. Rept. **405**, 279 (2005) [arXiv:hep-ph/0404175].
9. G. Jungman, M. Kamionkowski and K. Griest, Phys. Rept. **267**, 195 (1996) [arXiv:hep-ph/9506380].

INNOVATIVE EXPERIMENTAL PARTICLE PHYSICS THROUGH TECHNOLOGICAL ADVANCES — PAST, PRESENT AND FUTURE

H. W. K. CHEUNG

Fermi National Accelerator Laboratory,
P.O. Box 500,
Batavia, IL 60510-0500, USA
cheung@fnal.gov

This mini-course gives an introduction to the techniques used in experimental particle physics with an emphasis on the impact of technological advances. The basic detector types and particle accelerator facilities will be briefly covered with examples of their use and with comparisons. The mini-course ends with what can be expected in the near future from current technology advances. The mini-course is intended for graduate students and post-docs and as an introduction to experimental techniques for theorists.

1. Introduction

Despite the fancy title of this mini-course, the intention is to give a brief introduction to experimental particle physics. Since there are already some excellent introductions to this topic and some textbooks that cover various detectors in detail, a more informal approach to the topic is given in this mini-course. Some basic detector elements are covered while reviewing examples of real experiments, and experimental techniques are introduced by comparing competing experiments. Some aspects of experimental design are also briefly reviewed. Hopefully this will provide a more engaging introduction to the subject than a traditional textbook. This short mini-course cannot replace a real experimental physics course; the reader is just given a taste. Unfortunately the lack of space for this writeup means that not even the basic detection methods and detector types can be described. Instead detector types in *italics* will be briefly described in a glossary at the end of this writeup. For further reading, the reader can find the relevant physics of particle interactions and detailed descriptions of many different types of particle detectors in a number of textbooks and articles.[1]

Although this mini-course is devoted to the physics impact of some significant technological advances, it should be noted that the improvements in experimental techniques usually progress in steady steps. Quite often advances are linked to steady progress in the following areas:

- Higher energy available or/and higher production rate.
- Improvements in momentum or/and position resolution.
- Better particle identification methods.
- Increase in detector coverage or energy resolution.
- More powerful signal extraction from background.
- Higher accuracy (due to increase in data statistics, reduction of experimental systematic uncertainties, or reduction in theoretical uncertainties).

Discoveries are often made through a series of incremental steps, though of course the discoveries themselves can be in a surprising direction! The topics I have chosen for the two lectures of this mini-course is the discovery and subsequent study of the charm quark, and the future of bottom quark physics. The outlines for the two lectures are illustrated in Fig. 1.

2. Part I: Discovery of Charm

The discovery of the J/ψ meson is well documented by many books and articles[2] as well as in the Nobel lectures of Ting[3] and Richter.[4] Besides being a great classic story of discovery, we can also use it to illustrate some of the detection techniques and the physics and ideas behind the design of the experiments involved.

2.1. *A missed opportunity: Resolutions matter!*

Since the leptons, electrons and muons, are basically point-like, stable or long-lived, and interact primarily via the well understood and calculable electroweak force, they have served as the "eyes" in probing many experimental processes. One of those processes under study in the 1970's was hadron interactions. The interests in this study included the investigation of the electromagnetic structure of hadrons, the study of the then-called "Heavy photons" ρ, ω and ϕ and the search for additional ones, as well as the search for the neutral intermediate vector boson, the Z^0.

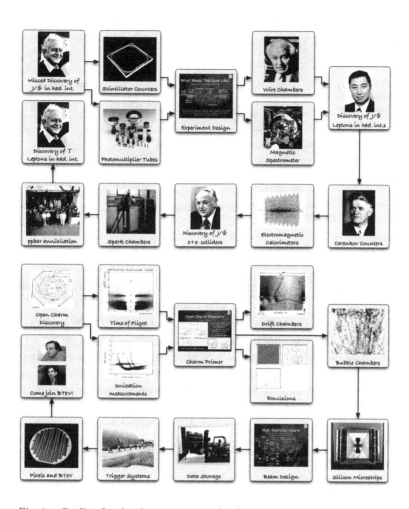

Fig. 1.   Outline for the the mini-course: (top) lecture 1; (bottom) lecture 2.

One experiment doing such a study offers a lesson on the importance of experimental resolution. This was an experiment at Brookhaven National Laboratory (BNL) using the Alternating-Gradient Synchrotron (AGS) carried out by Leon Lederman's group. They performed studies of $p + U \to \mu^+ \mu^- X$ and missed discovering the J/ψ in 1970, four years before the actual discovery.

A diagram showing Lederman's 1970 experiment is given in Fig. 2.[5,6] The experiment was to study the interaction of the 22-30 GeV proton beam on a Uranium target. The aim was to detect a pair of oppositely charged muons coming from the interaction.

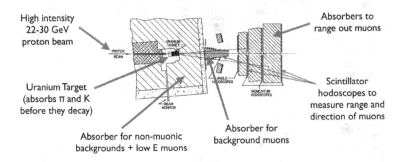

Fig. 2.   Diagram of the spectrometer for the Lederman 1970 experiment.[5]

The emphasis of this experiment was to get a clean signature for muons directly produced in the target. The main background to eliminate was muons from the decay of pions and kaons. A high atomic number target like Uranium has a short interaction length which serves to both cause a lot of the proton beam to interact and also to absorb pions and kaons produced in the interactions before they can decay. This is followed by additional material to absorb non-muonic backgrounds and low energy muons. Muons from hadron decay typically have lower energy than those directly produced in the primary proton-Uranium interaction. Another specially shaped heavy absorber serves to absorb more background muons while *scintillator hodoscopes* measure the direction of the surviving muons. The final material at the end of the detector serves to measure the range and therefore the energy of the muons.

Although all the absorber material helps to give a much cleaner sample of dimuon events, it also causes a lot of multiple Coulomb scattering (MCS), especially as the material is of high Z and therefore has short radiation length. This large MCS limited the dimuon mass resolution at 3 GeV/c^2 to about $\approx 13\%$, or ≈ 400 MeV/c^2. Even with all the absorber the signal-to-background (S/B) is relatively small. The S/B at low dimuon mass (≈ 2 GeV/c^2) was about 2%, increasing up to 50% at higher dimuon mass (≈ 5 GeV/c^2). This means relatively large background subtractions are needed. Another concern was the low acceptance and efficiency at low dimuon mass which therefore needed larger corrections.

The raw and corrected dimuon mass distributions are given in Fig. 3.

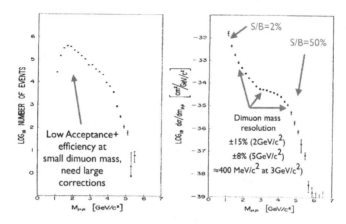

Fig. 3.   Results for the dimuon mass spectrum from the Lederman 1970 experiment.[5] (left) raw distribution; (right) corrected spectrum.

Although the large background subtraction and the uncertainty in the correction might have contributed to the missed discovery of a peak at the J/ψ mass, if the dimuon mass resolution were sufficiently better, the J/ψ peak would still have been observed. The experimenters did many tests and gave limits for a narrow state, but they had to conclude in the end that there was "no forcing evidence of resonant structure."

2.2. *Elements of experimental design*

With hindsight how would we change Lederman's 1970 experiment so we could observe the J/ψ? Instead of leaving it as a task to the reader, it is instructive to go through this in a little detail. The most obvious things to include in a redesign are the following:

(i) Improve the momentum resolution, which means having less material for MCS, using a magnet for momentum determination and using a finer spatial resolution detector than a scintillator hodoscope.

(ii) Increase the S/B, which means separating muons better from hadrons and enriching the sample of dimuons *vs.* single muons.

(iii) Achieve better acceptance and efficiency, which for a study of the dimuon mass spectrum means obtaining a flatter efficiency as a function of dimuon mass. A smooth efficiency across the dimuon mass is probably fine as long as the efficiency (correction) is well understood.

The average angular deflection due to MCS of directly produced (signal) muons is given by $\theta_{\text{MCS}} \sim (Z_{\text{Target}}/p_\mu)\sqrt{L_{\text{Target}}} \sim (1/p_\mu) \times (L_{\text{Target}}/\lambda_0)$. Where λ_0 is the radiation length. So to reduce the effects of MCS one should select a short target with long radiation length and use as high a beam energy as possible to produce more higher momentum signal muons. Targets with low Z/A will have longer radiation lengths but they also have lower density and thus a longer target would be needed to get the same number of inelastic proton interactions in the target. A large signal sample needs a target with high atomic number, since the dimuon signal rate $\sim A_{\text{Target}}$. Another consideration is that the absorption probability for pions and kaons $\sim A_{\text{Target}}^{0.7}$. Thus the S/B would increase with heavier targets and dense targets. One would need to do a Monte Carlo simulation to study what target material is optimal.

If the effects of MCS can be sufficiently reduced we would need to determine the momentum of the muon more precisely. This can be achieved with a *magnetic spectrometer* where the deflection in a known magnetic field can give the magnitude of the momentum. The angle of deflection can be obtained with low mass *Multiwire Proportional Chambers* (*MWPC's*) placed before any of the hadron absorbers.

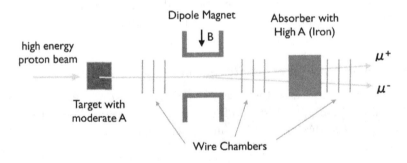

Fig. 4.   Schematic for one possible redesign of a detector studying dimuons where the main consideration was with improving the dimuon mass resolution.

An initial redesign of the detector might look something like the schematic in Fig. 4. However this does not have all the absorbing material of the 1970 Lederman detector to reduce background muons. For that experiment the S/B~ 0.04, while in our initial design it could be as small as 10^{-6}! To see how the S/B could be improved one has to consider the sources of background. The main ones are given below:

(i) Direct single muons – these should be relatively small at AGS energies since the production would be through electroweak processes. Production via the decay of τ leptons or charm particles is of the same level as the J/ψ, so getting an accidental dimuon pair through these decays should cause a negligible background.

(ii) Muons from decays of hadrons – these happen early due to an exponential decay and should therefore be absorbed early before they can decay. Also lower momentum hadrons will decay relatively sooner and thus make up a larger fraction of the decay muon background. One could try to reject softer muons from the data analysis. Also one could make multiple measurements of the momentum to reject muons from decay in flight.

(iii) Hadrons from "punch through" – a signal in a detector element placed after an absorber can arise due to the end of a hadronic shower leaking through the absorber. One can try to detect this by having multi-absorber/detection layers which can be used to recognize a hadronic shower signal from a typically minimum ionizing muon signature. One can also try to momentum analyze through a magnetic absorber.

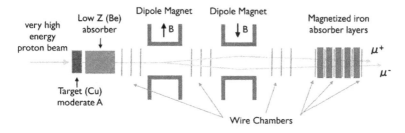

Fig. 5.   Schematic for a second possible redesign of a detector studying dimuons where S/B was included as a consideration as well as the dimuon mass resolution.

Another example of a revised design for the dimuon detector, this time also taking into consideration the S/B is shown in Fig. 5. It is seen that to do better than the 1970 Lederman experiment one needs a more complicated detector with considerably more advanced detectors. Even so one can see there is still a compromise made between getting the best S/B and the best dimuon mass resolution. One would need to do a serious Monte Carlo simulation to determine the optimal choices.

We have only really touched on the elements of experimental design. For example timing considerations have been completely ignored and we have assumed the wire chambers can handle the necessary rates. Instead of pursuing this further, and also before showing you Lederman's solution, we first turn to see how Ting solves this experimental design problem.

2.3. *Ting's solution*

It was recognized that the same physics could be studied by observing pairs of electrons instead of dimuons. Electrons can be produced by the same decays and have the same J^{PC} as muons, thus dielectrons should also be produced by the $J^P = 1^-$ ρ, ω, ϕ and J/ψ. However electrons differ in that they are about 200 times lighter than muons. This greatly changes the considerations for a detector designed to measure dielectron pairs compared to dimuon pairs. Although kaon and pion decays are no longer a serious source of background for a study of dielectrons, the electrons undergo much more scattering and absorption than muons. Thus the choice of materials and the detector types used to identify and track electrons is quite different.

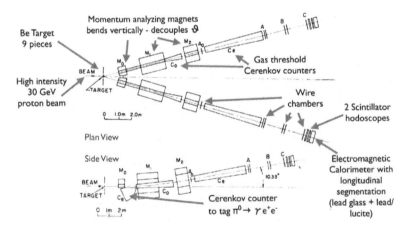

Fig. 6.   Schematic of the detector for Ting's J/ψ observation experiment.[3]

Figure 6 shows a schematic of the spectrometer used by Ting in his J/ψ observation experiment. It can be seen that the spectrometer is quite complex. Low Z beryllium targets and low mass *MWPC's* are used to avoid too many photons converting to e^+e^- pairs. The use of a *multi-magnet spectrometer* and *MWPC's* helped to achieve a very fine mass resolution of

≈ 5 MeV/c^2. Without electron identification the S/B would have been $\sim 10^{-6}$, thus a relative background rejection of 10^6–10^8 was needed. Typically a single particle identification detector can achieve a relative background rejection of 10^2–10^3 so multiple systems were combined. Both *Čerenkov counters* and *electromagnetic calorimeters* were used to identify electons. A special *Čerenkov counter* was used to specifically reject background from $\pi^0 \rightarrow \gamma e^+ e^-$. A relative background rejection of 10^8 was achieved and, together with a fine dielectron mass resolution, a spectacularly narrow and clean J/ψ signal was seen. The results and details of how this analysis was done are well documented in Ting's Nobel lecture[3] and the published papers.[7]

2.4. *Richter's solution*

There is another half of the J/ψ discovery story that cannot be covered in this writeup because of insufficient space. Revealed in that half would be additional important experimental techniques. For example Richter's observation of the J/ψ was made in an $e^+ e^-$ collider, a relatively new innovation at that time. A nearly 4π detector was used including *wire spark chambers* and *electromagnetic shower counters*. The J/ψ mass resolution was much better since it was governed by knowledge of the beam energy and thus the widths of states can be much better measured. That half of the story and the subsequent studies are well documented in Richter's Nobel lecture.[4]

2.5. *Improving charmonium spectroscopy*

An $e^+ e^-$ collider is an excellent study tool. This was recognized by Ting as well as by Richter. It is specially well suited to perform detailed studies of vector particles once their mass is known. This has been the case for charmonium, for bottomonium and for the Z^0. Narrow states with unknown masses are difficult to find. However special modifications were made to the SPEAR $e^+ e^-$ storage ring to enable scans in energy in a relatively short time. This enabled the discovery of the J/ψ by Richter's team as well as some of the charmonium excited states. A disadvantage is that the $e^+ e^-$ collisions can only directly produce states with $J^P = 1^-$, thus only these are measured with fine resolution. While some of the non-($J^P = 1^-$) charmonium states could be observed through the decays of the ψ', see Fig. 7, the measurements of their masses and widths can no longer be obtained with just the knowledge of the beam energies.

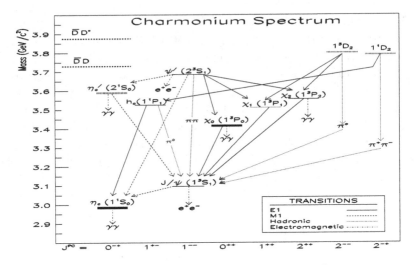

Fig. 7.   Schematic of the charmonium energy levels.

Further improvement in the knowledge of the charmonium spectrum has been achieved by using low energy $p\bar{p}$ collisions in an antiproton accumulator. The first of these experiments was done at the CERN ISR in R704, then in E760 and E835 at the Fermilab antiproton accumulator.[8] In the Fermilab experiments, a hydrogen gas-jet target is used and the antiproton beam is tuned to produce and precisely measure charmonium states of any J^P. The charmonium states are tagged by their electromagnetic decays using *lead glass shower counters* and *scintillating fibres*. All that is needed is to recognize signal from background. The actual mass and width measurements is determined with exquisite ($\sim 0.01\%$) resolution due to excellent knowledge of the beam energy.

2.6. *Lederman's two solutions*

We conclude the first part of the mini-course with two of Lederman's solutions to the dilepton experiment design problem. The first is his 1976 experiment that looked at e^+e^- pairs using a higher energy beam running at Fermilab.[9] This detector was a relatively simple experiment using a *magnetic spectrometer* for momentum determination and a *lead-glass calorimeter* for electron identification. Although the J/ψ was clearly visible in this experiment, the background was still relatively high. A cluster of events was observed at $M_{e^+e^-} \approx 6$ GeV/c^2 which lead to a claim of a possible observation of a narrow peak at this mass. What was observed was most likely a background fluctuation.[6] Lederman's second solution was a 1977 experiment to look at dimuons using a far more complicated detector and again running at Fermilab.[10] This was a far more successful experiment in which the first observation of the Υ was made, the first indication of a new fifth quark. Unfortunately observations of new quarks were apparently no longer deemed worthy of a Nobel prize by this time. However the reader need not feel too bad for Leon Lederman since he was awarded the Nobel prize anyway in 1988, sharing it with Melvin Schwartz and Jack Steinberger for their use of neutrino beams and discovering a second type of neutrino, the muon neutrino.[11]

3. Part II: More on Charm and Bottom Quarks

In Part I the discovery of charm was used to introduce some basic detectors components. *Scintillators*, *Photomultiplier Tubes*, *wire chambers*, *magnetic spectrometers*, *Čerenkov counters*, and *electromagnetic calorimeters* were mentioned. In Part II additional experimental topics are covered, namely the following: particle identification systems; the use of precision position detectors to observe detached vertices; the use of different beam types; and the evolution of trigger systems. The story for this part of the mini-course is the advancement of detection of particles containing charm and bottom quarks. The outline of this part is illustrated in the bottom section of Fig. 1.

3.1. *Open charm discovery*

In Sec. 2 we introduced the discovery of the charmonium ($c\bar{c}$) states where the charm quantum number is hidden. The charm quark explanation of the observed narrow states became universally accepted once states with open charm were discovered.

The two most commonly produced charm mesons are the D^0 ($c\bar{u}$) and the D^+ ($c\bar{d}$). The charm quark decays quickly to either a strange quark or a down quark. The ratio of the rates for these two decays is given by the ratio of the square of two CKM matrix elements: $\Gamma(c \to sW^*)/\Gamma(c \to dW^*) \sim |V_{cs}|^2/|V_{cd}|^2 \approx 20$. The $c \to sW^*$ decay is called Cabibbo favoured while the $c \to dW^*$ is Cabibbo suppressed. The virtual W can decay to either quarks or leptons. Thus most of the D^0 and D^+ mesons decay to states with a strange quark. The easiest decay modes to reconstruct are the all charged modes: $D^+ \to K^-\pi^+\pi^+$; $D^0 \to K^-\pi^+$; and $D^0 \to K^-\pi^+\pi^+\pi^-$. Since pions are the more copiously produced hadrons in an interaction, one needs to distinguish kaons from pions to observe these open charm signals.

Besides *Čerenkov counters* there are other particle identification methods for charged hadrons. One example is a *Time-of-Flight* (*TOF*) detector,
this was used in the discovery of open charm two years after the discovery
of the J/ψ. The discovery was made using the Mark I experiment at the
SPEAR e^+e^- collider, the same spectrometer which was used in the discovery of the J/ψ. The e^+e^- collider gives an inherently lower background
than hadron-hadron collisions since the electron and positron annihilate
completely. However the $D^+ \to K^-\pi^+\pi^+$ and $D^0 \to K^-\pi^+$, $K^-\pi^+\pi^+\pi^-$
were only discovered after the collection of additional data and using the
TOF system to separate kaons from pions.

A *TOF* system works by measuring the time it takes for a charged
particle to travel between two points. For particles of the same momentum,
this time difference depends on the particle's mass.

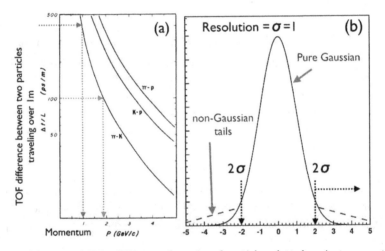

Fig. 8.   (a) Time-of-flight differences for pairs of particles plotted against momentum;
(b) Illustration of a Gaussian resolution function and example of non-Gaussian tails.

Figure 8(a) shows the difference in time-of-flight over one metre for pairs
of hadrons. The performance of a *TOF* system is given by the distance
(L) traveled between the two time measurements and the resolution ($\sigma_{\Delta t}$)
with which the time-of-flight measurement is made. Long distances and
fine resolution are needed. For example for Mark I $L \approx 2$ m and $\sigma_{\Delta t} \approx$
400 ps. This means that one can get $2\sigma_{\Delta t}$ separation between kaons and
pions for momenta < 1 GeV/c, *i.e.* at very low momentum. Even if the
time measurement resolution can be considerably reduced, *e.g.* to $\approx$100 ps

for the Fermilab CDF Run II experiment, it can be seen from Fig. 8(a) that with $L \approx 2$ m, a $2\sigma_{\Delta t}$ separation between kaons and pions is only achieved for momenta < 2 GeV/c. Even with only a $2\sigma_{\Delta t}$ separation at low momentum, the decays $D^0 \to K^-\pi^+$, $D^0 \to K^-\pi^+\pi^+\pi^-$ and $D^+ \to K^-\pi^+\pi^+$ could be isolated sufficiently from background at Mark I for them to make the discovery.[12] The mass plots are shown in Fig. 9, the $D^0 \to K^-\pi^+$ distribution is shown without and with a *TOF* kaon selection.

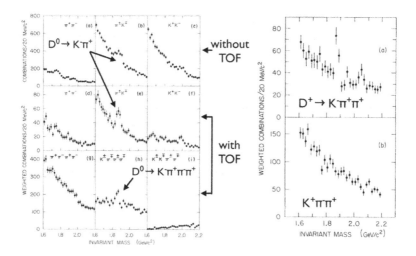

Fig. 9.   Mass plots from Mark I: (Left) D^0; (right) D^+.

The gas *Čerenkov counters* mentioned in Part I can be used to separate kaons and pions at higher momenta, but typically collider experiments like Mark I and CDF do not have the necessary space for them. There are a number of alternate particle identification systems.[1]

3.2. *Measurement uncertainties*

At this point it is worth making an aside about experimental resolution and the meaning of a $2\sigma_{\Delta t}$ separation. The importance of mass resolution was introduced in Sec. 2.1. Not only is the size of the resolution important, but the resolution function or shape matters also. When a quantity is measured experimentally one does not typically obtain an exact number, but there is some uncertainty. This uncertainty is normally separated into two components. One component is essentially statistical in nature and arise due to a lack of precision. The other is typically non-statistical and is due to our limited knowledge and affects the accuracy of the measurement. The former is called the statistical uncertainty or statistical error, while the latter is referred to as the systematic uncertainty.

A classic example of a statistical uncertainty is that due to limited statistics. *E.g.* when measuring the lifetime of a particle we have a limited number of particles to use, thus the lifetime distribution is measured with limited precision which leads to an uncertainty in the extracted lifetime. Another example is measuring a distance with a measuring tape. There is some uncertainty in positioning the tape at one end and in the reading and the precision of the scale at the other end. To reduce this uncertainty the measurement can be repeated many times and the average value used. We can illustrate the resolution with this simple example. If a histogram is made of these measurements (frequency *vs.* [value−nominal]) ideally the distribution is Gaussian as shown by the solid line in Fig. 8(b). The resolution is the sigma of the Gaussian distribution, and it gives the statistical uncertainty of any single measurement.

Imagine Fig. 8(b) shows the distribution of time-of-flight measurements for pions, where the nominal value is subtracted off and normalized to the resolution ($(\Delta t - \Delta t_0)/\sigma_{\Delta t}$). For any given pion the Δt measured can fall anywhere within the Gaussian distribution. In particular for a small fraction of the time it could be larger than $2\sigma_{\Delta t}$ from nominal, for a Gaussian distribution this probability is about 3%. Suppose that for a kaon the measured value of $(\Delta t - \Delta t_0)$ is greater than $2\sigma_{\Delta t}$. Then by requiring $(\Delta t - \Delta t_0) > 2\sigma_{\Delta t}$ we can select kaons and reject 97% of pions. This is for an ideal Gaussian distribution. The resolution function typically has non-Gaussian tails that go out much further as illustrated crudely by the dashed lines in Fig. 8(b). The rejection in this case would not be as good as 97%. Thus one needs to know the resolution function and must take care to try to avoid large non-Gaussian tails. For a real *TOF* system, non-Gaussian tails could arise from a number of sources, and the tails could also be asymmetric. Some of these sources include the following.

- The counter giving the time signal is finite in size and the measurement will depend on where the particle hits the counter.
- The system is made up of many counters whose relative timing and locations are not perfect.
- The calibration is not perfect, *e.g.* calibration tracks do not always come from exactly the same point, and the start time is not perfectly known.

- Some effects like MCS may affect the resolution and cause it to vary with the particle momentum.

Further coverage of statistical uncertainties and how to determine and handle them are beyond the scope of this mini-course, but there are many excellent books on this subject.[13]

The other component of a measurement uncertainty is called the systematic uncertainty and it is typically not statistical in nature. A classic example can again be illustrated by the case of measuring a distance with a measuring tape. If the scale of the measuring tape is wrong we would get a systematic error. Of course, if it were known that the scale was incorrect, we would correct the scale and the systematic error would be eliminated. Now let us assume that we must calibrate the measuring tape ourselves. We can only calibrate the scale within a certain accuracy, and this leads to a systematic uncertainty in the distance measured. For a more realistic example consider measuring the lifetime of a decaying particle. For a short lived particle like the D^0, the time is not directly measured. Instead, the distance (L) traveled between production and decay is measured and the momentum of the particle is also measured. The proper time for the decay is then $t = L m_{D^0}/p_{D^0}$. Besides the length and momentum scales, there are other potential sources of systematic uncertainties. The lifetime is extracted from a lifetime distribution containing many particle decays. This distribution may not be a pure exponential but could be modified due to detector acceptance and efficiency. The correction for acceptance and efficiency is typically determined using a Monte Carlo simulation. There are inherent uncertainties in the simulation that lead to an uncertainty in the correction function and thus to a systematic uncertainty in the lifetime. If the particle passes through matter before decaying or the daughter particles pass through matter, the lifetime distribution can also be affected by absorption of the parent or daughter particles. The cross sections for absorption may be poorly measured or not even known. This limited knowledge can also lead to a systematic uncertainty in the measurement. Finally, another source of systematic uncertainty could be backgrounds that mimic the signal but which are not properly accounted for. Typically, systematic uncertainties are not well defined and are not straightforward to determine. They are also usually not Gaussian distributed, and combining systematic uncertainties from different sources is problematic. Since even the meaning and definition of systematic uncertainties are difficult to quantify, ideally one should design an experiment to have a small systematic uncertainty (compared to the statistical uncertainty), so as not to have to worry about the details of the treatment and combining of systematic uncertainties. Further coverage of systematic uncertainties is beyond the scope of this mini-course. The understanding of systematics is beginning to be better understood and in some rare cases are even correctly taught.[14] However considerable disagreements are still common.

3.3. *Improving S/B for open charm*

Although the use of particle identification can be powerful in isolating a signal, it can be seen from Fig. 9 that there is considerable room for improvement. This is especially true in hadronic interactions which typically have higher backgrounds than in e^+e^- annihilations.

The lifetimes of the open charm particles are in the range 0.1–1 ps, which is small but finite and can be used to isolate a signal. Almost all the u- and d-quark backgrounds have essentially zero lifetime while the backgrounds from some strange particles decay after a long distance. Thus the signature of a charm particle is given by its decay a short distance away from the production point. For example, a 30 GeV D^0 travels an average length of about 2 mm, which is quite small but increases linearly with momentum.

To get a better sense of the scale involved, consider the decay of a a charm particle produced in a fixed-target experiment as illustrated in Fig. 10(a). The charm particle is produced and then decays after traveling a distance L_D. To separate the production and decay vertices we need to measure L_D with a resolution of $\sigma_{L_D} << L_D$. Since position detectors typically measure in the dimension transverse to the beam direction, it is more convenient to transform this essentially longitudinal resolution requirement into an transverse one. The typical angle that the charm particle is produced relative to the beam direction is $\theta \approx m_D/p_D$, where m_D and p_D are the mass and momentum of the charm particle respectively. The mean distance traveled by the charm particle is $L_D = \beta\gamma c\tau_D = c\tau_D p_D/m_D$ where τ_D is the lifetime of the charm particle. Thus to resolve the production and decay vertices we need $\sigma_{trans} << \theta L_D$, or $\sigma_{trans} << c\tau_D$, where σ_{trans} is the transverse position resolution of the detector (charged) tracking system. The values of $c\tau_D$ for the D^0, D^+ and Λ_c^+ are 123 μm, 312 μm and 60 μm respectively.

The resolution of the *MWPC's* depend on the wire spacing (s), and for a single detector plane is given by $\sigma_{trans} = s/\sqrt{12}$. The minimum wire spacings are in the range 1–2 mm depending on their cross sectional cover-

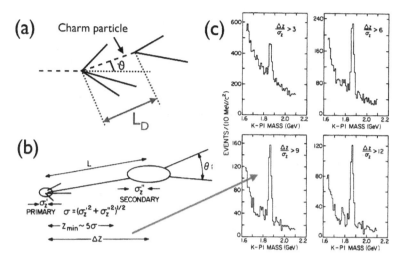

Fig. 10. (a),(b) Illustration of production and decay of a charm particle. (c) Invariant $K^-\pi^+$ mass plots from E691 showing the power of a detached vertex requirement.

age. For $s = 2$ mm, $\sigma_{trans} = 577$ μm, too large to resolve the production and decay vertices. The spatial resolution can be improved by measuring the time between a charged particle passing through the detector plane and when a signal is received in the wire closest to the point of passage. This is done in *Drift Chambers* and resolutions as low as $\sigma_{trans} \approx 100$ μm have been obtained in such a tracking system. This is still too large, especially considering that one typically needs better than 5–10σ vertex separation.

A tracking system with much better spatial resolution is needed. Historically, two detector technologies have been used that can give better resolutions: *photographic emulsions* and *Bubble Chambers*.

Detection using layers of photographic emulsions has been used for a long time and spatial resolutions of better than 10 μm have been obtained. Although these have been used relatively recently in DONUT to make the first direct observation of the ν_τ[15] they are not suitable for high rates.

Bubble Chambers have also been used historically to make important observations. Typically, the resolution of bubble chambers is not better than that for *Drift Chambers*. However, a sufficiently small bubble chamber, like the LEBC in the LEBC-EHS experiment, has achieved resolutions of ~10 μm, but again such bubble chambers are not suitable for high rates. The LEBC-EHS experiment reconstructed about 300-500 charm decays.[16]

What launched the high statistics studies of charm quark physics was the development and use of the *Silicon Microstrip Detector* (*SMD*). The Fermilab E691 photoproduction experiment included one of the first *SMD's* and collected a 10,000 sample of fully reconstructed charm decays, about two orders of magnitude more than other experiments of that time. Resolutions as good as $\sigma_{trans} \sim 10$ μm can be obtained and some data from E691 are shown in Fig. 10(c).[17] *SMD's* have now been used in many experiments including those studying bottom and top quarks.

3.4. *Going for higher statistics*

The road to higher statistics in charm studies is illustrated in Fig. 11, giving some selected milestones along the route.

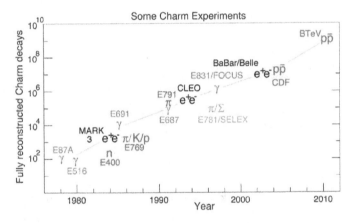

Fig. 11.   Number of fully reconstructed charm decays for different experiments as a function of time.

The charm quark experiments include e^+e^- and $p\bar{p}$ colliders as well as fixed-target experiments using photon and hadron beams. While larger charm data sets could be obtained in e^+e^- experiments by increasing the luminosity (*e.g.* CLEO), the fixed-target experiments needed additional technological advances – the 8 mm tape for data storage and high power commodity computing for data processing. For example, using these technologies and by building a more intense photon beam, the Fermilab FOCUS photoproduction experiment obtained a sample of 1 million fully reconstructed charm decays with published physics results one year after the end of data taking. Using 8 mm tapes with 30 times the capacity of 9-track tapes, the Fermilab E791 hadroproduction experiment could write out much more data and collected more charm than previous hadroproduction experiments. To do this, E791 used a wall of 42 8 mm tape drives in parallel to record data fast enough. To obtain substantially more statistics, a revolution in triggering is needed.

3.5. *The trigger system and the bottom quark*

Typically the particle interactions occur at a high rate and the S/B can be as low as 10^{-3}–10^{-8}. A trigger system is used to quickly decide whether an interaction contains signal and thus "trigger" the recording of the related data.

For a charm photoproduction experiment like FOCUS, the photon beam largely produces e^+e^- conversion pairs and only about 1 in every 500 photon interactions would produce hadrons. Only 1 in every 150 of those interactions producing hadrons contains a charm quark. It is relatively easy to recognize a photon conversion from an interaction producing hadrons. Since the fraction of hadron producing interactions containing charm is not too small, one just writes out all hadron producing interactions. For hadroproduction experiments on-the-other-hand, the fraction of charm is smaller by another factor of ten, thus either a lot more data must be written out and analyzed or a better, more intelligent trigger must be used.

Historically the trigger is a system of fast electronics that quickly processes special trigger signals produced by the detector and gives an electronic acceptance decision that is used to "trigger" the readout to save the data for that interaction. Long cables are used to delay the signals from the rest of the detector so they do not arrive at the readout before the trigger decision is made. In addition, the experiment is "dead" and unavailable to collect more data until the data readout is completed – this dead-time can be a significant fraction of the live-time. In the first stage of the FOCUS trigger, the decision must be made within about 370 ns from the time the interaction occurs.

Developments have made modern trigger systems much easier. Electronics are now faster, smaller and cheaper. Also, high speed data links and computing resources are more powerful. A large amount of memory is now affordable so that data from the detectors can be stored digitally while the trigger processing takes place. This eliminates the need for long signal cables which can degrade analog signals, and it also gives more time for trigger processing and virtually eliminates dead-time.

The ultimate trigger is if all the data could be recorded and analyzed before deciding which data to store. One can illustrate what is needed for such a trigger by using as an example the CDF or D0 experiments at Fermilab. In these experiments, protons and antiprotons cross every 396 ns, so there are about 2.5×10^6 crossing/s. If it took one second to fully analyze the data from one crossing in a single CPU, we would need 2.5 million CPU's to not lose data from any crossings. We would also need to temporarily store at least 2.5 million crossings worth of data. If one needs 300 KB/crossing, then 1 TB (10^3 GB) is needed. Since the processing time would have a long tail beyond one second, to be safe one would want about 1000 TB of RAM as well as the 2.5 million CPU's! Clearly a trigger that only partially processes the data is needed.

Even if the ultimate trigger cannot yet be realized, the developments mentioned above have provided the needed ingredients to separate out charm decays in hadronic collisions by looking for evidence of a detached decay vertex at the trigger level. Since recognizing bottom quark decays is similar to that for charm decays, this revolution in triggering has made possible an experiment that will reconstruct very large samples of charm and bottom decays.

Already large samples of charm and bottom quark decays are being collected using the BaBar and Belle e^+e^- experiments. To do better one must use hadronic collisions with sufficient energy like $p\bar{p}$ annihilations at the Fermilab Tevatron. The cross section for producing bottom quarks is much larger than in e^+e^- annihilations, *e.g.* about 100 μb compared to about 1.1 nb at the $\Upsilon(4S)$. The CDF and D0 $p\bar{p}$ experiments can collect sizable samples of charm and bottom decays, but to get 1000 times more rate than BaBar or Belle requires a specialized detector and data acquisition and trigger systems.

The BTeV experiment[18] is designed to study bottom and charm decays at the Tevatron. To maximize the yield for clean flavour-tagged B mesons for CP violation studies, the detector is placed in the forward direction allowing a *Ring Imaging Čerenkov Counter* (*RICH*) for excellent particle identification over a wide momentum range. A PbWO$_4$ crystal calorimeter provides efficient detection of photons and π^0's with excellent energy resolution. The BTeV experiment includes a *silicon pixel detector* that makes possible the recognition of detached vertices at the lowest trigger level.

Although a *SMD* can provide excellent spatial resolution, a lot of data processing is typically needed for interactions with many tracks due to the strip geometry. This is illustrated in Fig. 12(a). A single particle passing through a plane of strips will give a signal in one strip as illustrated in (i).

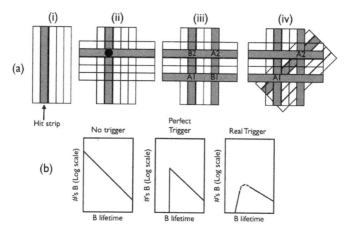

Fig. 12. (a) Illustration of pattern recognition in a *SMD*. (b) Illustration of the effect on the lifetime distribution of a detached vertex trigger.

The location along the hit strip can be determined by a second plane of strips oriented at 90° to the first plane as in (ii). However if, as illustrated in (iii), two particles pass through the two planes we would get four hit strips and one cannot tell if the two particles passed through points (A1, A2) or through (B1, B2). This ambiguity may be resolved by a third plane of strips at an angle as given in (iv). For a complex event with many particles the pattern recognition becomes quite complex and requires significant CPU power.

Ideally a trigger algorithm should be close to that used in the data analysis but with looser selection criteria. This is because a poorly chosen trigger algorithm can give rise to sizable systematic uncertainties. A simple example is illustrated in Fig. 12(b). The lifetime distribution of the B^0 meson is shown which is a pure exponential with the B^0 lifetime. Since backgrounds are typically at low lifetimes, an ideal trigger for collecting data to measure the lifetime would select decays with a large enough lifetime as illustrated in the middle distribution of Fig. 12(b). In a real trigger there is typically only time to do partial processing, and thus one might require only the presence of one or two detached tracks, instead of reconstructing the production and B^0 decay vertices. This could lead to a lifetime distribution illustrated by the right-most distribution of Fig. 12(b). Thus a correction function is needed to extract the correct B^0 lifetime.

The charged track pattern recognition is simplified in BTeV by the use of $400 \times 50 \ \mu m^2$ *silicon pixels* which can locate the position of a passing particle in 3-dimensions by a single hit pixel. Low momentum tracks undergo more MCS and can give rise to false detached tracks. In BTeV, the pixel detector is located in a dipole magnet so that tracks with low momentum can be rejected and not used at the trigger level.

Even with *silicon pixels* a full reconstruction cannot be done. Custom electronics using 500 FPGA's are used to help in processing the 500 GB/s data rate coming from the detectors. Further data processing and the pattern recognition is done on 500 commercial IBM-G5-equivalent processors. Two further levels of the trigger running on 1500 commodity CPU's reduce the data going to storage to a more manageable 200 MB/s.

The BTeV experiment nicely illustrates the convergence of a number of technological advances. Years of scientific progress have enabled such an experiment to be realized.

Glossary

Bubble chamber: A historic detector consisting of a cryogenic liquid maintained at a pressure above the equilibrium vapour pressure. The bubble chamber can be expanded to suddenly decrease the pressure so that charged particles passing through the liquid in a "superheated" condition will create a track of bubbles. Photographs are taken of the bubbles in multiple views to reconstruct the particle trajectories.

Calorimeter: A device to measure the energy of particles. The two distinct types are *electromagnetic calorimeters* and hadronic calorimeters. They work by completely absorbing the shower produced by a particle and producing a signal proportional to its energy. Calorimeters must be calibrated to give the absolute particle energy.

Čerenkov counter: A detector based on the Čerenkov effect (for which Čerenkov shared the 1958 Nobel prize). Particles traveling faster than light in a given medium emits a cone of (Čerenkov) light. A threshold Čerenkov counter contains a gas, for example, with a well chosen refractive index so that for a given particle momentum one particle type (*e.g.* pions) will emit light while another (*e.g.* kaons) will not. The angle of the cone of light also depends on the particle velocity which is used in other forms of *Čerenkov counters* like the *RICH*. The amount of Čerenkov light emitted is typically low, about 100 times less intense than scintillation light in a scintillator.

Drift chamber: A *wire chamber* where one measures the time between when a charged particle passes through and when a signal in the nearest signal wire is received. Typically many wires are used to form drift cells where the electric field is tailored to obtain a fairly uniform drift velocity across the cell. The spatial resolution is better than a *MWPC* but a drift chamber is more complex and typically cannot handle as high a rate of particles.

Electromagnetic calorimeter: A *calorimeter* for measuring the energies of photons and $e^{\pm}$ through their electromagnetic interactions. These calorimeters can be made from dense crystals like $PbWO_4$ or *lead-glass*, or can be sandwiches made of multiple layers of dense absorber and detection material.

Emulsions: Usually a layer of photographic emulsion several hundred μm thick in which a traversing charged particle causes the nearest silver halide grains to develop. Each grain is typically 0.2 μm in diameter with about 270 developed grains/mm. The emulsion must be scanned to reconstruct the particle trajectories.

Lead-glass shower counters: A dense glass used to detect photons and $e^{\pm}$ and for *electromagnetic calorimeters*. The detection is based on Čerenkov light.

Magnetic Spectrometer: A detector system used to determine the momentum of charged particles by measuring the deflection of the particles in a known magnetic field. Various magnetic field configurations can be used *e.g.* dipole, solenoid, or toroid. Deflection of particles are measured using position detectors, usually *wire chambers*, but can be *e.g. scintillator hodoscopes, scintillating fibres* or a *SMD*.

Multiwire proportional chamber (MWPC): A *wire chamber* where the location of a passing charged particle is determined by the location of the wire closest to it.

Photomultiplier Tube (PMT): A device to detect a small quantity of light using the photoelectric effect (for which Einstein received the 1921 Nobel prize). The maximum sensitivity of the photocathode in a typical *PMT* is for blue light.

Ring imaging Čerenkov counter (RICH): A *Čerenkov counter* where the angle of the emitted Čerenkov light is measured to enable the identification of particles over a wide momentum range.

Scintillator: A material that produces light through fluorescence when a charged particle passes through it. Scintillators used include inorganic crystals like $PbWO_4$, organic liquids and plastic. A classic plastic scintillator is made of polystyrene that produces light in the UV. The UV light is shifted to blue with a tiny doping of primary and secondary fluors to better match the photosensitivity of a *PMT*. Most of the light comes in a fast component (few ns) and strong signals are possible with sufficient scintillator thickness.

Scintillator fibres: *Scintillator* in the form of long flexible fibres with an outer acrylic sleeve so that the scintillation light is isolated to the fibre, but still totally internally reflected along the fibre to the ends. Typically a few mm in diameter they are used for position detectors or in *calorimeters* as either the detection material or as a mechanism for readout.

Scintillator hodoscope: A single detector plane made of strips of *scintillator*. Used to detect the position of a charge particle. Two planes can be overlapped with the strips in one plane oriented at 90° to the other to locate the particle in both transverse dimensions.

Silicon microstrip detector (SMD): Detection is based on essentially a silicon semiconductor p-n junction where the depletion region is enlarged by a bias voltage. The depletion layer can be considered as a solid state ionization chamber. A charged particle passing through the depletion region liberates electron-hole pairs which create signals on very thin, closely spaced readout strips. *SMD's* have the detection regions arranged as long uniformly separated strips. The strip separation can be in the range 10–300 μm.

Silicon pixels: Similar to the *SMD* but the active region is in the form of rectangles so that a "hit" pixel locates a passing particle in both transverse dimensions. The readout is however more complicated.

Spark wire chamber: A parallel-plate gas chamber in which a high voltage pulse is applied immediately after the passage of a passing charged particle. Sparks form along the trail of ions caused by the charged particle passing through the gas. This can provide a visualization of the track useful for public demonstration. High speed readout is typically done magnetostrictively or capacitively.

Time-of-flight detector: A system for identifying charged particles based on measuring their velocity between two points. The time-of-flight between two points is usually measured using *scintillator* counters possibly in conjunction with a measurement of the time of an interaction. The particle momentum is also measured giving a velocity that can distinguish particle types through their differing masses. See Sec. 3.1.

Wire chamber: For detection of charged particles through their ionization of usually noble gas atoms. A high voltage causes ionized electrons to accelerate and create an avalanche of electrons and positive ions. Detection of the avalanches in a plane of wires can give the position of the passing particle. Many types of wire chambers have been used, the original type (*MWPC*) was invented by Charpak for which he received the 1992 Nobel prize.

Acknowledgments

My thanks to Jeff Appel for some helpful suggestions for these lectures and for a careful reading of this writeup. This work was supported by the Universities Research Association Inc. under Contract No. DE-AC02-76CH03000 with the U. S. Department of Energy.

References

1. Some examples of textbooks are: R. Fernow, *Introduction to experimental particle physics*, CUP 1986; K. Kleinknecht, *Detectors for particle radiation*, 2nd Ed., CUP 1998. There are also some excellent articles, *e.g.* in F. Sauli (Ed.), *Instrumentation in High Energy Physics*, World Scientific, 1992.
2. R. N. Cahn and G. Goldhaber, *The experimental foundations of particle physics*, CUP 1989.
3. S. C. C. Ting, Nobel Lecture, 11 Dec. 1976, the full text is available at `http://nobelprize.org/physics/laureates/1976/ting-lecture.pdf`.
4. B. Richter, Nobel Lecture, 11 Dec. 1976, the full text is available at `http://nobelprize.org/physics/laureates/1976/richter-lecture.pdf`.
5. J. H.Christenson *et al.*, *Phys. Rev. Lett.* **21**, 1523 (1970).
6. L. M Lederman, Nobel Lecture, 8 Dec. 1988, the full text is available at `http://nobelprize.org/physics/laureates/1988/lederman-lecture.pdf`.
7. J. J. Aubert *et al.*, *Phys. Rev. Lett.* **33**, 1404 (1974); *Nucl. Phys.* **B89**, 1 (1975).
8. M. Ambrogiani *et al.*, *Phys. Rev.* **D64**, 052003 (2001).
9. D. C. Horn *et al.*, *Phys. Rev. Lett.* **36**, 1236 (1976).
10. S. W. Herb *et al.*, *Phys. Rev. Lett.* **39**, 252 (1977).
11. G. Danby *et al.*, *Phys. Rev. Lett.* **9**, 36 (1962). See also the Nobel lectures at `http://nobelprize.org/physics/laureates/1988/`
12. G. Goldhaber *et al.*, *Phys. Rev. Lett.* **37**, 255 (1976); I. Peruzzi *et al.*, *Phys. Rev. Lett.* **37**, 569 (1976).
13. Some examples of statistics textbooks are: P. R. Bevington and D. K. Robinson, "Data Reduction and Error Analysis for the Physical Sciences", 2nd

Ed., McGraw-Hill 1992; G. Cowan, "Statistical Data Analysis", OUP 1998; D. S. Sivia, "Data Analysis: A Bayesian Tutorial", OUP 1996.

14. R. Barlow, "Systematic Errors: Facts and Fictions", hep-ex/0207026.

15. K. Kodama *et al.*, *Phys. Lett.* **B504**, 218 (2001).

16. M. Aguilar-Benitez *et al.*, *Z. Phys.* **C40**, 321 (1988).

17. J. C. Anjos *et al.*, *Phys. Rev. Lett.* **58**, 311 (1987); K. Sliwa *et al.*, *Phys. Rev.* **D 32**, 1053 (1985); J. R. Rabb *et al.* *Phys. Rev.* **D 37**, 2391 (1988).

18. http://www-btev.fnal.gov/

SEMINARS

GRAND UNIFICATION AND PHYSICS BEYOND THE STANDARD MODEL

ERNEST MA

Physics Department,
University of California,
Riverside, CA 92521, USA
ernest.ma@ucr.edu

Recent progress in some selected areas of grand unification and physics beyond the standard model is reviewed. Topics include gauge coupling unification, $SU(5)$, $SO(10)$, symmetry breaking mechanisms, finite field theory: $SU(3)^3$, leptonic color: $SU(3)^4$, chiral color and quark-lepton nonuniversality: $SU(3)^6$.

1. Introduction

Up to the energy scale of 10^2 GeV, we are confident that the fundamental gauge symmetry of particle physics is that of the Standard Model (SM), i.e. $SU(3)_C \times SU(2)_L \times U(1)_Y$. New physics may appear just above this scale, but there may also be a much higher energy scale where the three gauge groups of the SM become unified into some larger symmetry. This is the notion of grand unification and depends crucially on the values of the three observed gauge couplings at the electroweak scale, as well as the particle content of the assumed theory from that scale to the unification scale.

2. Gauge Coupling Unification

The basic tool for exploring the possibility of grand unification is the renormalization-group evolution of the gauge couplings as a function of energy scale, given in one loop by

$$\alpha_i^{-1}(M_Z) = \alpha_i^{-1}(M_U) + (b_i/2\pi)\ln(M_U/M_Z), \tag{1}$$

with the experimentally determined values $\alpha_3(M_Z) = 0.1183(26)$, $\sin^2\theta_W(M_Z) = 0.23136(16)$, $\alpha^{-1}(M_Z) = 127.931(42)$, where $\alpha_2^{-1} =$

$\alpha^{-1} \sin^2 \theta_W$, and $\alpha_1^{-1} = (3/5)\alpha^{-1} \cos^2 \theta_W$ (assuming $\sin^2 \theta_W(M_U) = 3/8$). The coefficients b_i are obtained from the assumed particle content of the theory between M_Z and M_U. It is well-known that the three gauge coupings do not meet if only the particles of the SM are included. However, if the SM is extended to include supersymmetry (MSSM) thereby increasing the particle content, they do meet at around 10^{16} GeV.

A recent detailed analysis[1] using the more accurate two-loop analogs of Eq. (1) shows that the MSSM does allow the unification of gauge couplings but there remains a possible discrepancy, depending on the choice of inputs at the electroweak scale. In fact, this small discrepancy is taken seriously by proponents of specific models of grand unification, and has been the subject of debate in the past two years or so.

3. $SU(5)$

Consider the particle content of the MSSM. There are three copies of quark and lepton superfields:

$$(u, d) \sim (3, 2, 1/6), \quad u^c \sim (3^*, 1, -2/3), \quad d^c \sim (3^*, 1, 1/3), \tag{2}$$
$$(\nu, e) \sim (1, 2, -1/2), \quad e^c \sim (1, 1, 1), \tag{3}$$

and one copy of the two Higgs superfields:

$$(\phi_1^0, \phi_1^-) \sim (1, 2, -1/2), \quad (\phi_2^+, \phi_2^0) \sim (1, 2, 1/2). \tag{4}$$

The quarks and leptons can be embedded into $SU(5)$ as follows:

$$5^* = (3^*, 1, 1/3) + (1, 2, -1/2), \tag{5}$$
$$10 = (3, 2, 1/6) + (3^*, 1, -2/3) + (1, 1, 1), \tag{6}$$

but the Higgs superfields do not form complete multiplets: $\Phi_1 \subset 5^*$, $\Phi_2 \subset 5$. Their missing partners are $(3^*, 1, 1/3)$, $(3, 1, -1/3)$ respectively and they mediate proton decay. In the MSSM, such effective operators are dimension-five, i.e. they are suppressed by only one power of M_U in the denominator and can easily contribute to a proton decay lifetime below the experimental lower bound.

Recalling that there is a small discrepancy in the unification of gauge couplings. This can be fixed by threshold corrections due to heavy particles at M_U. Using these heavy color triplet Higgs superfields to obtain exact unification, it was shown that[2] their masses must lie in the range 3.5×10^{14} to 3.6×10^{15} GeV. However, the experimental lower bound on the decay lifetime of $p \rightarrow K^+ \bar{\nu}$ is 6.7×10^{32} years, which requires this mass to be greater than 7.6×10^{16} GeV. This contradiction is then used to rule out minimal $SU(5)$ as a candidate model of grand unification.

The above analysis assumes that the sparticle mass matrices are related to the particle mass matrices in a simple natural way. However, proton decay in the MSSM through the above-mentioned dimension-five operators depends on how sparticles turn into particles. It has been pointed out[3] that if the most general sparticle mass matrices are used, these operators may be sufficiently suppressed to avoid any contradiction with proton decay.

Instead of adjusting the color triplet masses to obtain exact unification, a new and popular way is to invoke extra space dimensions. For example, in a five-dimensional theory, if Higgs fields exist in the bulk, then there can be finite threshold corrections from summing over Kaluza-Klein modes. A specific successful $SU(5)$ model[4] was proposed using the Kawamura mechanism[5] of symmetry breaking by boundary conditions.

4. $SO(10)$

The power of $SO(10)$ is historically well-known. A single spinor representation, i.e. **16**, contains the **5*** and **10** of $SU(5)$ as well as a singlet N, which may be identified as the right-handed neutrino. The existence of three heavy singlets allows the three known neutrinos to acquire naturally small Majorana masses through the famous seesaw mechanism, and the decay of the lightest of them may also generate a lepton asymmetry in the early Universe which gets converted by sphalerons during the electroweak phase transition to the present observed baryon asymmetry of the Universe.

What is new in the past two years is the realization of the importance of the electroweak Higgs triplet contained in the **126** of $SO(10)$. Whereas the Higgs triplet under $SU(2)_R$ provides N with a heavy Majorana mass, the Higgs triplet under $SU(2)_L$ provides ν with a small Majorana mass.[6] This latter mechanism is also seesaw in character and may in fact be the dominant contribution to the observed neutrino mass. For a more complete discussion of this and other important recent developments in $SO(10)$, see the talk by Alejandra Melfo in these proceedings.

5. Symmetry Breaking Mechanisms

The breaking of a gauge symmetry through the nonzero vacuum expectation value of a scalar field is the canonical method to obtain a renormalizable field theory. If fermions have interactions which allow them to pair up to form a condensate with $\langle \bar{f} f \rangle \neq 0$, then the symmetry is also broken, but now dynamically. With extra dimensions, a recent discovery is that it is possible in some cases for a theory without Higgs fields (in the bulk or on our brane) to be recast into one with dynamical symmetry breaking on our brane. It is of course known for a long time that the components of gauge fields in extra dimensions may also be integrated over the nontrivial compactified manifold so that

$$\int A_i dx^i \neq 0, \tag{7}$$

thereby breaking the gauge symmetry.[7] More recently, bulk scalar field boundary conditions in a compact fifth dimension, using $S^1/Z_2 \times Z_2'$ for example,[5] have become the mechanism of choice for breaking $SU(5)$ and other grand unified groups to the MSSM. This method can also be applied to breaking supersymmetry itself.[8]

6. Finite Field Theory: $SU(3)^3$

If $\beta_i = 0$ and $\gamma_i = 0$ in an $N = 1$ supersymmetric field theory, then it is also finite to all orders in perturbation theory if an isolated solution exists for the unique reduction of all couplings.[9] This is an attractive possibility for a grand unified theory between the unification scale and the Planck scale. The conditions for finiteness are then boundary conditions on all the couplings of the theory at the unification scale where the symmetry is broken, and the renormalization-group running of these couplings down to the electroweak scale will make predictions which can be compared to experimental data. In particular, the mass of the top quark and that of the Higgs boson may be derived. Successful examples using $SU(5)$ already exist.[10,11] Recently, an $SU(3)^3$ example has also been obtained.[12]

Consider the product group $SU(N)_1 \times ... \times SU(N)_k$ with n_f copies of matter superfields $(N, N^*, 1, ..., 1) + ... + (N^*, 1, 1, ..., N)$ in a "moose" chain. Assume Z_k cyclic symmetry on this chain, then

$$b = \left(-\frac{11}{3} + \frac{2}{3}\right) N + n_f \left(\frac{2}{3} + \frac{1}{3}\right)\left(\frac{1}{2}\right) N = -3N + n_f N. \tag{8}$$

Therefore, $b = 0$ if $n_f = 3$ independent of N and k.

Choose $N = 3$, $k = 3$, then we have the trinification model,[13] i.e. $SU(3)^3$ which is the maximal subgroup of E_6. The quarks and leptons are given by $q \sim (3, 3^*, 1)$, $q^c \sim (3^*, 1, 3)$, and $\lambda \sim (1, 3, 3^*)$, denoted in matrix notation respectively as

$$
\begin{pmatrix} d\ u\ h \\ d\ u\ h \\ d\ u\ h \end{pmatrix}, \quad \begin{pmatrix} d^c\ u^c\ h^c \\ d^c\ u^c\ h^c \\ d^c\ u^c\ h^c \end{pmatrix}, \quad \begin{pmatrix} N\ E^c\ \nu \\ E\ N^c\ e \\ \nu^c\ e^c\ S \end{pmatrix}. \tag{9}
$$

With three families, there are 11 invariant f couplings of the form $\lambda q^c q$ and 10 invariant f' couplings of the form $\det q + \det q^c + \det \lambda$. An isolated solution of $\gamma_i = 0$ is

$$
f_{iii}^2 = \frac{16}{9} g^2, \tag{10}
$$

and all other couplings $= 0$. Assuming that $SU(3)^3$ breaks down to the MSSM at M_U, this predicts $m_t \sim 183$ GeV, in good agreement with the present experimental value of $178.0 \pm 2.7 \pm 3.3$ GeV.

7. Leptonic Color: $SU(3)^4$

Because of the empirical evidence of gauge coupling unification, almost all models of grand unification have the same low-energy particle content of the MSSM, including all models discussed so far. However, this does not rule out the possibility of new physics (beyond the MSSM) at the TeV energy scale, without spoiling unification. I discuss two recent examples. The first[14] is nonsupersymmetric $SU(3)^4$ and the second[15] is supersymmetric $SU(3)^6$.

In trinification, quarks and leptons are assigned asymmetrically in Eq. (9). To restore complete quark-lepton interchangeability at high energy, an $SU(3)^4$ model of quartification[14] has been proposed. The idea is to add leptonic color[16] $SU(3)_l$ which breaks down to $SU(2)_l \times U(1)_{Y_l}$, with the charge operator given by

$$
Q = I_{3L} + I_{3R} - \frac{1}{2} Y_L - \frac{1}{2} Y_R - \frac{1}{2} Y_l. \tag{11}
$$

The leptons are now $(3, 3^*)$ under $SU(3)_L \times SU(3)_l$ and $(3, 3^*)$ under $SU(3)_l \times SU(3)_R$, i.e.

$$
l \sim \begin{pmatrix} x_1\ x_2\ \nu \\ y_1\ y_2\ e \\ z_1\ z_2\ N \end{pmatrix}, \quad l^c \sim \begin{pmatrix} x_1^c\ y_1^c\ z_1^c \\ x_2^c\ y_2^c\ z_2^c \\ \nu^c\ e^c\ N^c \end{pmatrix}. \tag{12}
$$

The exotic particles x, y, z and x^c, y^c, z^c have half-integral charges: $Q_x = Q_z = Q_{y^c} = 1/2$ and $Q_{x^c} = Q_{z^c} = Q_y = -1/2$, hence they are called "hemions". They are confined by the $SU(2)_l$ "stickons" to form integrally charged particles, just as the fractionally charged quarks are confined by the $SU(3)_q$ gluons to form integrally charged hadrons.

The particle content of $SU(3)^4$ immediately tells us that if unification occurs, then $\sin^2 \theta_W(M_U) = \sum I_{3L}^2 / \sum Q^2 = 1/3$ instead of the canonical $3/8$ in $SU(5)$, $SU(3)^3$, etc. This means that it cannot be that of the MSSM at low energy. Instead the SM is extended to include 3 copies of hemions at the TeV scale:

$$(x,y) \sim (1,2,0,2), \quad x^c \sim (1,1,-1/2,2), \quad y^c \sim (1,1,1/2,2), \qquad (13)$$

under $SU(3)_C \times SU(2)_L \times U(1)_Y \times SU(2)_l$, without supersymmetry. In that case, it was shown[14] that the gauge couplings do meet, but at a much lower unification scale $M_U \sim 4 \times 10^{11}$ GeV. However, proton decay is suppressed by effective higher-dimensional Yukawa couplings with $\tau_p \sim 10^{35}$ years. Also, the exotic hemions at the TeV scale have $SU(2)_L \times U(1)_Y$ invariant masses such as $x_1 y_2 - y_1 x_2$, so that their contributions to the S, T, U oblique parameters are suppressed and do not spoil the agreement of the SM with precision electroweak measurements.

8. Chiral Color and Quark-Lepton Nonuniversality: $SU(3)^6$

Each of the $SU(3)$ factors in supersymmetric unification may be extended:

$$SU(3)_C \to SU(3)_{CL} \times SU(3)_{CR}, \qquad (14)$$

which is the notion of chiral color;[17]

$$SU(3)_L \to SU(3)_{qL} \times SU(3)_{lL}, \qquad (15)$$

which is the notion of quark-lepton nonuniversality;[18,19] and

$$SU(3)_R \to SU(3)_{qR} \times SU(3)_{lR}, \qquad (16)$$

which is needed to preserve left-right symmetry. Quarks and leptons are now $(3,3^*)$ under $SU(3)_{CL} \times SU(3)_{qL}$, $SU(3)_{qR} \times SU(3)_{CR}$, and $SU(3)_{lL} \times SU(3)_{lR}$. The three extra $(3,3^*)$ multiplets x, x^c, η transform under $SU(3)_{qL} \times SU(3)_{lL}$, $SU(3)_{lR} \times SU(3)_{qR}$, $SU(3)_{CR} \times SU(3)_{CL}$ respectively, with x, x^c having the same charges as λ and zero charge for η. With this assignment, $\sin^2 \theta_W(M_U) = 3/8$.

Because all the fermions are arranged in a moose chain, this model is automatically free of anomalies, in contrast to the case of chiral color by itself or quark-lepton nonuniversality by itself, where anomalies exist and have to be canceled in some *ad hoc* way. At the TeV scale, the gauge group is assumed to be $SU(3)_{CL} \times SU(3)_{CR} \times SU(2)_{qL} \times SU(2)_{lL} \times U(1)_Y$ with the following 3 copies of new supermultiplets:

$$h \sim (3, 1, 1, 1, -1/3), \quad h^c \sim (1, 3^*, 1, 1, 1/3), \quad \eta \sim (3^*, 3, 1, 1, 0); \quad (17)$$

$$(\nu_x, e_x) \sim (1, 1, 2, 1, -1/2), \quad (e_x^c, \nu_x^c) \sim (1, 1, 1, 2, 1/2); \quad (18)$$

$$\begin{pmatrix} N_x & E_x^c \\ E_x & N_x^c \end{pmatrix} \sim (1, 1, 2, 2, 0). \quad (19)$$

Again they all have $SU(2)_L \times U(1)_Y$ invariant masses. With this particle content, it was shown[15] that unification indeed occurs at around 10^{16} GeV. What sets this model apart from the MSSM is the rich new physics populating the TeV landscape. In addition to the particles and sparticles listed above, the heavy gauge bosons and fermions corresponding to the breaking of chiral color to QCD as well as quark-lepton nonuniversality to the usual $SU(2)_L$ should also be manifest, with unmistakable experimental signatures.

The consequences of $SU(2)_{qL} \times SU(2)_{lL} \rightarrow SU(2)_L$ have been discussed[19] in some detail. They include the prediction $(G_F)_{lq} < (G_F)_{ll}$, which may be interpreted as the apparent violation of unitarity in the quark mixing matrix, i.e. $|V_{ud}|^2 + |V_{us}|^2 + |V_{ub}|^2 < 1$, as well as effective $\sin^2 \theta_W$ corrections in processes such as $\nu q \rightarrow \nu q$, polarized $e^- e^- \rightarrow e^- e^-$, and the weak charge of the proton, etc. However, the constraints from Z^0 data imply that these effects are very small and not likely to be measurable within the context of this model. On the hand, since the new particles of this model are required to be present at the TeV scale, they should be observable at the Large Hadron Collider (LHC) when it becomes operational in a few years.

9. Conclusion

Assuming a grand desert from just above the electroweak scale to 10^{16} GeV, the particle content of the MSSM allows the unification of the three known gauge couplings. If studied closely, taking into account proton decay and neutrino masses, etc., this appears to favor $SO(10)$ as the grand unified symmetry over $SU(5)$ but the latter is still viable, especially if a fifth dimension is invoked for example.

Instead of a single simple group, the product $SU(N)^k$ supplemented by a cyclic Z_k discrete symmetry is an interesting alternative. Using a moose chain in assigning the particle content of such a supersymmetric theory, a necessary condition for it to be finite is to have 3 copies of this chain, i.e.

3 families of quarks and leptons. A realistic example has been obtained[12] using $N = k = 3$.

For $N = 3$, $k = 4$ without supersymmetry, the notion of leptonic color which has a residual unbroken $SU(2)_l$ gauge group can be implemented in a model of $SU(3)^4$ quartification[14]. This model allows unification at 10^{11} GeV without conflicting with proton decay, and predicts new half-integrally charged particles (hemions) at the TeV scale.

For $N = 3$, $k = 6$ with supersymmetry, the notions of chiral color and quark-lepton nonuniversality can be implemented[15], which cooperate to make the theory anomaly-free and be observable at the TeV scale, without spoiling unification.

In a few years, data from the LHC will tell us if the MSSM is corrrect [as predicted for example by $SU(5)$ and $SO(10)$], or perhaps that supersymmetry is not present but other new particles exist [as predicted for example by $SU(3)^4$], or that there are particles beyond those of the MSSM as well [as predicted for example by $SU(3)^6$]. Excitement awaits.

Acknowledgments

I thank Javier Solano and all the other organizers of V-SILAFAE for their great hospitality and a stimulating meeting in Peru. This work was supported in part by the U. S. Department of Energy under Grant No. DE-FG03-94ER40837.

References

1. W. de Boer and C. Sander, Phys. Lett. **B585**, 276 (2004).
2. H. Murayama and A. Pierce, Phys. Rev. **D65**, 055009 (2002).
3. B. Bajc, P. Fileviez Perez, and G. Senjanovic, Phys. Rev. **D66**, 075005 (2002).
4. L. J. Hall and Y. Nomura, Phys. Rev. **D66**, 075004 (2002).
5. Y. Kawamura, Prog. Theor. Phys. **103**, 613 (2000); **105**, 691, 999 (2001).
6. E. Ma and U. Sarkar, Phys. Rev. Lett. **80**, 5716 (1998); E. Ma, Phys. Rev. Lett. **81**, 1171 (1998).
7. Y. Hosotani, Phys. Lett. **B126**, 309 (1983).
8. J. Scherk and J. H. Schwarz, Phys. Lett. **B82**, 60 (1979).
9. C. Lucchesi, O. Piguet, and K. Sibold, Helv. Phys. Acta **61**, 321 (1988).
10. D. Kapetanakis, M. Mondragon, and G. Zoupanos, Z. Phys. **C60**, 181 (1993).
11. T. Kobayashi, J. Kubo, M. Mondragon, and G. Zoupanos, Nucl. Phys. **B511**, 45 (1998).
12. E. Ma, M. Mondragon, and G. Zoupanos, hep-ph/0407236.
13. A. De Rujula, H. Georgi, and S. L. Glashow, in *Fifth Workshop on Grand Unification*, ed. by K. Kang, H. Fried, and P. Frampton (World Scientific, Singapore, 1984), p. 88.

14. K. S. Babu, E. Ma, and S. Willenbrock, Phys. Rev. **D69**, 051301(R) (2004).
15. E. Ma, Phys. Lett. **B593**, 198 (2004).
16. R. Foot and H. Lew, Phys. Rev. **D41**, 3502 (1990).
17. P. H. Frampton and S. L. Glashow, Phys. Lett. **B190**, 157 (1987).
18. H. Georgi, E. E. Jenkins, and E. H. Simmons, Phys. Rev. Lett. **62**, 2789 (1989).
19. X.-Y. Li and E. Ma, Mod. Phys. Lett. **A18**, 1367 (2003).

QCD EVOLUTION IN DENSE MEDIUM*

M. B. GAY DUCATI

Instituto de Física
Universidade Federal do Rio Grande do Sul,
Caixa Postal 15051, CEP 91501-970, Porto Alegre, Brazil
gay@if.ufrgs.br

The dynamics of the partonic distribution is a main concern in high energy physics, once it provides the initial condition for the Heavy Ion colliders. The determination of the evolution equation which drives the partonic behavior is subject of great interest since is connected to the observables. This lecture aims to present a brief review of the evolution equations that describe the partonic dynamics at high energies. First the linear evolution equations (DGLAP and BFKL) are presented. Then, the formulations developed to deal with the high density effects, which originate the non-linear evolution equations (GLR, AGL, BK, JIMWLK) are discussed, as well as an example of related phenomenology.

1. Introduction

The knowledge of the dynamics of the high density Quantum Chromodynamics (hdQCD) is one of the most challenging problems to solve nowadays in high energy physics. In this lecture I intend to review the linear approaches for the evolution of the parton distributions and the formalisms developed to deal with the high density QCD systems, including non-linear effects, required to restore unitarity.

The increasing of the parton density requires a formulation of the QCD at high energies, that assures the unitarity limit of the cross section. A reliable treatment should evolve both, in the linear and non-linear regimes. The linear regime is valid in a system where the parton evolution should be described only by emission diagrams and the non-linear regime is reached

---

*Work partially supported by CNPq, Brazil.

when the density of the partonic system is so high such that the recombination among partons (nonlinearity) should be important in the evolution.

2. The Linear Evolution Equations

In this section it will be shortly presented the DGLAP and BFKL dynamics and their descriptions concerning the Deep Inelastic Scattering at high energies. The DIS is the process of interaction of a lepton and a nucleon exchanging an electroweak boson, producing many particles at the final state X [1]. The process is $l(k)\ N(p) \to l'(k')\ X(p')$, where k, k', p and p' are the four momenta of the initial and final lepton, incident nucleon and final hadronic system, respectively. The main variables for this process are $Q^2 = -q^2 = -(k-k')^2$, which is the square of the transfered momentum, $s = (k+p)^2$, the square of the center of mass energy, $W^2 = (q+p)^2 = (p')^2$, the square of the center of mass energy of the virtual boson-nucleon system. The hard scale is given by q^2 (< 0), corresponding to the process resolution, and $x = Q^2/2p.q$, represents the momentum fraction of the hadron carried by the parton in the interaction. In terms of the partonic content of the nucleon the structure function, which reflets its overall distribution, is given by

$$F_2(x, Q^2) = \sum_i e_i^2 q_S^i(x, Q^2), \quad \text{with } q_S^i(x, Q^2) \equiv [q^i(x, Q^2) + \bar{q}^i(x, Q^2)], \quad (1)$$

where q_S is the singlet quark distribution. It is the F_2 function that is object of main experimental studies, specially at HERA, at small x. This observable presents a direct dependence on the parton evolution. Concerning the DGLAP evolution equation, the quark distribution function can be shown to evolve as [2]

$$\frac{\partial q_S^i(x, Q^2)}{\partial \ln Q^2} = \frac{\alpha_s(Q^2)}{2\pi} \left[\int_x^1 \left(P_{qq}(\frac{x}{x_1}) q_S^i(x_1, Q^2) + P_{qg}(\frac{x}{x_1}) g(x, Q^2) \right) \right], \quad (2)$$

where P_{qq} is one of the splitting functions P_{ij}, describing the transition between the parton (quarks or gluons) j to the parton i, from fraction of momentum x_1 to x. This evoution is represented diagramatically in Fig. 1

The Eq. (2) and the correspondent equation for gluon distribution were independently derived by Dokshitzer, Gribov and Lipatov, and by Altarelli and Parisi, known as DGLAP equations in leading order. The approach is based on perturbative QCD, however a suitable non perturbative initial condition, extracted from the experiment for a given boson virtuality, is required.

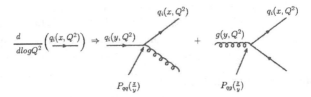

Fig. 1.   Representation of the DGLAP evolution.

It can be shown by successive derivations that $q_S(x, \ln Q^2) \sim \sum_n (\alpha_s \varepsilon)^n$, which corresponds to the emission of n gluons, showing that the DGLAP equations resum the leading $\ln Q^2$. This can be understood as ladder diagrams with a strong ordering in transverse momenta $k_\perp$, i.e., $Q_0^2 \ll k_{\perp_1}^2 \ll ... \ll k_{\perp_n}^2 \ll Q^2$. The scale Q_0^2 is the cut, or transition value between perturbative and non perturbative physics.

At small-x the gluons dominate, since $P_{gg}^{(0)}(z) \sim \frac{2N_c}{z}$, and the parton distributions have the general behavior $xp_i(x, Q^2) \sim x^{-\lambda}$, $\lambda > 0$. More likely for initial condition $xq_i(x, Q_0^2) \sim Const$ results in $xq_i(x, Q^2) \sim exp\sqrt{\ln(\ln Q^2)\ln 1/x}$, known as double leading logarithm approximation (DLA), having as region of validity $\alpha_s \ll 1$, $\alpha_s \ln(1/x) \ll 1$ and $\alpha_s \ln(1/x) \ln Q^2 \approx 1$.

For very low x values the $\ln s$ becomes large and $\alpha_s \ln 1/x \approx 1$ and the DLA is not valid. Then, Balitsky-Fadin-Kuraev-Lipatov (BFKL) proposed the resum of all leading logarithms of Bjorken x, or the energy [3]. The BFKL evolution equation is proposed for an unintegrated gluon distribution function in the transverse momentum variable. The differential form of the BFKL equation for the non-integrated gluon distribution is

$$\frac{\partial \phi(x, k_\perp^2)}{\partial \ln(1/x)} = \frac{3\alpha_s}{\pi} k_\perp^2 \int_0^\infty \frac{dk_\perp'^2}{k_\perp'^2} \left\{ \frac{\phi(x, k_\perp'^2) + \phi(x, k_\perp^2)}{|k_\perp'^2 - k_\perp^2|} + \frac{\phi(x, k_\perp^2)}{\sqrt{4k_\perp'^4 + k_\perp^4}} \right\}, \quad (3)$$

where the non-integrated gluon function $\phi(x, k_\perp^2)$ is related to the usual gluon distribution function by

$$xg(x, Q^2) = \int^{Q^2} \frac{dk_\perp^2}{k_\perp^2} \phi(x, k_\perp^2). \quad (4)$$

The initial condition to the equation (3) should be considered at a sufficient small value of x_0, such that the following condition should be satisfied $\alpha_s \ll 1$, $\alpha_s \ln(Q^2/Q_0^2) \ll 1$, $\alpha_s \ln(1/x) \approx 1$. At high energy limit, the BFKL equation should be represented by a ladder diagram , with longitudinal momenta strongly ordered $x \ll x_{i+1} \ll ... \ll x_1 \ll 1$, and $Q^2 \approx k_{\perp i+1} \approx ... \approx k_{\perp 1} \approx Q_0^2$.

Its solution grows as a power of the center of mass energy s with the consequent violation of the Froissart bound [4] ($\sigma \leq \ln^2 s$) at very high energies. The cure for this problem was not reached in the next to leading order calculation [5], and is still under research. Both BFKL and DGLAP predict a cross section that violates the Froissart limit at high energies, consequently limitating the use of such evolution equations in the region of not very small x.

From this very brief discussion on the main issues of the linear formalisms for the dynamics of the parton distributions, it gets clear the need to improve the formulation in order to include the unitarity corrections preserving the Froissart limit.

3. The Nonlinear Evolution Equations

The DGLAP and BFKL solutions predict a strong growth of the cross section $\sigma(\gamma^* N)$ at high energies. Such behavior shows that the perturbative QCD, in these formalisms, does not provide a limit to the growth of the cross section, requiring some dynamical effects, not originally present in the DGLAP and BFKL approaches.

Intuitively, we can associate $xg(x, Q^2)$ to the number of gluons into the nucleon, n_g, per rapidity unity, $y = \ln(1/x)$, with transverse size of order $1/Q^2$. Approaching the high density regime, the gluons may begin to superpose spatially in the transverse direction and to interact, behaving no more as free partons. These interactions should slower, or even stop, the intense growing of the cross section, fixing the limit πR^2_{HAD} in the small x regime. Introducing the function κ, with probabilistic interpretation

$$\kappa = \sigma_0 \frac{xg(x, Q^2)}{\pi R^2} , \tag{5}$$

it is possible to estimate in which kinematical region one can expect modifications in the usual evolution equations. So to say, for $\kappa << 1$ the system obeys the linear evolution equations, governed by individual partonic cascades, without interactions among the cascades. As $\kappa \approx \alpha_s$, partons from distinct cascades begin to interact due to spatial superposition. This specific kinematical regime, or the onset of the recombination mechanism, was first studied by Gribov, Levin and Ryskin [6] around twenty years ago, proposing the introduction of non linear terms into the evolution equation.

4. The GLR Approach

In 1983, Gribov, Levin and Ryskin [6] introduced the mechanism of parton recombination in perturbative QCD for high density systems, expressing this as unitarity corrections included in a new evolution equation known as GLR equation. In terms of diagrams it considers the dominant non-ladder contributions, or multi-ladder graphs, also denoted fan diagrams.

Following DGLAP, the number of partons of low fraction of momentum increases very rapidly, in contrast with a more diluted system at intermediate values of x. The transition between these regimes should be characterized by a critical value x. The same can be argued through BFKL formalism, with the difference that in this case the increasing of the partonic distributions, takes place at a fixed transverse scale, although the evolution presents the fluctuations in the transverse plane due to the characteristic diffusion in BFKL.

It is important to emphasize that in both linear dynamics only the decay processes are considered in the partonic evolution, however we expect that the anihilation mechanism should contribute in the low x regime, providing some control of the increasing of the partons distribution functions. To express the recombination mechanism it is needed a formulation in terms of the probability to recombine two incident partons. As a first approximation one considers the anihilation probability as proportional to the square of the probability to find one incident parton, introducing a non-linear behavior.

Taking $\rho = \frac{xg(x,Q^2)}{\pi R^2}$ as the gluon density in the transverse plane, one has the general behavior: for splitting $1 \to 2$, the probability is proportional to $\alpha_s \rho$, for anihilation $2 \to 1$, the probability is proportional to $\alpha_s^2 \rho^2/Q^2$; where $1/Q^2$ stands for the size of the produced parton. For $x \to 0$, ρ increases and the anihilation process becomes relevant. Considering a cell of volume $\Delta \ln Q^2 \Delta \ln(1/x)$ in the phase space allows one to write the modification of the partonic density in terms of the gluon distribution as,

$$\frac{\partial^2 xg(x,Q^2)}{\partial \ln Q^2 \partial \ln 1/x} = \frac{\alpha_s N_c}{\pi} xg - \frac{\alpha_s^2 \gamma}{Q^2 R^2} [xg]^2 . \tag{6}$$

This equation is the GLR equation [6]. The work of Mueller and Qiu [7] gives $\gamma = 81/16$ for $N_c = 3$.

The Eq. (6), provides the reduction of the growing of $xg(x,Q^2)$ at low x, in comparison with the linear equations. It is also predicted for the asymptotic region $x \to 0$ the saturation of the gluon distribution, with a critical line between the perturbative region and saturation region, setting its region of validity.

In the asymptotic limit one obtains $xg(x, Q^2)\big|_{SAT}^{GLR} = \frac{16}{27\pi\alpha_s}Q^2 R^2$. Since the GLR only includes the first non-linear term, although it predicts saturation in the asymptotic regime its region of validity does not extend to very high density where higher order terms should contribute significantly.

5. The AGL Approach

This approach was developed by Ayala, Gay Ducati and Levin (AGL) [8]to extend the perturbative treatment of QCD up to the onset of high density partons regime, through the calculation of the gluon distribution which is the solution of a non-linear equation that resums the multiple exchange of gluon ladders, in double leading logarithm approximation (DLA).

It is based on the development of the Glauber formalism for perturbative QCD [9], considering the interaction of the fastest partons of the ladders with the target, nucleon or nucleus. In this formalism a virtual probe G^* that interacts with the target in the rest frame, through multiple rescatterings with the nucleons is considered. In this reference frame the virtual probe can be interpreted following the decomposition of the Fock states, and its interaction with the target occurs by the decay of the component gg, as represented in the Fig. 2.

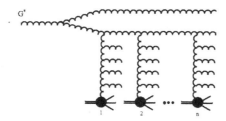

Fig. 2.   Representation of the gg interaction with the nucleus.

For small-x this pair has a lifetime much bigger than the nucleus (nucleon) radius and the pair is separated by the fixed transverse separation r_t during the interaction, which is represented by the exchange of a ladder of gluons strongly ordered in transverse momentum. Then, the cross section for this process is calculated through the dipole picture and knowing that for the virtual probe with virtuality Q^2 the relation $\sigma^{G^*A}(x, Q^2) = (\frac{4\pi^2\alpha_s}{Q^2})xg_A(x, Q^2)$ is valid, the nuclear gluon distribution including Unitarity Corrections (UC) is reached in the form,

$$xg_A(x, Q^2) = \frac{2R_A^2}{\pi^2} \int_x^1 \frac{dx'}{x'} \int_{1/Q^2}^{1/Q_0^2} \frac{d^2r_t}{\pi r_t^4} \left[C + \ln(\kappa_G) + E_1(\kappa_G)\right], \quad (7)$$

where C is the Euler constant, E_1 is the exponential function and where the κ_G function was introduced as $\kappa_G(x, r_t^2) = \frac{3\alpha_s}{2R_A^2}\pi r_t^2 xg(x, 1/r_t^2)$. Equation (7) is the master equation for the interaction of the gg pair with the target, and is known as the Glauber-Mueller formula. The expansion of Eq. (7) in terms of κ_G gives as the Born term the DGLAP equation in the small x region, the higher order terms corresponding to the unitarity corrections naturally implemented in this formalism.

In this approach the gluon pair emission is described in DLA of perturbative QCD and from the Feynman diagrams of order α_s^n, it should be extracted only the terms that contribute with a factor of order $(\alpha_s \ln 1/x \ln Q^2/Q_0^2)^n$. The interaction of the gluon pair with the target operates through the exchange of a ladder which satisfies the DGLAP evolution equation in the DLA limit.

With the aim to obtain a non-linear evolution equation containing the unitarity corrections through the inclusion of all the interactions besides the fastest parton from the ladder, the master equation for the gluon is differentiated in $y = \ln 1/x$ and $\varepsilon = \ln Q^2$, obtaining in terms of κ_G the main evolution equation,

$$\frac{\partial^2 \kappa(y, \varepsilon)}{\partial y \partial \varepsilon} + \frac{\partial \kappa(y, \varepsilon)}{\partial y} = \frac{N_c \alpha_s}{\pi} \left[C + \ln(\kappa_G) + E_1(\kappa_G)\right]. \quad (8)$$

It should be mentioned that large distance effects are absorved in the initial condition for the evolution, and situating in a convenient region of Q^2 only short distance effects are present, meaning a perturbative calculation is reliable.

Equation (8) was derived in Ref. [8], and refered for simplicity as AGL equation. In this approach all contributions from the diagrams of order $(\alpha_s y \ln Q^2)^2$ are resummed; in the limit $\kappa \to 0$ the DGLAP evolution in DLA is fully recovered; for $\kappa < 1$, and not large, the GLR equation is recovered and for $\alpha_s y \ln Q^2 \approx 1$ the equation is equivalent to the Glauber approach. A comprehensive phenomenology was developed with the AGL high density approach [8,10,11,12,13].

Non-perturbative effects are not explicitly described in the approaches studied up to here and this is the object of a distinct formalism JIMWLK [14] that we will briefly comment in a next subsection. In Fig. 3 the comparison

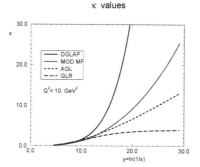

Fig. 3. Comparison among perturbative approaches.

between the solutions of the equations AGL, GLR, DGLAP and Glauber-Mueller (MOD MF) formula is presented, where the control of the growing of the gluon distribution in the non linear approaches is very evident.

6. The Balitsky-Kovchegov Equation

The unitarization problem in QCD was addressed as an extension of the dipoles formalism for the BFKL equation by Kovchegov [15]. This work proposes a non linear generalization of BFKL equation, also addressed previously in Ref. [16] by the use of OPE to QCD obtaining the evolution of Wilson line operators. The scattering of a dipole (onium - $q\bar{q}$) with the nucleon is described by a cascade evolution corresponding to the successive subdivision of dipoles from the father dipole. Each dipole has multiple scatterings with the nucleons of the target, implying multiple ladders exchange to be resummed in order to obtain the cross section of the interaction of the dipole with the nucleus. As a result it is derived the evolution equation having the unitarized BFKL Pomeron as solution, in the $LL(1/x)$ approximation.

The scattering of the onium $q\bar{q}$ (dipole) with the nucleus in the rest frame, takes place through a cascade of soft gluons, which once taken in the $N_c \rightarrow \infty$ limit is simplified by the suppression of non-planar diagrams. The gluons are replaced by $q\bar{q}$ pairs and the dipole Mueller's technique for the perturbative cascade can be employed [17]. The Balitsky-Kovchegov (BK) formulation, as the AGL, is a perturbative QCD calculation and the considered dipoles from the cascade interact independently with the nucleus.

Assuming no correlation among the dipoles, the forward scattering amplitude for the interaction, onium nucleus $N(\vec{x_{01}}, \vec{b_0}, Y)$ is then given

by [15],

$$N(\vec{x_{01}}, \vec{b_0}, Y) = -\sum_i^{\infty} \int n_i(x_{01}, Y, \vec{b_1}, \vec{x_1}, \dots, \vec{b_i}, \vec{x_i}) \prod_i \gamma(\vec{x_i}, \vec{b_i}) \frac{d^2 x_i}{2\pi x_i^2} d^2 b_i \quad (9)$$

where b is the impact parameter, $\vec{x_0}$ and $\vec{x_1}$ being the spatial position of the quark and antiquark pair, $\vec{x_{01}} = \vec{x_1} - \vec{x_0}$ and n represents the dipole density, $\gamma(\vec{x_{01}}, \vec{b_0})$ is the propagator of the pair $q\bar{q}$ through the nucleus, describing the multiple rescattering of the dipole with the nucleons within the nucleus. We denote this equation as the BK equation.

The physical representation is comparable with the approach Glauber-Mueller since the incident photon generates a $q\bar{q}$ that subsequently emits a gluon cascade further interacting with the nucleus, as represented in the Fig. 4. Although beginning the formulations with distinct degrees of freedom both BK and AGL resum the multiple rescatterings in their respectives degrees of freedom, which allows to consider they should coincide in a suitable common kinematical limit [18]. There is interesting phenomenological works with BK equation, for instance see the Refs. [19,20].

Fig. 4.   Dipole cascate in the BK equation.

In DLA, where the photon scale of momentum Q^2 is bigger than Λ^2_{QCD}, the K equation, deriving in $\ln(1/x_{01}^2 \Lambda^2_{QCD})$ results in

$$\frac{\partial^2 N(\vec{x_{01}}, \vec{b_0}, Y)}{\partial Y \partial \ln(1/x_{01}^2 \Lambda^2_{QCD})} = \frac{\alpha_s C_F}{\pi} \left[2 - N(\vec{x_{01}}, \vec{b_0}, Y) \right] N(\vec{x_{01}}, \vec{b_0}, Y), \quad (10)$$

which is the evolution in transverse size of the dipoles from x_{01} up to $1/\Lambda_{QCD}$, setting that the successive emission of dipoles generates larger transverse size for each higher generation.

The linear term reproduces BFKL at low density, and the quadratic term introduces UC and the equation reproduces GLR once we assume N directly related to the gluon distribution.

7. The JIMWLK Formulation

At very high energies the growth of the parton distribution is expected to saturate below a specific momentum scale Q_s, forming a Color Glass

Condensate (CGC)[21,22,23,24]. This saturated field, meaning the dominant field or gluons, has a large occupation number and allows the use of semi-classical methods. These methods provide the description of the small x gluons as being radiated from fast moving color sources (parton with higher values of x), being described by a color source density ρ_a, with internal dynamics frozen by Lorentz time dilatation, thus forming a color glass. The small x gluons saturate at a value of order $xG(x, Q^2) \sim 1/\alpha_s >> 1$ for $Q^2 \lesssim Q_s^2$, corresponding to a multi-particle Bose condensate state. The color fields are driven by the classical Yang-Mills equation of motion with the sources given by the large x partons. The large x partons move nearly at the speed of light in the positive z direction. In the CGC approach the light cone variables are employed, where, $x^\mu \equiv (x^+, x^-, x_\perp)$. The large x partons (fast) have momentum p^+, emitting (or absorbing) soft gluons with momentum $k^+ << p^+$, generating a color current with the + component $J_a^+ = \delta(x^-)\rho_a$. In this framework there is a separation between fast and soft partons, implying that the former have large lifetime while soft partons have a short lifetime.

In order to have a gauge-invariant formulation, the source ρ_a must be treated as a stochastic variable with zero expectation value, and an average over all configurations is required and it is performed through a weight function $W_{\Lambda^+}[\rho]$, which depends upon the dynamics of the fast modes, and upon the intermediate scale Λ^+, which defines fast $(p^+ > \Lambda^+)$ and soft $(p^+ < \Lambda^+)$ modes. The effective theory is valid only at soft momenta of order Λ^+. Indeed, going to a much softer scale, there are large radiative corrections which invalidate the classical approximation. The modifications to the effective classical theory is governed by a functional, nonlinear, evolution equation, derived by Jalilian-Marian, Iancu, McLerran, Kovner, Leonidov and Weigert (JIMWLK) [22,23,24] for the statistical weight function $W_{\Lambda^+}[\rho]$ associated with the random variable $\rho_a(x)$.

The functional evolution equation to the statistical weight function in a most condensed form, can be written as

$$\frac{\partial W_\tau[\rho]}{\partial \tau} = \frac{1}{2} \int_{x_\perp y_\perp} \frac{\delta}{\delta \rho_\tau^a(x_\perp)} \chi_{ab}(x_\perp, y_\perp)[\rho] \frac{\delta}{\delta \rho_{tau}^b(y_\perp)} W_\tau[\rho]. \qquad (11)$$

The dependece of the effective theory upon the separation scale is employed with the rapidity variable $\tau = \ln(1/x) = \ln(P^+/\Lambda^+)$. The kernel χ_{ab} takes into account all changes in the source color correlation due to new source modes, which are the gluons with $x < \frac{\Lambda^+}{P^+}$. Then, the quantum evolution consists in adding new correlations to the source density ρ_a.

However, a complete solution of the cited evolution equation was not yet obtained. Some approximated solutions have been studied in the literature [25], as well as numerical solutions, for instance, by lattice calculations (for a brief review about solutions see Ref. [26]).

In order to make predictions or comparison with data, some phenomenological treatment should be given to the weight function. For example, an approximation to the weight function which is reasonable when we have large nuclei is used and consists in taking a Gaussian form [27,28,29], which can accommodate both the BFKL evolution of the gluon distribution at high transverse momenta, and the gluon saturation phenomenon at low transverse momenta [29]. A non-local Gaussian distribution of color sources has been predicted in Ref. [25] as a mean-field asymptotic solution for the JIMWLK equation and provides some modifications concerning phenomenological properties of the observables [30,31]. The local Gaussian weight function assures that the color sources are correlated locally, on the other side, the non-local Gaussian allows correlations over large distances, with repercussion on the observables.

8. Phenomenology

A large contribution on phenomenology has been realized in the last years using linear and non-linear equations, however I intend to focus in this section in one phenomenological issue of the JIMWLK evolution equation and this implies to treat some observable considering the Color Glass Condensate. The dilepton production at RHIC and LHC energies will be the observable used to investigate such properties of the JIMWLK equation, however, the motivation for the choice of such observable is based on the Cronin effect [32].

The Cronin effect was discovered in the late's 70s and is related to the enhancement of the hadron transverse momentum spectra at moderated p_T (2-5 GeV) in comparison with the proton-proton collisions. The effect should be interpreted as being originated by the multiple scatterings of the partons from the proton propagating through the nucleus, resulting in a broadening of the transverse momentum of the initial partons. The Cronin effect was recently measured by the RHIC experiments, in $Au - Au$ and $d - Au$ collisions, and presents a distinct behavior: the suppression of the peak is founded and is claimed to be due to final state interactions [33]. Although the Cronin effect was observed in the hadron transverse momentum spectra, it can also be analyzed in the dilepton transverse momentum spectra, since multiple scatterings are expected as an initial state effect. The dilepton p_T spectra should clarify the properties of the Cronin effect, once no final state interactions occur in this observable.

In Ref. [31] is has been investigated the phenomenological aspects of the JIMWLK in the dilepton production at RHIC and LHC. The correlation function (which is the object of evolution of the JIMWLK equation) should be taken as a local or non-local Gaussian. The dilepton production in the forward rapidity using the CGC approach was investigated, more precisely the ratio between the proton-nucleus and proton-proton differential cross section for RHIC and LHC,

$$R_{pA} = \frac{\frac{d\sigma(pA)}{\pi R_A^2 \, dM \, dx_F \, dp_t^2}}{A^{1/3} \frac{d\sigma(pp)}{\pi R_p^2 \, dM \, dx_F \, dp_t^2}}. \tag{12}$$

The ratio in the Eq. (12) is similar to the one used to investigate the Cronin effect.

The BRAHMS experiment at forward rapidities [34,35] shows experimental data concerning the Cronin effect, and a suppression of the Cronin peak is found. Using the CGC the Ref. [30] studies the Cronin effect considering a local and non-local Gaussian distribution for the weight function and the Cronin peak suppression is reached with the non-local Gaussian, emphasizing this form of the weight function could be the right physics at forward rapidities.

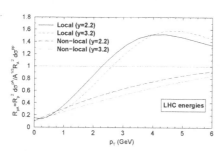

Fig. 5.   Ratio between proton-nucleus and proton-proton at LHC energies.

In the Fig. 5 we present the result for the ratio R_{pA} and we verify the peak and the suppression, if the local or non-local Gaussian are used, the later brings the same behavior presented by the Cronin study in the CGC. This implies that the dilepton p_T distribution used for comparison between pp and pA cross sections, provides a powerful tool to study the Cronin effect and the dynamics of the Color Glass Condensate in a high density system.

9. Summary

As a summary of the formulations for hdQCD at present, the Fig. 6 presents their different regions of applicability in the $\tau = \ln(1/x)$ versus Q^2 plane.

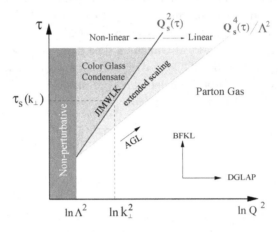

Fig. 6.   A map of the quantum evolution in the $\tau \times \ln Q^2$ plane.

The linear evolution equations drive the evolution at $\kappa \ll 1$, in a region identified as parton gas, where the parton density is not very high.

As $\kappa \approx \alpha_s$, is the kinematical regime that requires the QCD dynamics for high partons density, first suggested in the GLR evolution equation and in the map is shown as the region of application of the AGL equation.

The region of very high density parton systems provides the use of semi-classical techniques, identified in the phase space diagram (Fig. 6) by the CGC formalism, where the JIMWLK evolution equation drives the partonic evolution.

High density QCD is an exciting subject of great interest and under development in particle physics. The knowledge of the solution of evolution equations is an open subject and is very important to provide more application and predictibility of the QCD theory.

Acknowledgments

MBGD thanks the kind invitation for this lecture in the V SILAFAE and would like to point out the huge effort of the organizers to achieve this important meeting among Latino-American particle physicists.

References

1. A. M. Cooper-Sarkar, R.C.E. Devenish and A. De Roeck, *Int. J. Mod. Phys* **A13**, 3385 (1998).
2. V. N. Gribov and L. N. Lipatov,*Sov. Journ. Nucl. Phys.* **15**, 438 (1972); Yu. L. Dokshitzer, *Sov. Phys. JETP* **46**, 641 (1977); G. Altarelli and G. Parisi, *Nucl. Phys.* **B126**, 298(1977).
3. E.A. Kuraev, L.N. Lipatov and V.S. Fadin, *Phys. Lett.* **B60**, 50 (1975); Sov. Phys. JETP **44**, 443 (1976); *Sov. Phys. JETP* **45**, 199 (1977); Ya. Balitsky and L.N. Lipatov, *Sov. J. Nucl. Phys.* **28**, 822 (1978).
4. M. Froissart,*Phys. Rev.* **123**, 1053 (1961). A. Martin, *Phys. Rev.* **129**, 1462 (1963).
5. V.S. Fadin, L.N. Lipatov, *Phys. Lett.* **B429**, 127 (1998); M. Ciafaloni, D. Colferai, G.P. Salam, *Phys. Rev.* **D60**, 114036 (1999).
6. L. V. Gribov, E. M. Levin, M. G. Ryskin, *Phys. Rep.* **100**, 1 (1983).
7. A. H. Mueller, J. Qiu, *Nucl. Phys.* **B268**, 427 (1986).
8. A. L. Ayala, M. B. Gay Ducati and E. M. Levin, *Nucl. Phys.* **B493**, 305 (1997); *ibid***511**, 355 (1998).
9. A. H. Mueller, *Nucl. Phys.* **B335**, 115 (1990); ibid. **335**, 335 (1990).
10. M. B. Gay Ducati, V. P. Gonçalves, *Nucl. Phys.* **A680**, 141C (2001). *Phys. Lett.* **B487**, 110 (2000); Erratum *ibid.* **491**, 375 (2000). *Phys. Rev.* **C60**, 058201 (1999). *Phys. Lett.* **B466**, 375 (1999).
11. E. Gotsman *et al.*, *Nucl. Phys.* **B539**, 535 (1999).
12. M. B. Gay Ducati, *Braz. J. Phys.* **31**, 115 (2001); *Rev. Mex. Fis.* **48** Suplemento 2, 26 (2002).
13. E. Gotsman, E. Levin and U. Maor, Phys. Lett. B **425**, 369 (1998)
14. L. McLerran, R. Venugopalan, *Phys. Rev.* **D49**, 2233 (1994); ibid. 3352, **D50**, 2225 (1994); **D53**, 458 (1996).
15. Y. Kovchegov, *Phys Rev.* **D60**, 034008 (1999).
16. Ya. Balitsky,*Nucl. Phys.* **B463**, 99 (1996)
17. A.H. Mueller and B. Patel, *Nucl. Phys.* **B425**, 471 (1994).
18. M.B. Gay Ducati, V.P. Gonçalves, *Nucl. Phys.* **B557**, 296 (1999).
19. E. Gotsman, E. Levin, M. Lublinsky, U. Maor, E. Naftali and K. Tuchin, J. Phys. G **27**, 2297 (2001).
20. E. Levin and K. Tuchin, Nucl. Phys. B **573**, 833 (2000)
21. L. McLerran, R. Venugopalan, *Phys. Rev.* **D49**, 2233 (1994); *ibid.* 49, 3352 (1994).
22. J. Jalilian-Marian., A. Kovner, A. Leonidov, H. Weigert, *Nucl. Phys.* **B504**, 415 (1997); *Phys. Rev.* **D59**, 014014 (1999).
23. E. Iancu, A. Leonidov, L. D. McLerran, *Nucl. Phys.* **A692**, 583 (2001); *Phys. Lett.* **B510**, 133 (2001);
24. E. Ferreiro, E. Iancu, A. Leonidov, L. McLerran, *Nucl. Phys.* **A703**, 489 (2002).
25. E. Iancu, A. Leonidov and L. D. McLerran, *Phys. Lett.* **B510**, 133 (2001).
26. H. Weigert, ePrint arXiv:hep-ph/0501087.
27. E. Iancu and R. Venugopalan, ePrint arXiv:hep-ph/0303204.

28. F. Gelis, A. Peshier, *Nucl. Phys.* **A697**, 879 (2002).
29. E. Iancu, K. Itakura, L. McLerran, *Nucl. Phys.* **A724**, 181 (2003).
30. J.P. Blaizot, F. Gelis, R. Venugopalan, *Nucl. Phys.* **A743**, 13 (2004).
31. M. A. Betemps, M.B. Gay Ducati,*Phys. Rev.* **D70**, 116005 (2004).
32. J. W. Cronin *et al.*, *Phys. Rev.* **D11**, 3105 (1975).
33. A. Accardi, ePrint arXiv:nucl-th/0405046.
34. R. Debbe, by the BRAHMS collaboration, *J. Phys.* **G30**, S759 (2004).
35. I. Arsene *et al.* [BRAHMS Collaboration], *Phys. Rev. Lett.* **93**, 242303 (2004).

FUTURE EXPERIMENTS — GRID AND LHC*

ALBERTO SANTORO

DFNAE-IF-UERJ
Rua So Francisco Xavier,524
20550-013 Rio de Janeiro - RJ -Brazil
Alberto.Santoro@cern.ch

The next generation of Experiments in High Energy Physics at CERN will soon start in the largest Particles accelerator, the Large Hadron Collider (LHC). All the experiments together will produce an amount of data totally out of scale for a traditional data management. A new computing architecture has been created to face this challenge. In this talk we show that it will be almost impossible to do High Energy Physics in the era of LHC without using Grid the Grid computing.

1. Introduction — LHC and Future Experiments

The LHC was developed as a consequence of the evolution of High Energy Physics (HEP) ideas, new developments of the Theory of Particles, new development of Technologies on behalf of the advancement of the Science. Proton-proton collider at the energy of 14 TeV is a significant progress compared to the energies of other experiments. As a consequence four collaborations with four big Detectors has been proposed and are being built at CERN employing a big number of world institutions, physicists, engineers, computing scientists and technicians.

The four Detectors are the ALICE[1], LHCb[2], ATLAS[3] and CMS[4]. (For details of each collaboration/experiment go to the references of the Internet). Figure 1 shows the map of (Geneva) Switzerland/France where CERN/LHC and the four experiments are situated.

The physics main purpose of the experiments and Detectors at LHC is the search for the Higgs boson. The main parameters for LHC and the

---

*This work is supported by FAPERJ and CNPQ.

Large Hadron Collider (LHC) @ CERN

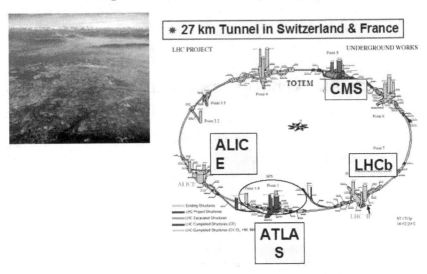

Fig. 1.   This is the LHC site where we can see the four places for the four Detectors.

experiments, are Beam Energy of $\sqrt{s} = 14.TeV$, and a luminosity limit at $L = 10^{-34}cm^{-2}s^{-1}$. Approximately 2000 physicists from about 150 Institutions and 50 countries are currently involved at CMS, one of the experiments of LHC. These numbers will have an impact on our computing strategy and the number and the size of the events to be produced. If only 1/3 of the physicists start working we will get serious problem with the present bandwidth. The international computing traffic will be really big.

2. New World — New Parameters — New Challenges

We would like to call attention to these new parameters imposed by the new technologies what we are calling of a *New World for Science*. They will impose a new way of work and put new challenges for the whole community. Some of these parameters oblige us to cross the frontiers of the possible in science and technology. Each one of the experiments at LHC will produce about 5 Petabytes of data per year. Then the four experiments will produce about 20 Petabytes per year. To have an idea about this amount of data we compare with the popular CD. If we use the commercial CD to store these data and pile them up we will have a 20 Km high tower of CDs.

The *new world* is characterized by the expected new science to advance the knowledge of the fundamental interactions, discovering the Higgs, expected new phenomena to integrate our scientific world. All these topics become very important for Particle Physics in general. The list of topics is long and offers a very exciting research opportunity. By 2010, or even before, we expect, results that will show this new scientific era.

While we need 10^{12} events to have the probability of observing one single Higgs, many other topics can be exploited comfortably with the amount of data available.

With this amount of data it will be impossible to work as we have worked so far a new architecture of computing is mandatory. This architecture will be the GRID. Let us move on the next chapter to describe this new wave to work on HEP.

3. GRID Projects

Recently we have seen other developments of computing, both in hardware and software. The success of the PC is so great that it becomes an option to be used in the new clusters. The complimentary advantage of using PC to build cluster for supercomputing is the easy technology to build and to fix the machines. This allows people to use their creativity to work with big amount of data. Video conference is been used in daily work among high energy physicists of several countries. The progress of the groups is much more visible and the development of each work in a collaboration is much better controlled. We can follow almost everything without being in the site of the experiment. Linux as the basic software, is very friendly and has become for all of us an option to work with the clusters without requiring an enormous amount of money for software. Many different versions exists in the market and HEP ommunity has choose the Red Hat as the basic flavor. Networks have experienced a big development and fast links have shown the possibility of building GRID with high speed lines.

For a GRID we want the best combination of CPU and storage devices, shared and used on a very fast network. We can also add, as part of a realistic definition, a subject as the center of interest of the GRID group which, in our case, is High Energy Physics. Technologies are still not ready to think about a "general" GRID, accepting all type of jobs. We are working in this direction. We will describe a bit the present development of the HEP projects. The main current activity is related to the software and the appropriated architecture for each cluster.

3.1. *iVDGL + GriPhyN + PPDG = Trilling*

In the beginning each project make an effort to develop its own software and exercise their ideas in a free way. Nevertheless the key of the development is to keep the collaboration among the groups. So, soon it appears the Trilling Project which is a coordination between iVDGL[5] (International Virtual Data Grid Laboratory), GriPhyN[6] (Grid Physics Network) and PPDG[7]. These three projects are dedicated to some set of experiments. Let us talk a bit about of each one of them.

3.2. *GRID for Alice, Atlas, LHCb, CMS*

ALICE

In this talk we will just give an idea of what is each experiment of the LHC. ALICE is one of them dedicated mainly to plasma physics using heavy ions to collide in the center of mass with big energies. For many details about the ALICE detector and the whole experiment the interested reader has should go to the web page of Alice[1]. The collaboration is made of 1000 physicists from almost one hundred institutions all over the world. The ALIEN (ALIce ENvironment) was created as a grid-like system for job submission and data management. This system intend to help ALICE to build a computing model soon to be defined by collaboration taking into account Functionatility, Interoperability, Performance, Scalability, Standards. These are not so different of each one of the experiments at LHC.

ATLAS

ATLAS is a collaboration of approximately 2000 physicists and engineers from about 150 Institutes and 34 Countries. For details of the Detector ATLAS of 7000 Tons, 25 m of diameter and 46 m long the reader should go to the web page of Atlas [3]. Many technologies have been developed by ATLAS collaboration to build one of the most interesting general purpose detector. The aims of the experiment are: Measure the Standard Model Higgs Boson, detect Supersymmetric states, study Standard Model QCD (Chromodinamics), EW (Electroweek), HQ (Heavy Quark) Physics, and New Physics (?). The Collaboration has been in intensive collaborative developments of software for GRID to be used in its experiment. Some of the tools developed with LHCb are: GridView (Simple tool to monitor status of testbed), Gripe (unified user accounts), Magda (Manager for Grid Data), Pacman (package management and distribution tool), Grappa (web portal using active notebook technology), GRAT (Grid Application Toolkit), Grdsearcher (browser), GridExpert (Knowledge Database), VO-

Toolkit (Site Authentication,Authorization), and so on.

LHCb

Like ALICE, LHCb (The Large Hadron Collider Beauty experiment) it is a detector dedicated mainly to b physics. The Collaboration has 563 physicists from 12 Countries and 50 Institutes. The experiment expect, to get $10^{12}b\bar{b}$ pairs per year and this is a much higher statistics than the current B factories. Another set of numbers expected for LHCb experiments is: (i) 200,000 reconstructed $B^0 \rightarrow J/\psi K_s$ events per year; (ii) 26,000 reconstructed $B^0 \rightarrow \pi^+\pi^-$; (iii) all B Mesons and Barions, and so on. The expected competition comes from BTeV experiment at Tevatron/Fermilab. LHCb collaboration has been produced many useful software for analysis in the near future in cooperation with Atlas. Two examples, beyond those pointed out above on Atlas subsection, GANGA (Gaudi ANd Grid) as an user interface for Grid and DIRAC (Distributed Infrastructure with Remote Agent Control) for Monte Carlos event production. Much details of this experiment we can find on the web page of LHCb [2].

CMS

I have choose the CMS detector to show in a picture in figure 2 to give an idea of these LHC detectors . CMS collaboration has approximately 2000 physicists, 160 Institutions and about 40 countries. The weight is 12,500 Tons, having 15 m of diameter, 22 m length and will use a Magnet field of 4 Tesla. The Detector is composed by a Silicon Microstrips as a central Tracker; Electromagnetic and Hadronic Calorimeters; a Superconducting Coil; Iron Yoke; Muon Barrel and Muon Endcaps. These are the main components shown in the figure 2. CMS will explore a big number of physics topics. In the case of Higgs, the collaboration intend to explore the full range of 100 - 1000 GeV as the allowed region of Higgs masses. Topics as QCD, Heavy Flavor Physics, SUSY, New Phenomena are part of the list of the future analysis.

The Caltech group, and many other groups at CMS have developed many softwares for the collaboration (see reference[9] to have an idea about this development). The Monte Carlo Production (MOP) is one of the important tool to help the collaborators use it in GRID for the production of Monte Carlo events, taking into account all the power installed on the Grid CMS sites. CLARENS is a software that consists of a server communication among several clients. One impressive, interesting and useful software is MonAlisa for monitoring sites and machines in a world level.

The Brazilian group constituted by physicists, engineers and computer scientists, from several institutions CBPF (Centro Brasileiro de Pesquisas

Físicas), UERJ (Universidade do Estado do Rio de Jeneiro), UFBA (Universidade Federal da Bahia),UFRJ (Universidade Federal do Rio de Janeiro), UNESP (Universidade Estadual Paulista), UFRGS (Universidade Federal do Rio Grande do Sul) is a consortium as a National and Regional collaboration. The group has succeed to approve a project to build a tiers 2 which will evolve to a tiers1 in the future. This site is called T2-HEPGRID BRASIL [10] and is being built with collaboration with Caltech group. This is being built to be part of the CMS GRID.

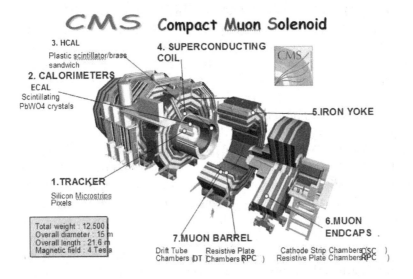

Fig. 2.   CMS Detector.

3.3. *Other GRID projects*

There are a permanent re-organization of this new system created for HEP. At the beginning many projects started to work isolated with poor communication and after they started to work together, in a narrow collaboration. They are the iVDGL [5], the GriPhyN [6], and the PPDG [7]. Now, the Trillium organizer project joins the 3 independent projects. Other organizations appear in several levels: the LCG [11] is the Large Hadron Collider Computing Grid Project. It is dedicated to the development of the European projects for GRID associated to LHC. Similar organization is the GRID3 [12] in United States organizing the GRID projects in USA dedicated to HEP experiments. Two other organizations are also appearing. The EGEE [13] (Enabling Grids for E-science in Europe) is organized to take into account all GRID projects for Science in Europe. The corresponding organization in USA is the OSG [14] (the Open Science Grid). All these organizations are collaborating for the progress of the HEPGRID.

Outside of the HEP there are many other organizations for GRID with specific purpose like the Mamogrid for Cancer of breast, Grid for Biology, Astronomy and others. All of them are in the Internet. There are also many sites for Grid softwares, the most popular are Globus[16] and the Condor [17]. For more information look at the books on reference [18]

4. Digital Divide and GRID

GRID is a great invention and a new Internet revolution. There is a lot of work to do before it become a real and useful tool. Many communities are doing a big effort to get it as soon as possible. One benefit for every-body came with the GRID: it shows explicitly the problem of the Digital Divide[19].

Digital Divide is the unequal distribution of the network bandwidth. There are cases where it is very hard to think of a solution at short time and there are some cases where the solution is available and depends only of political decisions from local administrations. The case of the Rede Rio is an example of this last case. UERJ is the biggest University supported by the State of Rio de Janciro, that gives also the financial support to Rede Rio, but it is in a very bad situation. In 2002 the bandwidth of the University was 2 Mbps while other Federal institutions were at 155 Mbps and there is a project with exclusive dedicated 155 Mbps. In 2004 we achieved to 24 Mbps while the other institutions duplicate their capacity and are preparing to go to 1 Gbps while there is no news to upgrade the network of UERJ. This is one example of Digital Divide: UERJ is explicitly excluded of the "club" defined by the administration of Rede Rio. We find many examples like that in Brazil, in Latin America and in the rest of the world. The Digital Divide characterize a situation of the difference between institutions, Cities and Countries. This means that there are Institutions, Cities and Countries that are the "natural" places to install facilities and others that are not. This is the perverse side of the Digital Divide: to condemn some institutions, cities and countries to living in the backward computing era.

ICFA has understood the serious risk of having the whole program of LHC condemned to be only between USA and Europe, with a few exceptions and has created a SCIC/DD Standing Committee on Interregional Connectivity/Digital Divide and a interested reader can go to the web page to get all reports about the problem at the world level[15]. There are other initiatives like ALICE (America Latina Interconectada com a Europa or Latin America Interconnectivity with Europe) which is a project to create a regional Latin American Research networking infrastructure interconnected with GEANT in Europe. This project is co-funded by the European Commission and create CLARA as its network.

5. Latin America and New Technologies — Conclusion

As we are in Latin America we have to pay attention to our region and work hard to upgrade the technologies on the continent in order to participate of the development of science in this century. GRID is a real possibility to start our development in collaboration with CERN. It will be certainly also a possibility for other developments in science. As request by my colleagues I have presented a proposal to the assembly of SILAFAE in Lima/Peru, as shown in the figure 3 below.

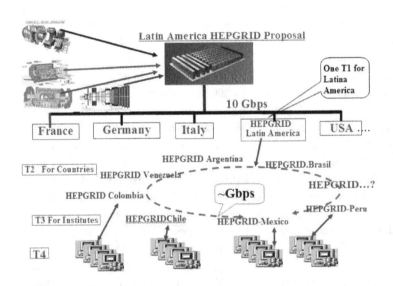

Fig. 3.   This was the draft-proposal presented during the meeting in Lima/Peru.

Acknowledgments

I would like to thank CNPq and FAPERJ for the financial support for the development of GRID and the Organizers of this SILAFAE for have invited me to give this talk. It was very useful for me to attend this meeting and I have attended very good talks from several Latin American colleagues.

References

1. http://www.alice.cern.ch/
2. http://www.lhcb.cern.ch/
3. http://www.atlas.cern.ch/
4. http://www.cms.cern.ch/
5. http://www.ivdgl.org/
6. http://www.griphyn.org/
7. http://www.ppdg.net/
8. http://www.opensciencegrid.org/
9. http://ultralight.caltech.edu/gaeweb/
10. http://www.hepgridbrasil.uerj.br/
11. http://lcg.web.cern.ch/LCG/
12. http://www.grid3.com/
13. http://public.eu-egee.org/
14. http://www.opensciencegrid.org/
15. http://icfa-scic.web.cern.ch/ICFA-SCIC/
16. http://www.globus.org
17. http://www.cs.wisc.edu/condor/
18. (i) The Grid Blueprint for a New computing Infrastructure, Edited by Ian Foster and Carl Kesselman, and the second volume too; (ii) Grid Computing Making the Global Infrastructure a Reality — Fram Berman , Anthony J.G. Hey and Geoffrey C. Fox
19. The Digital Divide — Facing a Crisis or Creating a Myth? — Edited by Benjamin M. Compaine

BTEV: USING HEAVY QUARK DECAYS TO TEST THE STANDARD MODEL[*]

M. SHEAFF

Physics Department, University of Wisconsin,
1150 University Avenue,
Madison, WI 53706, USA
sheaff@fnal.gov

The decays of hadrons containing heavy quarks provide a multitude of ways in which to search for physics effects outside the standard model. The BTeV experiment at Fermilab has been designed to exploit this. Detailed plans for the measurements BTeV will make including the anticipated precision will be discussed. Comparisons will be made to the results expected both from the B factories currently taking data and from the LHCb experiment now under construction at CERN.

1. Introduction

The primary physics goal of the BTeV experiment at Fermilab[1] is to search for physics beyond the Standard Model (SM). The design of the experiment allows us to do this in a number of ways, including: precise measurements of CP violation parameters in $B_{u,d,s}$ decays to look for deviations from those expected in the SM; measurement of the B_s mixing parameter, which could be the first measurement if it turns out to be well above the value predicted by the SM; the search for CP violation and mixing in the charm sector, which are predicted by the SM to be very small; and searches for b and c decays that are predicted to be very rare or even forbidden in the SM. If we do find inconsistencies with the SM, our broad physics program allows multiple means by which we can elucidate the underlying physics.

---

[*]Work partially supported by INT-0086456 and INT-0072436 of the U.S. National Science Foundation.

BTeV is a crucial component of the experimental program that is needed to answer outstanding fundamental questions in flavor physics. This program involves present and future experiments with high statistics samples of K, D, and B mesons as well as neutrino oscillation and astrophysics experiments[2]. For a more complete understanding of flavor physics we need to answer such questions as: Why are there three families of fermions in both the quark and lepton sectors? In fact, why are there families at all? What about the mass hierarchy? There are mass differences of many orders of magnitude between the lightest and heaviest fermions in both families. Is this related in some way to the mixing angles between the states? Are the quark mixing angles fully described by the SM, or do they arise at least in part from new physics? Are there similarities as well as differences in the mass hierarchies and mixing angles for the families of quarks and leptons? Is CPT violated, and if so, how? What interactions in the early universe caused the complete matter/antimatter asymmetry that we see today?

We know from previous measurements that there is CP violation in the decays of K and B mesons. This is predicted in the SM and attributed in that model to the complex phase terms in the Cabibbo-Kobayashi-Maskawa (CKM) matrix describing the mixing between the quark mass states and their weak eigenstates. This effect is believed to be too small to account for the complete matter/antimatter asymmetry we now observe. Precise measurements of CP violation in B meson decays which can test the CP violating terms in the CKM matrix elements therefore provide a fertile ground for searches for physics beyond the SM.

The BTeV experiment will take place at the Tevatron Collider and in the *forward* region, $1.9 < \eta < 5.3$. The main reason for pursuing these studies in a hadron machine, i.e., the Tevatron, is the very large production cross section for $b\bar{b}$ pairs, 20,000 per second at the design luminosity of $2 \times 10^{32} cm^{-2} s^{-1}$ and energy of 2 TeV, to be compared to 10 per second in an $e^+ e^-$ collider operating at the $\Upsilon(4S)$ and with a luminosity of $2 \times 10^{34} cm^{-2} s^{-1}$. Another compelling reason is that a number of additional states, e.g., B_s, B_c, b-baryons, Ξ_{cc}, Ω_{ccc}, and Ξ_{bc}, are produced in the hadron collider.

The experiment will take place in the forward region along the direction of the incident anti-proton beam. The layout resembles a typical fixed target experiment. The advantage of this is that the B's are produced at higher momentum on average than in the central region and thus travel farther on average before they decay. This leads to a larger average separation between primary and secondary vertices. Thus we will be able to select events based on the vertex separation we measure divided by the error with which we measure it with good acceptance. This selection criterion was found to be the most effective in selecting charm events over background for the fixed target charm experiments. While this selection was made offline for those experiments, BTeV is designed to make this selection online in the first level trigger. The importance of this capability can hardly be overstated. It allows us to select very high statistics samples of b decays including many final states, hadronic as well as semi-leptonic, with high efficiency. This comprehensive data set provides a plethora of ways by which we can challenge the SM.

Because of the large boost the lifetimes of the b states can be measured with high accuracy. The mass resolution of these states will be better, too, because the decay tracks will be higher in momentum and will therefore suffer less multiple scattering. Another advantage of this configuration is that there is sufficient space for several particle identification detectors. These are essential for the separation of different b decay states which overlap when decay tracks are misidentified and for tagging.

2. Description of the BTeV Spectrometer

The BTeV Spectrometer is shown in Figure 1. The spectrometer as originally proposed had two identical arms, one placed along the proton direction, the other along the antiproton direction. While Stage 1 approval was given for the original proposal in June 2000, funding constraints at Fermilab led to a request for a proposal for a descoped detector during the Fall of 2001. Stage 1 approval for the one-arm spectrometer was received in May 2002. The change resulted in a significant reduction in cost, since it is now necessary to build one each of the forward tracking system, Ring Imaging Cerenkov Counter, electromagnetic calorimeter, and muon system rather than two. However, we have retained the full pixel detector system and data acquisition bandwidth. This and a revised design for the RICH mean that we can expect close to the same, and, in some states, even an improved performance from the single-arm spectrometer than from the two-arm spectrometer in the original proposal. This is despite the loss of a factor of two in our geometric acceptance. Fortunately, in almost all events, both b and $\bar{b}$ are boosted into the same arm so that the loss of B decays is not more than a factor of two.

The away side toroids will still be placed along the proton direction. They are needed for shielding and for constant floor loading in case we do

BTeV Detector Layout

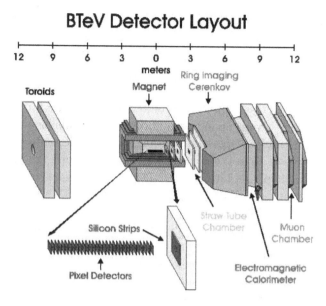

Fig. 1. The BTeV Spectrometer.

eventually build a second spectrometer. They also support the compensating magnet, as do the toroids in the spectrometer. Compensating magnets are needed for BTeV because the experiment magnet is a dipole rather than a solenoid, and it therefore perturbs the particle beams.

The different detectors that make up the spectrometer are labeled on the figure. A short description of each follows. Further information is available on the Web pages[1].

There are thirty stations of silicon pixel detectors lined up perpendicular to the beams inside the dipole magnet surrounding the interaction region. Each station contains two planes 10 cm by 10 cm in size. Each plane is divided into two half-planes so that the detectors can be moved out of the beams when they are being injected into the Tevatron and ramped up to full energy. The full system contains a total of 23 million pixels. The pixels themselves are small in one direction, 50μm, and relatively large in the other, 400μm. The two planes in each station are oriented such that one plane has the narrow dimension in x and the other in y. Fast electronics chips which contain front-end circuits to amplify, digitize, and read out the pixel hits are bump-bonded to each pixel chip. These assemblies are then mounted on a substrate which supports them and also provides cooling.

A high-density interface circuit provides the electrical connections between the assemblies as well as between the assemblies and the outside world.

There are many good reasons for choosing to use pixel detectors rather than strip detectors close to the interaction point. First, the x and y association is made directly for each hit eliminating the need for a time-consuming calculation to do this. Second, since each pixel is small in area it has a small capacitance. This results in a much better signal to noise ratio for the hits. Pixels also have low occupancy and are very fast as well as radiation hard. They provide excellent spatial resolution, $\sim$5-10μm depending on track angle. This is better than the nominal $50\mu m/\sqrt{12}$ because the pixel data are not recorded simply as hit (or not). The pixel pulse heights are digitized with a three-bit flash analog-to-digital converter (FADC) to allow interpolation for improved spatial resolution.

All pixel hits will be read out from every beam crossing and used to form a first level trigger based on the number of tracks which do not point to the production vertex. A rough momentum measurement is possible so that only tracks above a minimum momentum will be used in the selection. Simulations indicate that we can achieve trigger efficiencies of better than 50% for most reconstructible B decays of interest while rejecting approximately 99% of minimum bias events. A second trigger based on tracks that are seen in the muon chambers will be included at level one and will be used to check the efficiency of the pixel vertex trigger.

The muon chambers are made from gas-filled stainless steel tubes 1 cm in diameter with 30 μm gold-plated tungsten anode wires running down the center of the tubes. Two planes of tubes, the one offset from the other by one-half cell, make up each of the four views r,u,v,r (u, v rotated by $\pm 22.5^{\circ}$ relative to r) that form an octant. The offset resolves the left-right ambiguity for hits in each cell thereby improving the resolution from 1 cm/$\sqrt{12}$ to 0.5 cm/$\sqrt{12}$. There are three stations of muon detectors made up from these octants. Toroidal steel magnets lie in front of the first station of planes and between the first and second stations. Unmagnetized iron is placed between the second and third chamber stations. The hits from high transverse momentum muons from B decays in the three doublets of planes in any one view will lie on a plane. By making a cut on the distance of closest approach of a muon to this plane for each crossing/octant/view, trigger efficiencies of better than 50% can be achieved for reconstructible $J/\psi \rightarrow \mu^{+}\mu^{-}$ with a rejection factor of 100 for minimum bias background.

There are forward tracking chambers just inside and downstream of the dipole magnet for improved momentum measurement compared to what can be achieved using the pixels alone. They will also be used to identify vees from the decays of K mesons and Lambda baryons which occur downstream of the pixel detectors. The tracking chambers are comprised of gas-filled straw tubes at larger angles and silicon strips closer to the beam for improved resolution where both the track momenta and the occupancy are higher. They are expected to give momentum accuracy of 1% or better over the full range of momenta and angles they cover.

The electromagnetic calorimeter will be built from lead tungstate ($PbWO_4$) crystals each shaped as a "tower" facing the interaction point. This material is also being used by the CMS experiment at CERN and has been chosen because it is relatively radiation hard. Simulations as well as test beam data indicate that this detector will provide CLEO/BaBar/Belle-like performance both in terms of efficiency and energy resolution but now in a hadron collider.

The Ring Imaging Cerenkov Detector (RICH) in the new design includes both a gas and a liquid radiator, each with its own detector system. The liquid RICH extends to higher angles than either the electromagnetic calorimeter or the muon system. Its performance is well-matched to the low momentum tracks found in this angular region. The detector is able to distinguish both muons and electrons from pions at these momenta. The result is a factor of 2.4 (3.9) improvement in efficiency for the identification of single (double) leptons for B decays with lepton daughters. The new RICH provides good separation of π's, K's, and p's through a wide range of momenta. This is important when reconstructing a B decay of interest, since other decays may reconstruct incorrectly in the same mass region if one or more tracks are incorrectly identified.

The muon system, the electromagnetic calorimeter, and the RICH all contribute to "tagging" the B decay being studied either by identifying it or the "other" B in the event as B or $\bar{B}$. This is crucial to many CP violation measurements. Methods that can be used are: identifying the kaon, muon, or electron from decay of the "other" B; identifying the kaon (for B_s) or pion (for B^o) produced closest to the B decay of interest in phase space; or overall B jet charge. Simulations indicate that the combined efficiency for tagging (including *dilution* by events that have been incorrectly identified) that we can expect is 10% for B_d decays and 13% for B_s decays. This has been included in our sensitivity projections for the various angle measurements in the cases where they require a "tagged" sample.

3. Measuring the Parameters of the CKM Matrix

Because the CKM matrix is unitary, the product of any two rows or columns must add to zero. The six products so formed can be thought of as triangles in the complex plane whose angles must add to $180°$ if the SM is correct. The one of interest to us is the triangle formed by the product of the b and d columns. It has angles that are expected to be relatively large, which we refer to as β, α, and γ, where[3]

$$\beta = arg\left(-\frac{V_{tb}V_{td}^*}{V_{cb}V_{cd}^*}\right), \gamma = arg\left(-\frac{V_{ub}^*V_{ud}}{V_{cb}^*V_{cd}}\right), \alpha = \pi - (\beta + \gamma). \qquad (1)$$

These angles can be determined by measuring CP violation in the B system.

Two angles from others of the six triangles are also of special interest because they provide important tests of the SM. These are χ and χ', where

$$\chi = arg\left(-\frac{V_{cs}^*V_{cb}}{V_{ts}^*V_{tb}}\right), \chi' = arg\left(-\frac{V_{ud}^*V_{us}}{V_{cd}^*V_{cs}}\right). \qquad (2)$$

We plan to measure these as well, although it will take somewhat longer, since they are expected to be very small, $\chi \sim 0.03$ and χ' smaller still.

3.0.1. Measuring the angle β

The sine of 2β has already been measured by BaBar and BELLE with good precision, giving a current average of 0.731 ± 0.056[4]. We expect that the two experiments will improve this number by the time BTeV starts running, but we believe we will be able to improve it still further. Our expected sensitivity to $sin(2\beta)$ using the decay $B^o \to J/\psi K_s$ at the design luminosity in only one year of running is ± 0.017.

This measurement still leaves a fourfold ambiguity in the determination of the angle β. This can partially be resolved by performing a measurement of the time evolution versus the K decay time integrated over all B decay times of the four decay rates for $B^o \to J/\psi K^o, K^o \to \pi l \nu$ for tagged B decays[5]. These are the four decay rates that arise from combinations of B and $\bar{B}$ with K and $\bar{K}$. The rates differ depending on whether $cos(2\beta)$ is positive or negative. Differences in the rates are larger for smaller $sin(2\beta)$. Using this technique we expect to be able to pin down the sign of β in about one year since we expect to reconstruct approximately 1700 untagged decays of this type and we need on the order of 100 low-background, tagged B decays for the measurement, which should be achievable given our expected effective tagging efficiency of 10% for B^o mesons.

3.1. *Measuring the angle α*

Using the decay $B^o \to \pi^+\pi^-$, our sensitivity to A_{cp} in one year is expected to be ± 0.030. However, unlike the decay $B^o \to J/\psi K_s$ used in the measurement of the angle β, the contribution from Penguin diagrams is predicted to be large in this case, making it difficult to extract the angle α. In order to unravel the various contributions we would have to measure the decays with π^o daughters as well, $B^- \to \pi^-\pi^o$ and $B^o \to \pi^o\pi^o$. The second decay is extremely difficult to detect. A more promising alternative, suggested by Snyder and Quinn [6], appears to be to extract the decay amplitudes for B^o and $\bar{B}^o$ to $\pi^+\pi^-\pi^o$ final states via the three $\rho\pi$ intermediate states which contribute using the Dalitz plot distributions. The angle between the sum of the amplitudes which contribute for B^o and the sum which contribute for $\bar{B}^o$ is precisely α.

We have performed a mini-Monte Carlo study of this technique using GEANT to simulate the acceptance and smearing of the BTeV detector. We input the signal according to the Snyder and Quinn matrix element and superposed this on an incoherent background with signal to noise of 4:1. The results of trials using 1000-event signals showed that sensitivity to α in two years of running will be $< 4^o$ for α between 77 and 111 degrees.

3.2. *Measuring the angle γ*

There are two model independent ways that we can use to measure the angle γ. The first is to perform a time-dependent, flavor-tagged analysis of the decays $B_s \to D_s^\pm K^\mp$ [7] with expected sensitivity in one year of running of $\pm 8^o$. The second is to measure the rate difference between $B^- \to D^o K^-$ and $B^+ \to D^o K^+$ [8] for which the sensitivity in one year of running is expected to be $\pm 13^o$. Other measurements which are expected to have better experimental sensitivities, $\sim 4^o$, but which have theoretical errors associated with them are to perform rate measurements of $K^o\pi^\pm$ and $K^\pm\pi^\mp$ [9], or to perform rate measurements in $K^o\pi^\pm$ and asymmetry measurements in $K^\pm\pi^o$. See Ref.[10] for a discussion of this. Another suggestion[11] is to exploit U-spin symmetry ($u \leftrightarrow d$), by measuring time-dependent asymmetries in both $B^o \to \pi^+\pi^-$ and $B_s^o \to K^+K^-$.

3.3. *Measuring the angle χ*

Asymmetry measurements using the decays of B_s to $J/\psi\eta$ and $J/\psi\eta'$ will yield an expected sensitivity of ± 0.024 for χ in one year of running. Since the angle itself is expected to be of this order it will take a number of years to get the required precision using these modes alone. The measurement using these decay modes is made possible only because of the excellent resolution provided by the PbWO$_4$ crystal calorimeter. The expected signal to noise is 15:1 for the mode containing η and 30:1 for the mode containing η'. Both muon and electron decay modes for the J/ψ are included in our estimates.

3.4. x_s reach

We have run mini-Monte Carlo studies to estimate our reach in measuring the B_s mixing parameter using the decay $B_s \rightarrow D_s^- \pi^+$, which we have found gives us the best sensitivity. From these studies we estimate that it will take us one year to measure it if it is around 70 and a little over three years if it is near 80. If it is much smaller, as predicted by the SM, we expect that it will have already been measured by CDF.

4. Other Searches for Physics Beyond the SM

We have a number of means by which we can search for physics beyond the SM other than to measure and attempt to overconstrain the CP violation terms in the CKM matrix. One is to look for decays that it predicts to be forbidden or at least very rare. Using b and c decays mediated by loop diagrams, the experiment is sensitive to mass scales of up to a few TeV. These decays may provide the *only* means by which we can distinguish among the various extensions of the SM. We also plan precise measurements of CKM parameters other than those involved in CP violation to look for inconsistencies. The large samples of charm decays we expect, e.g., of order 10^8 tagged $D^o \rightarrow K\pi$ in the first 10^7 seconds of running, should allow us to look for evidence of mixing or CP violation at the level of the SM predictions, 10^{-6} and 10^{-3}, respectively. There is a wide window in which to search for new physics between these values and the current experimental limits.

5. Comparison to Other Experiments

An e^+e^- collider operating at the $\Upsilon(4S)$ with a luminosity of $2 \times 10^{36} cm^{-2}s^{-1}$, which is two orders of magnitude higher than present luminosities, would be required to match the sensitivity of BTeV. We therefore expect the main competition for heavy quark physics in the BTeV era to come from the LHCb experiment at CERN, which, like BTeV, resembles a typical fixed target experiment and is constructed in the forward region of a hadron collider. LHCb will benefit from the higher energy of the LHC both because the production of $b\bar{b}$ pairs is five times higher and because the b production cross section relative to the total cross section is higher by a factor of about 1.6. Also, because of the shorter time between beam crossings, there will be fewer interactions in each crossing to sort through. LHCb will also come online sooner, although the machine in which it operates will be new and is likely to require some time to reach design luminosity. BTeV will have a larger and more comprehensive data set because of our ability to select events with a detached vertex using the low occupancy pixel detector placed close to the beam ($\sim$ 6mm) within a magnetic field. The large bandwidth of the data acquisition system means that we can write five times as many b's/sec to tape as can LHCb. We will gain another factor of two from longer running times each year. Also, the smaller overall size of BTeV allows use a crystal calorimeter with which we can reconstruct states containing π^o's and γ's with excellent mass resolution and thus high S/B.

The heavy quark experimental programs planned by BTeV and LHCb are very rich indeed. The two will provide leading edge physics results starting near the end of this decade.

References

1. The BTeV Collaboration - the proposal, the update to the proposal and related documents are available on the Web at
   http://www-btev.fnal.gov/public/hep/general/proposal/index.shtml.
2. U. Nierste, Fermilab, "B Physics and CP Violation", talk presented at the *PHENO 2002 Symposium*, Madison, Wisconsin, April 2002.
3. R. Aleksan, B. Kayser and D. London, *Phys. Rev. Lett.* **73**, 18 (1994) (hep-ph/9403341).
4. S. Eidelman *et al.* (Particle Data Group), Phys. Lett. **B592**, 1 (2004).
5. B. Kayser, "Cascade Mixing and the CP-Violating Angle Beta", *Les Arcs 1997, Electroweak Interactions and Unified Theories*, p389, hep-ph/9709382. Previous work in this area was done by Y. Aimov, *Phys. Rev.* **D42** 3705 (1990).
6. A. E. Snyder and H. R. Quinn, *Phys. Rev.* **D48**, 2139 (1993).
7. D. Du, I. Dunietz, and Dan-di Wu, *Phys. Rev.* **D34**, 3414 (1986). R. Aleksan, I. Dunietz, and B. Kayser, *Z. Phys.* **C54**, 653 (1992). R. Aleksan, A. Le Yaouanc, L. Oliver, O. Pène, and J.-C. Raynal, *Z. Phys.* **C67**, 251 (1995).
8. D. Atwood, I. Dunietz, and A. Soni, *Phys. Rev. Lett.* **78**, 3257 (1997). M. Gronau and D. Wyler, *Phys. Lett* **B265**, 172 (1991).
9. R. Fleischer and T. Mannel, *Phys. Rev.* **D57**, 2752 (1998).
10. M. Neubert, *JHEP* **9902**, 14 (1999), (hep-ph/9812396).
11. R. Fleischer, *Phys. Lett.* **B459**, 306 (1999).

RECENT RESULTS FROM CDF AND D0 EXPERIMENTS

C. AVILA

(For the CDF and D0 Collaborations)

Universidad de Los Andes,
Cra. 1E No. 18 A 10,
Bogota, Colombia
cavila@uniandes.edu.co

Run II preliminary physics results obtained by CDF and D0 experiments by the summer of 2004 are presented. We also discuss the status of other physics analysis as well as the search for the standard model Higgs boson and searches for new physics.

1. Introduction

Tevatron run II data taking for physics analysis started at the beginning of 2002, for both CDF and D0 experiments, and it is expected to finish in 2009. The goals of the collider experiments at Fermilab are: 1)To perform precision measurements to test the standard model (SM) and search for the SM Higgs boson; and 2) Search for new physics beyond the SM. The key parameter to accomplish those goals is the amount of integrated luminosity the experiments can record on tape. To be able to increase the luminosity delivered to the experiments, the Tevatron went trough an upgrade after the conclusion of run I. Some of the accelerator parameters that have changed from run I to run II are: a) The proton antiproton collision energy was raised from $\sqrt{s}$=1.8 TeV to $\sqrt{s}$=1.96 TeV; b) The bunch crossing time was reduced from 3.6 μs (6x6 bunches colliding) to 396 ns (36x36 bunches colliding); c) A new 150 GeV proton storage ring, the Main Injector, is now used in run II to inject proton and antiprotons to the Tevatron and also to provide higher proton intensities to the antiproton target. d) A new antiproton storage ring, the Recycler (which operates in the same Main Injector tunnel), allows to reuse antiprotons at the end of each store.

There are more few upgrades planned for the Tevatron in the period 2004-2007 aiming to increase the number of antiprotons per bunch, which is the most critical parameter at the Tevatron to obtain higher luminosities. The expected integrated luminosity by the end of run II is between 4.4 fb^{-1} to 8.6 fb^{-1} [1]. At the end of summer 2004 the integrated Luminosity already recorded on tape is about 0.5fb^{-1}, per experiment, which is a factor of 5 more statistics compared to run I. In this note we present the status of different data analysis that are being carried out by the two experiments, because there are so many analysis underway, we only select some topics to discuss here: Electroweak measurements, top quark measurements, search for SM Higgs bosons and some searches for new physics beyond the SM.

2. The Detectors

The CDF and D0 detectors are multipurpose detectors used to identify and measure the momentum and energy of electrons, muons, photons, jets and to establish missing transverse energy that is the signature of neutral particles that do not interact with the detector. There are three main systems in each collider detector: a) A tracking system within a superconducting solenoid (1.4 T at CDF and 2.0 T at D0); b) A calorimeter system; and c) A muon detection system. The tracking system consists of a silicon microstrip tracker and a central tracking system. The silicon microstrip tracker is the inner most part of the detector and cosists of different layers of silicon strip detectors surrounding the interaction point and covering a seudorapidity range of $|\eta| < 3.0$, the design of the silicon tracker for both experiments has been motivated by the large interaction region that has an extension of about $\sigma_z = 25$ cm which is due to the structure of the colliding bunches. The silicon microstrip tracker allows 3D track reconstruction with a resolution of about $10\mu m$ which is useful to identify secondary vertices from b quark decays (B hadrons travel in average about 3mm before decaying), the tagging of the b decays is crucial for many areas of physics like the top quark physics, search for higgs boson and searches for new physics. The total number of silicon readout channels is about 720,000 and 790,000 for CDF and D0 experiments, respectively. Surrounding the Silicon tracker the experiments have the central tracking system that is used to measure the momentum of charged particles by measuring the radius of curvature of their trajectories bent by the magnetic field produced by the superconducting solenoid. The CDF Central tracker is based on drift chambers with 8 concentric layers of sense wires (about 32000 wires) and a time of flight detector, while D0 uses 8 concentric layers of scintillating fibers that are readout with VLPCs, the total number of scintillating fibers at D0 is about 78,000. Outside the superconducting solenoid there is the calorimeter which is used to identify and measure the energy of electrons, photons and jets and to establish energy balance in the transverse plane to the beam. Both CDF and D0 experiments use sampling calorimeters that consist of layers of high dense inert materials inmersed in a sensitive material. CDF uses plates of lead and stainless steel sandwiched with scintillation blocks; D0 uses Uranium plate absorbers embeded in liquid Argon. The outermost part of the detectors corresponds to the muon tracking system that is composed of a magnetised toroid and few layers of drift chambers and scintillation blocks, the drift chambers allow a measurement of the momentum of the muons from the bending of their trajectory in the toroid and the scintillation blocks are used for muon trigger and measurement of time of flight which is useful to reject cosmic ray muons. Apart from the three main systems, each experiment uses roman pots that allow to tag and measure the momentum of protons and/or antiprotons after a diffractive collision. The roman pots are located very far from the interaction point in order to track protons and/or antiprotons scattered at very small angles. A more detailed information about the CDF and the D0 detectors can be found in the references [2,3].

3. Electroweak Physics

Until LHC starts operation, the Tevatron is the only place where W and Z bosons can be produced, due to the high energy collisions and high luminosities at the Tevatron, W and Z bosons are produced copiously which allows different precision measurements of their properties, among them: masses of W and Z, W decay width, W charge asimmetry, W, Z cross sections, diboson cross sections, Z-quark couplings etc. W, Z boson production at the tevatron is dominated by quark-antiquak annihilation, we only concentrate our analysis in the leptonic decays of the vector bosons, the cleanest channels are: $W \to e\nu, \mu\nu, Z \to e^+e^-, \mu^+\mu^-$. The signature for $Z \to l^+l^-$ is 2 high P_T leptons ($P_T > 15$ GeV) with opposite charge. The signature for $W \to l\nu$ is large missing transverse energy ($\not{E}_T > 25$ GeV) and a central isolated lepton with large P_T. The selected W and Z samples have been used to measure the cross section times branching ratio. Table 1 shows a summary of these measurements performed by both CDF and D0 experiments, in the table we have also included measurements on the channel $Z \rightarrow \tau^+\tau^-$ which is useful to check lepton universality. The τ decay measurements are in agreement with the measurements on the electron and muon channels.

Table 1. Run II measurements of W, Z production. The uncertainties are statistical, systematic and from luminosity measurement.

	CDF		D0	
	$L(pb^{-1})$	σ x B(Z$\rightarrow$ l$^+$l$^-$)pb	$L(pb^{-1})$	σ x B(Z$\rightarrow$ l$^+$l$^-$) pb
e	71	$255.2 \pm 3.9 \pm 5.5 \pm 15.3$	41	$275 \pm 9 \pm 9 \pm 28$
μ	72	$248.9 \pm 5.9 \pm 7.0 \pm 14.9$	117	$261.8 \pm 5.0 \pm 8.9 \pm 26.2$
τ	72	$242 \pm 48 \pm 26 \pm 15$	68	$222 \pm 36 \pm 57 \pm 22$
	$L(pb^{-1})$	σ x B(W$\rightarrow$ lν)pb	$L(pb^{-1})$	σ x B(W$\rightarrow$ lν) pb
e	72	$2782 \pm 14 \pm 61 \pm 167$	42	$2844 \pm 21 \pm 128 \pm 284$
μ	72	$2772 \pm 16 \pm 64 \pm 166$	17	$3226 \pm 128 \pm 100 \pm 322$

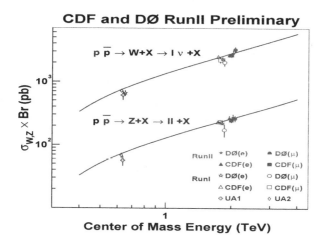

Fig. 1.   Measurements of the W and Z boson cross sections at the Tevatron as a function of center of mass energy. The solid lines are NLO predictions

Figure 1 shows a comparison of sigma x branching ratio measurements performed in run II with measurements at lower energies and with NNLO calculations. In the near future both experiments will improve the precision of their measurements by analysing data samples with higher luminosities that the ones presented here. Also measurements of W mass, W width, W charge asymmetry, W polarization, inclusive W,Z cross sections are being obtained [4] which are not discussed here. At the Tevatron we have also the chance to study diboson production (WW, WZ, ZZ) which can be used to probe the gauge structure of electroweak physics and search for anomalous couplings. The understanding of WW and WZ production is important to understand backgrounds in searches for Higgs production.

4. Top Quark Physics

The top quark, with its unusually high mass, plays a special role in the understanding of electroweak symmetry breaking. Also, indications for non-SM physics may be found as anomalies in top quark production and decay. The only source of top quarks, until LHC starts taking data, is the Tevatron. The top quark physics program at the Tevatron in run II comprises a variety of precision measurements of top quark properties that are very important to test the SM, for example, a precise measurement of the mass of the top can help to constraint the mass of the Higgs boson through electroweak fits; perturbative QCD can be tested through a precise measurement of the $t\bar{t}$ production cross section, etc.

Top quark at the Tevatron has been observed to be produced in $t\bar{t}$ pairs through $q\bar{q}$ annihilation ($\approx$ 85 %) and gg fusion ($\approx$ 15 %). Single top production is also expected via electroweak interaction with about 40 % of the $t\bar{t}$ cross section, but has not been observed yet due to the high backgrounds that are associated to the final states. Observation of single top production is a major goal for the Tevatron run II. The two leading electroweak processes that can produce a single top are: 1) s channel W* production decaying into t and b quarks ($p\bar{p} \rightarrow$ t $\bar{b}$ + X); and 2) t-channel Wg fusion ($p\bar{p} \rightarrow$ tq$\bar{b}$ + X).

The SM predicts that the top quark decays almost 100 % of the times into Wb (t$\rightarrow$ Wb), the different W decay modes produce three distinctive topologies to study final states of $t\bar{t}$ decays: 1) The dilepton channel($t\bar{t} \rightarrow W^+bW^-\bar{b} \rightarrow l\nu bl\nu b$): both W bosons decay into leptons (e, μ), the branching ratio for this topology is about 7%, the signature for this channel is two high P_T leptons, large $\not{E}_T$ from the neutrinos and two high P_T jets from the fragmentation of the two b quarks, the main backgrounds are Drell Yan production, diboson production and fake leptons; 2) The Lepton +jets channel ($t\bar{t} \rightarrow W^+bW^-\bar{b} \rightarrow l\nu$ b jjb) : one W decays leptonically and the other hadronically with a branching ratio of about 35 %, the signature for this cahnnel is one large P_T lepton, large $\not{E}_T$ and 4 jets (2 from the b quarks and 2 from the hadronic decay of the W), the main background is production of W bosons with associated jets which can be reduced with b tagging ; 3) All jets channel($t\bar{t} \rightarrow W^+ b W^- \bar{b} \rightarrow$ jjbjjb): both W's decay hadronically with a branching ratio of about 44 %, the signature is 6 jets in the final state and therefore suffers from high QCD multijet background. The standard procedure to measure the cross sections is very much the same for both experiments: count events with the expected signature, subtract SM backgrounds, apply corrections for expected acceptance, reconstruction inefficiencies and other biases and divide by the integrated luminosity that corresponds to the data sample analyzed. Table 2 shows the measurements performed by each experiment in run 1 for the $t\bar{t}$ production cross sections in each channel and the combined channels. Figure 2 shows the preliminary results that have been obtained with run II data, where the amount of luminosity used for the different analysis is shown.

Table 2. Run I measurements of $t\bar{t}$ production cross section.

	CDF	D0
channel	$\sigma_{t\bar{t}}$ (pb)	$\sigma_{t\bar{t}}$ (pb)
dilepton	$8.4^{+4.5}_{-3.5}$	6.4 ± 3.4
lepton+jets	$5.7^{+1.9}_{-1.5}$	5.2 ± 1.8
all jets	$7.6^{+3.5}_{-2.7}$	7.1 ± 3.2
combined	$6.5^{1.7}_{1.4}$	5.9 ± 1.7

The experimental signature for single top productions is one high p_T isolated lepton, high $\not{E}_T$ and two or more jets, at least one being from a b quark. The dominant backgrounds are W+jets, $t\bar{t}$ and multijets. Cross section limits at 95 % CL from run I data are: σ_t <13(22) pb and σ_{s+t} < 14 pb from CDF(D0) experiment. In run II with 162 pb^{-1} and searching in the lepton+jets channel CDF has obtained the following limits at 95% CL: σ_t <8.5 pb and σ_{s+t} <13.7 pb. D0 has also looked in the lepton+jets channel and with 164 pb^{-1} of data has obtained the following preliminary values: σ_t <19.8 pb and σ_{s+t} < 15.8 pb.

The mass of the top was measured in run 1 by both experiments in different channels as it is shown in table 3, where the kinematics of the event is used in a variety of fitting techniques that take into account the combinations of how to assigned the physical objects to the observed objects. The most precise measurement from run I comes from a recent analysis performed by D0 in the lepton + jets channel [5] where each event has an associated probability to be signal or background based on the matrix element information, this probability is convoluted with a transfer function that re-

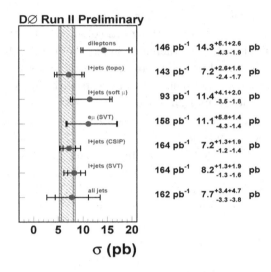

Fig. 2.   Preliminary measurements of $t\bar{t}$ cross section in run II

lates the object at the parton level to the object after reconstruction. The world average of the mass of the top from run I is now M_{top}=178.0±2.7±3.3 GeV/c^2 [6]. There are already preliminary results of analysis of the mass of the top from run II, [7] which are in agreement with results from run I but with higher errors specially due to systematic uncertainties that need better understanding.

Table 3. Run I measurements of the top quark mass.

	CDF	D0
channel	M_{top} (GeV/c^2)	M_{top} (GeV/c^2)
dilepton	167.4 ±10.3±4.8	168.4±12.3±3.6
lepton+jets	176.1 ± 4.8 ±5.3	180.1 ±3.6±3.9
all jets	186.0 ±10.0±5.7	

There are other analysis using top events to measure W helicity [8] properties of top decays [9] that we don't discuss here.

5. Search for the SM Higgs Boson

The Higgs boson is the only particle predicted by the SM that has not been observed yet. Electroweak simmetry breaking is explained, within the

SM, by the Higgs mechanism where the W and Z bosons interact with a scalar field (the Higgs field), the coupling of the vector bosons to the Higgs field explains the masses of those bosons. If fermions are also coupled to the Higgs field that would also explain the origin of their masses. The discovery of the Higgs boson (the quantum of the Higgs field) is one of the major goals in high energy physics. At the Tevatron, the production cross section for the Higgs boson, as expected from the standard model, is low (of the order of 0.1pb to 1pb depending on its mass and the production mechanism) and the backgrounds are very high, therefore high integrated luminosities are needed to discover the Higgs boson. Precision electroweak measurements, including the mass of the top, can be used to estimate the mass of the Higgs boson. Using the recent world average for the mass of the top, the most probable value obtained for the mass of the Higgs is 117 GeV with an upper limit of 251 GeV [5]. Direct searches from LEP have excluded the region of masses below 114.4 GeV [10]. Given that the electroweak fits favor a light higgs boson and that the Tevatron performance has drastically improved during 2004, there are still good chances for the Fermilab experiments to discover the Higgs.

The main production mechanism is gluon fusion (gg → H). For a low mass Higgs ($M_{higgs} < 135 GeV$) the dominant decay mode is H→ $b\bar{b}$, however it is almost imposible to extract the Higgs signal from this channel due to the high backgrounds of 2 b jets from QCD. To search for a Higgs boson in this mass range, we have to look for a Higss produced associated with a W or Z (HW, HZ production) and with the vector bosons dacaying leptonically. For a Higgs with high mass ($M_{higgs} > 135 GeV$) the dominant decay mode is H→ $WW^{(*)}$, and then look for the leptonic decays of the W's. Those are the modes being exploited by CDF and D0 experiments. Some extensions of the standard model predict an enhancement of the decay H→ $\gamma\gamma$ which is also being searched as a possible evidence of physics beyond the SM [11].

A recent study of the Higgs discovery potential of the Tevatron collider experiments in run II [12] shows that to make a 5 σ discovery of a Higgs with a mass of 120 GeV each experiment would need to accumulate about 10 fb^{-1} of data, and a 3 σ discovery of a Higgs with that mass would require about 4 fb^{-1} of data per experiment. The luminosity expectations for run II are within that range. With the amount of luminosity recorded at the moment the experiments are calibrating their detetectors and tuning their algorithms to understand background processes and establishing new limits on the Higgs boson production cross section. Figure 3 shows the new limit established by the D0 experiment in the cross section x branching ratio for the channel H→ $WW^{(*)}$ using a luminosity of about 174 pb^{-}1 [13].

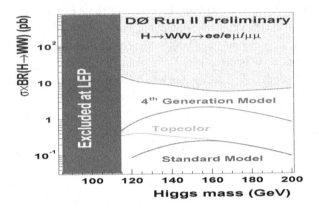

Fig. 3.   D0 limits on H→ WW

6. Searches for New Physics

The standard model has been very successful in predicting many phenomena in high energy physics that have been tested experimentally with great accuracy, but still there are many questions that are left out without an answer, among some of them: we still do not have an explanation for the mechanism of electroweak symmetry breaking, we don't understand why the planck scale is so many orders of magnitudes higher than electroweak scale ($M_{planck}/M_{EW} \approx 10^{16}$) , etc. New theoretical models have been proposed to answer some of these questions: Large extra dimensions, supersimmetry, leptoquarks, Higgs bosons beyond SM, Technicolor, etc. Those models predict new signatures beyond the SM that can be tested by the collider experiments at Fermilab.

Models predicting extra dimensions postulate that gravity propagates freely in n extra, compact spatial dimensions, which explains why in our 3+1 dimensions gravity is so weaker than the other fundamental forces. There are two experimental signatures that the experiments look for: 1) in hadron-hadron collisions gravitons can be produced together with a jet or a photon, the graviton then propagates in the extra dimensions, that are invisible to our detectors, producing as a signature missing transverse energy ($\not{E}_T$) associated to a jet or a photon. 2) Virtual graviton exchange will modify the di-photon and di-lepton production cross sections which can be written as:

$$\frac{d^2\sigma}{dMdcos\theta^*} = f_{SM} + f_{int}\eta_G + f_{KK}\eta_G^2 \qquad (1)$$

Where f_{SM} accounts for the SM cross section, f_{int} is the interference of the graviton induced amplitude and the SM amplitude, f_{KK} is the amplitude for graviton exchange and η_G is the parameter that accounts for extra dimensions, $\eta_G = F/M_{planck}^4$, different formalisms give different values for F [14], [15], [16]; No excess of events over background have been observed yet. New limits on the value of the planck scale have been established by both experiments for each formalism [17], within the GRW formalism [14] D0 has established a 95 % CL of $M_{plank} > 1.43$ TeV, CDF has established a slightly lower limit.

Supersymmetry (SUSY) is the most popular extension of the SM that postulates that every known particle should have a more massive partner which has the same quantum numbers except for the spin. When looking for signatures of supersymmetry the trilepton signature is considered the golden signature for associated production of charginos and neutralinos because the SM backgrounds are very small. D0 has been searching for 4 different signatures within this channel: two electrons + an isolated track, two muons + an isolated track, one electron + one muon + an isolated track, two muons with the same charge. The data samples used correspond to an integrated luminosity of 147 - 249 pb^{-1}. After all cuts have been applied there are 1,1,0,1 events remaining with expected SM backgrounds of 0.7 $\pm$0.5,1.8$\pm$0.5,0.3$\pm$0.3 and 0.1$\pm$0.1 events for the four channels mentioned above. Many other searches for supersimmetry are being carried out [18] which are not discussed here.

Another search that is being carried out if for leptoquarks. Leptoquarks are color triplet bosons that carry both lepton and baryon numbers which are predicted in models like Technicolor, GUT, SUSY with R parity violation. At the Tevatron leptoquarks are expected to be produced in pairs either by gluon gluon fusion or quark-antiquark annihilation and expected to decay in a quark and a lepton. Leptoquarks can be scalar (spin=0) or vector (spin=1). We consider three generations of leptoquarks according to the type of quarks they decay. Searches for first generation leptoquarks with 175 pb^{-1} by D0 have produced lower 95 % CL limit in the mass of the first generation leptoquarks, in the channel $LQ\overline{LQ} \rightarrow e^{\pm}e^{\mp}q\overline{q}$ $M_{LQ} > 238$ GeV and for $LQ\overline{LQ} \rightarrow e^{\pm}\nu q\overline{q}$ the limit is $M_{LQ} > 194$ GeV. Searches for second generation leptoquarks with 198 pb^{-1} in the channel $LQ\overline{LQ} \rightarrow \mu^{\pm}\mu^{\mp}q\overline{q}$ D0 has obtained a lower limit (95 % Cl) of M_{LQ} >240 GeV [19].

Many other searches for new physics are being carried out by both experiments which are not discussed here [20].

Acknowledgments

The beams division at Fermilab is making a tremendous effort to provide higher luminosities to both collider experiments. CDF and D0 collaborations are producing new measurements that are very important for the high energy physics community, thanks to all the people involved in this huge effort. I also want to thank the V Silafae organizing commitee for their hospitality and their efforts to get a well organised symposium.

References

1. P. Garbincius, *hep-ex*/ **0406013**.
2. F. Abe, et al *Nucl. Instrum. Meth.* **A271**, 387 (1988)
3. S. Abachi, et al,*FERMILAB-PUB*-**96-357-E** (1996).
4. S. Mattingly, et al, *hep-ex*/**0409024**
5. V. M. Abazov et al, *Nature* **429**, 638 (2004)
6. CDF collaboration, D0 Collaboration, Tevatron Electroweak working group *hep-ex* **0404010**.
7. L. Cerrito, *hep-ex* **0405046**.
8. V. M. Abazov et al, *hep-ex*/**0404040**.
9. K. Sliwa, *hep-ex*/**0406061**.
10. LEP collaborations, *Phys. Lett.* **B 565**, 61 (2003).
11. A. Melnitchouk,*Fermilab-thesis* **2003-23** (2003).
12. CDF and D0 collaborations,*FERMILAB-PUB* **03/320-E** (2003).
13. D0 collaboration,*hep-ex*/-**0410062**.
14. G. Giudice, R. Ratazzi, *J. Wells, Nucl. Phys.* **B544**, 3 (1999).
15. T. Hann, J. Lykken, R. Zhang, *Phys. Rev. D* **59**, 105006 (1999).
16. J. Hewett, *Phys. Rev. Lett.* **82**, 4765 (1999).
17. A. Pompos, *hep-ex*/ **0408004**.
18. A. Meyer, *hep-ex*/ **0411072**.
19. S. Ming, *hep-ex*/ **0405075**.
20. J. Linnemann, *hep-ex*/ **0405079**.

MATTER UNDER EXTREME CONDITIONS THE ALICE EXPERIMENT*

G. HERRERA CORRAL

and

ALICE-MEXICO†

Physics Department, CINVESTAV, P.O. Box 14 740
Mexico, D.F. 07300 , Mexico
gherrera@fis.cinvestav.mx

The ALICE experiment will study heavy ion collisions at the highest energy ever achieved. The main goal of ALICE is the observation of a transition of ordinary matter into a plasma of quarks and gluons as well as the investigation of this new state of matter. An important part of the ALICE detector is the V0 system, which consists of two sub-detectors: the V0A located at 3.4 mts. from the interaction point on one side and the V0C located at 0.9 mts. on the other side. These detectors are discs of scintillator plastic segmented in cells which are read with optical fibers. The response of the detector should be fast to provide the signal for the level 0 trigger. The cosmic ray trigger is also an important subsystem of ALICE. The detector consists of an array of scintillator panels which will cover the magnet. It will provide a veto signal and will collect interesting cosmic ray events when ALICE does not take data with the collider. The system will be also a calibration tool for other devices of ALICE. We focus on the description of these systems in which several mexican institutions are participating.

1. Introduction

The Large Hadron Collider (LHC) is now in construction at the European Center for Nuclear Research (CERN). In year 2007 it will provide beams of protons and heavy ions at an unprecedented energy in the center of mass. Four experiments will be performed at the LHC. The ALICE experiment (A Large Ion Collider Experiment) will be dedicated to the study of heavy

---

*This work is supported by CONACYT.
†The ALICE-Mexico group is listed at the end of the contribution.

ion collisions. CMS (Compact Muon Solenoid) and ATLAS (A Toroidal LHC AparatuS) will focus on the physics of proton -proton interactions. The LHCb experiment (Large Hadron Collider beauty experiment) will do precision measurements of CP violation and rare decays.

The accelerator will run in proton-proton mode 90% of the time and will deliverheavy ions the remaining 10% of the run time, this means 10^7s effective time in the proton-proton mode and 10^6s in heavy ion. In lead-lead collisions the buch crossing will be 125 ns while in the proton-proton mode it will be 25 ns.

The LHC will accelerate lead ions and make them collide with an energy of 5.5 TeV in center of mass per nucleon pair.

The ALICE experiment will study the high density matter created in Pb-Pb collisions also known as plasma of quarks and gluons. At the energies achieved by the LHC, the density, the size and the lifetime of the excited quark matter will be high enough as to allow a careful investigation of the properties of this new state of matter. The temperature will exceed by much the critical value predicted for the transition to take place. The particle production will be dominated by hard processes providing a mean to study the system in the earliest stage. The ALICE detector will have tracking and particle identification system over a wide range of transverse momentum which goes from 100 MeV/c to 100 GeV/c. Topics like parton energy loss in a dense medium, heavy quark and jet production, prompt photons etc will be addressed providing important information in that energy regime.

Here we will not give a description of the physics that will be studied with the ALICE detector. We refer the interested reader to ref. [1].

A longitudinal view of the ALICE detector is shown in Fig. 1. A detailed description of the ALICE detector can be found in ref. [2].

In the forward direction a set of tracking chambers inside a dipole magnet will measure muons while electrons and photons are measured in the central region. Photons will be measured in a high resolution calorimeter 5 m below from the interaction point in the central part. The PHOS is built from which has a high light output. The track measurement is performed with a set of six barrels of silicon detectors and a large Time Projection Chamber. The Time Projection Chamber has an effective volume of 88 . It is the largest TPC ever built. These detectors will make available information on the energy loss allowing particle identification too. In addition to this a Transition Radiation Detector and a Time of Flight system will provide excellent electron identification and particle separation for pions, kaons and protons at intermediate momentum respectively. The Time of

Flight system uses Multi-gap Resistive Plate Chambers (MRPC) with a total of 160,000 readout channels. A Ring Imaging Cherenkov will extend the particle identification capability to higher momentum particles. It covers 15% of the acceptance in the central area and will separate pions from kaons with momenta up to 3 GeV/c and kaons from protons with momenta up to 5 GeV/c. A Forward Multiplicity Detector consisting of silicon strip detectors and a Zero Degree Calorimeter will cover the very forward region providing information on the charge multiplicity and energy flow. An array of Cherenkov counters will give the precise time of the interaction. The V0 system consisting of two scintillation counters on each side of the interaction point will be used as the main interaction trigger. The central part of the detector will have a moderate magnetic field. In the top of the magnet a Cosmic Ray Detector will signal cosmic muons arrival.

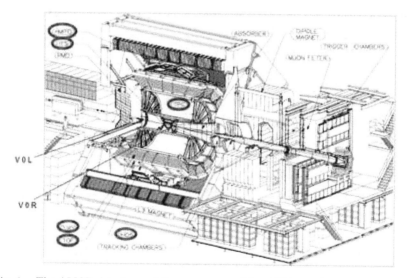

Fig. 1. The ALICE detector and its components. The ACORDE (A Cosmic Ray Detector) on the top of the magnet and the V0A are constructed in Mexico.

2. The V0 Detector

The V0 system consists of two detectors (V0A and V0C) which will be located in the central part of ALICE. The V0A will be installed at a distance of 340 cm from the interaction point as shown in Fig. 2, mounted in two rigid half boxes around the beam pipe. Each detector is an array of 32 cells of scintillator plastic, distributed in 4 rings forming a disc with 8 sectors as shown in Fig. 3. For the V0C the cells of rings 3 and 4 are divided into two identical pieces that will be read with a single photo-multiplier. This is done to achieve uniformity of detection and a small time fluctuation. The light produced in the scintillator plastic is collected by wavelength shifting fibers and transported to photomultipliers by clear optical fibers. This setup is based on experimental tests and simulations.

The V0 detector system has several functions: 1. On line vertex determination. With the use of the timing information one should be able to locate the main vertex. 2. Track distribution consistent with beam-beam interaction. This will allow a strong suppression of undesirable beam-gas events. 3. On-line centrality measurement. With a rough multiplicity measurement one may discriminate central from non central collisions. 4. It will provide a wake up signal to the electronics of the Transition Radiation Detector within 100 ns after the interaction. 5. Luminosity measurement by counting triggered events.

In the pp mode the mean number of charged particles within 0.5 units of rapidity is about 3. Note that each ring covers approximately 0.5 units of rapidity (Table 1). The particles coming from the main vertex will interact with other components of the detector generating secondary particles. In general, each cell of the V0 detector will on the average register one hit. For this reason the detector should have a very high efficiency.

Table 1.   Pseudo-rapidity (η) and angular acceptances (degrees) of the V0A detector rings.

Ring	η_{max}	η_{min}	ang.accep.max	ang.accep.min
1	5.1	4.5	0.7	1.3
2	4.5	3.9	1.3	2.3
3	3.9	3.4	2.3	3.8
4	3.4	2.8	3.8	6.9

In Pb-Pb collisions the number of particles in a similar pseudo-rapidity range could be up to 4000 once secondary particles are included. Comparing the number of hits in the detector in the pp and Pb-Pb mode we can see that the required dynamic range will be 1–500 minimum ionizing particles. The segments of the V0A detector will be obtained with a megatile construction (see ref. [3]). This technique consists of machining the scintillator plastic and filling the grooves with loaded epoxy in order to separate one sector from the other. Since this is done all the way through the plastic the epoxy restores mechanical strength. The surface of the slices are then wrapped with teflon tape. The wavelength shifting fibers that collect the light are then embedded in the plastic in 3 mm deep grooves on both sides of the cell. These fibers bring the light to the edge of the disc where they are coupled to clear fibers with optical connectors. The construction is made of 2 cm thick scintillator plastic from Bicron (BC404). The wavelength shifting fibers WLS (BC9929AMC) are 1 mm diameter and the clear fibers (BCF 98 MC) are 1.1 mm diameter also from Bicron. At the end of the WLS inside the plastic, aluminum coating is used as a reflector. To achieve optimal light transmission from the WLS to clear fibers, silicone optical grease from Bicron is used. A single connector will joint the WLS fibers from a cell to a clear fiber. The photomultiplier tubes (PMT) will be installed inside the magnet not far from the detector. In order to tolerate the magnetic field, fine mesh tubes have been chosen. The final decision on the PMT to be used is still under study.

In order to fulfill its functions the V0 electronics should be compatible with the proton-proton and heavy ion modes. The system will provide the following information: 6. Minimum-bias trigger: this signal is generated if at least one channel of V0A and one of V0C is fired. 7. Beam-gas trigger: one signal is generated by V0A in case a beam gas occurs on its side and the other by the V0C in case it occurs on the opposite side. 8. Centrality triggers: these are generated if one of the following two conditions or the two of them are fulfilled: — the charged seen in V0A and V0C after a collisions is larger than a threshold — the number of channels after the collision is larger than an expected value. 9. A measure of the multiplicity. This is obtained measuring the total charge by analyzing the anode pulse from the PMTs. 10. A measure of the time difference between particles detected and the beam crossing signal. 11. A quick wake-up signal for the Transition Radiation Detector.

A more detailed description of the V0 system can be found in ref. [4].

3. The Cosmic Ray Detector for ALICE

The cosmic array on top of the ALICE magnet will serve two purposes: 12. Performing as a cosmic ray trigger for the TRD and TPC detectors in the calibration phases. 13. Study of rare cosmic ray events in conjunction with the ALICE tracking detectors. Together with other ALICE devices it will

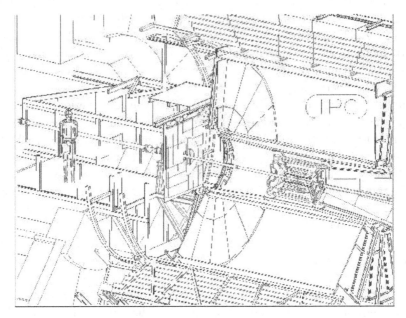

Fig. 2.   Location of the V0A detector in front of the Photon Multiplicity Detector. Here only one half of the detector in a D shaped box is shown.

provide interesting information about cosmic rays originated by primaries with an energy of eV. In particular multi-muon events can be studied with the use of the Time Projection Chamber. The Cosmic Ray Detector consists of an array of 60 scintillator counters located in the upper part of the ALICE magnet. Fig. 4 shows the distribution of the panels in its final position. The plastic used for the construction of the detector was part of the DELPHI detector [5]. The material was carefully studied and the design of the detector was done according to the capabilities of the plastic available. The material was transported to Mexico where the construction is taking place. Each module has a sensitive area of and is built with two superimposed plastics. The doublet has an efficiency of 90 % or more along the module.

The electronics design is also made in Mexico [6]. The requirements are as follows: 14. It should generate a single muon trigger to calibrate the Time Projection Chamber and other components of ALICE. 15. It should generate a multi-muon trigger to study cosmic rays. 16. It will provide a wake-up signal for the Transition Radiation Detector. 17. It should include a calibration system for periodical testing. 18. It will scan 120 channels (but

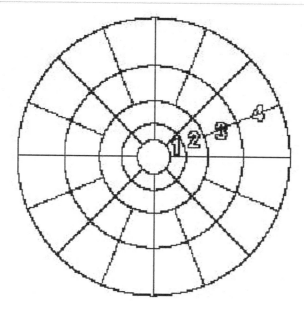

Fig. 3. Segmentation of the V0 detectors. On the left a prototype of V0 rings 3 and 4 is shown.

it is designed to scan up to 200 channels for possible enlargement of the surface covered) in synchronization with the LHC clock. — It will produce a coincidence signal when two overlapped plastics are fired. — It generates a trigger signal when cosmic muons are detected.

The electronics should generate the signal in 100 ns and will provide spatial information of the scintillator fired. It will also store the information on this spatial location. Fig. 5 shows a box diagram of the coincidence module for the cosmic ray detector. As can be seen, sets of two plastics are put in coincidence after a discriminator.

In the beginning of 2005, 20 panels will be sent to CERN to be mounted at the top of the Time Projection Chamber before the latter is lowered in the ALICE cavern. Later on, the remaining 40 panels will be deliver and installed on the top of the magnet.

Acknowledgments

I would like to thank the organizers of the V SILAFAE for the invitation to deliver this talk . The financial support provided by them made possible my participation in the event.

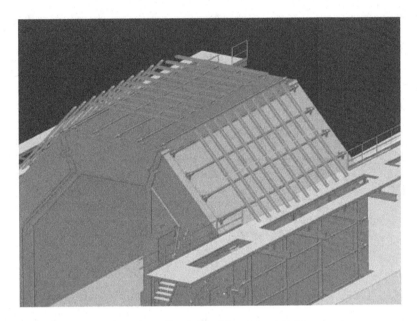

Fig. 4.   Array of panels on the top of the ALICE magnet.

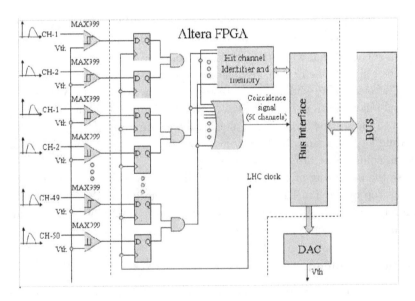

Fig. 5.   Electronic coincidence module for the cosmic ray detector.

ALICE-Mexico

G. Contreras (Departamento de Física Aplicada, CINVESTAV), J. García García, A. Cerna, L. Montaño, G. Herrera Corral, A. Zepeda (Departamento de Física, CINVESTAV), A. Ayala, E. Cuautle, J.C. D Olivo, S. Vergara, M. I. Martínez, L. Nellen, G. Paic (Instituto de Ciencias Nucleares, UNAM), R. Alfaro, E. Belmont, V. Grabski, A. Martínez, A. Menchaca (Instituto de Física, UNAM), L. Villaseñor (Universidad de Michoacán), A. Fernández, R. López, M. A. Vargas (Universidad de Puebla).

References

1. Journal Phys. G: Nucl. Part. Phys 30 (2004)1517
2. ALICE Technical Proposal, Ahmad et al. CERN/LHCC/ 95-71, 1995
3. S. Kim, et al. Nucl. Inst. Meth. Phys. Res. A306(1995)206
4. ALICE Technical Design Report, Forward Detectors FMD, T0, V0 CERN/LHCC-2004-025, 2004.
5. R.I. Dzhelyadin, et al., DELPHI Internal Note 86-108, TRACK 42 , CERN 1986
6. S. Vergara, private communication. ALICE Internal Note in preparation.

RECENT RESULTS FROM PHOBOS AT RHIC

EDMUNDO GARCIA[6]*

for the PHOBOS Collaboration

B.B.Back[1], M.D.Baker[2], M.Ballintijn[4], D.S.Barton[2], R.R.Betts[6], A.A.Bickley[7],
R.Bindel[7], W.Busza[4], A.Carroll[2], Z.Chai[2], M.P.Decowski[4], E.Garcia[6], T.Gburek[3],
N.George[2], K.Gulbrandsen[4], C.Halliwell[6], J.Hamblen[8], M.Hauer[2], C.Henderson[4],
D.J.Hofman[6], R.S.Hollis[6], R.Hołyński[3], B.Holzman[2], A.Iordanova[6], E.Johnson[8],
J.L.Kane[4], N.Khan[8], P.Kulinich[4], C.M.Kuo[5], W.T.Lin[5], S.Manly[8], A.C.Mignerey[7],
R.Nouicer[2,6], A.Olszewski[3], R.Pak[2], C.Reed[4], C.Roland[4], G.Roland[4], J.Sagerer[6],
H.Seals[2], I.Sedykh[2], C.E.Smith[6], M.A.Stankiewicz[2], P.Steinberg[2], G.S.F.Stephans[4],
A.Sukhanov[2], M.B.Tonjes[7], A.Trzupek[3], C.Vale[4], G.J.van Nieuwenhuizen[4],
S.S.Vaurynovich[4], R.Verdier[4], G.I.Veres[4], E.Wenger[4], F.L.H.Wolfs[8], B.Wosiek[3],
K.Woźniak[3], B.Wysłouch[4]

[1] Argonne National Laboratory, Argonne, IL 60439-4843, USA
[2] Brookhaven National Laboratory, Upton, NY 11973-5000, USA
[3] Institute of Nuclear Physics PAN, Kraków, Poland
[4] Massachusetts Institute of Technology, Cambridge, MA 02139-4307, USA
[5] National Central University, Chung-Li, Taiwan
[6] University of Illinois at Chicago, Chicago, IL 60607-7059, USA
[7] University of Maryland, College Park, MD 20742, USA
[8] University of Rochester, Rochester, NY 14627, USA

The PHOBOS detector is one of four heavy-ion experiments at the Relativistic
Heavy Ion Collider (RHIC) at Brookhaven National Laboratory. In this paper
we will review some of the results of PHOBOS from the data collected in p+p,
d+Au and Au+Au collisions at nucleon-nucleon center-of-mass energies up to
200 GeV. In the most central Au+Au collisions at the highest energy, evidence
is found for the formation of a very high energy density and highly interactive
system, which can not be described in terms of hadrons, and which has a
relatively low baryon density.

---

*Corresponding author. E-mail: ejgarcia@uic.edu

1. Introduction

The bulk of the hadronic matter is made of quarks and gluons (partons), bound into neutrons and protons. These hadrons then form nuclear structures held together by the strong interactions mediated by the gluons. The fundamental interactions between partons are well described by the theory of Quantum Chromodynamics (QCD). However, due to the non-commutative nature of the parton interactions, the phase structure of strongly interacting matter is only partially understood in terms of QCD. The knowledge we have of the properties of strongly interacting matter comes from experimental data.

At distances close to the hadronic size, the QCD coupling constant decreases with decreasing distance between the partons. As a consequence it is expected that at high temperatures a parton conglomerate should have properties of an ideal relativistic gas [1], traditionally designated as the Quark-Gluon Plasma (QGP). The transition between the QGP and the strongly interacting matter that exist at normal temperature and density (protons and neutrons) happens, according to QCD, with the spontaneous breaking of "chiral symmetry" and "deconfinement" [2].

Lattice gauge calculations suggest that at low baryon densities there is a phase differentiation in a highly interactive matter below and above a critical temperature $T_c \sim 170$ MeV, or energy density $\epsilon \sim 1$ GeV/fm^3 [3]. The closest approach to the creation of matter under these conditions is achieved in relativistic heavy-ion collisions. The most recent facility for the study of these collisions is the Relativistic Heavy Ion Collider (RHIC). Four experiments at RHIC have studied the collisions of p+p, Au+Au and d+Au systems at nucleon-nucleon center-of-mass energies, $\sqrt{s_{NN}}$, from 19.6 GeV to 200 GeV. Data from these experiments are being studied to get a better understanding of the physics of heavy-ion collisions and in particular to search for the evidence of the phase transition of the highly interactive parton matter or QCD matter. This paper summarizes some of the results obtained by the PHOBOS collaboration.

The PHOBOS apparatus is composed of three major subsystems; a charged particle multiplicity detector covering almost the entire solid angle; a two arm magnetic spectrometer with particle identification capability, and a suite of detectors used for triggering and centrality determination. A full description of the PHOBOS detector and its properties can be found in Ref. [4]. Also, a description of some of the techniques used for the PHOBOS event selection (triggering) and event characterization (vertex position and centrality of the collision) can be found in Ref. [5].

2. The Formation of a Very High Energy Density State at RHIC

Figure 1 shows the evolution of the midrapidity charged particle density $dN_{ch}/d\eta]_{|\eta|\leq 1}$, per participating nucleon pair, $N_{part}/2$, as a function of $\sqrt{s_{NN}}$ [6]. The data are consistent with a logarithmic extrapolation from lower energies as shown by the solid line drawn in the plot. The mid rapidity particle density at $\sqrt{s_{NN}}=$ 200 GeV is almost a factor of two higher than the value observed at the maximum SPS energy. The details of the evolution of the charged particle pseudorapidity density are shown in Fig. 2, where $dN_{ch}/d\eta$ is presented for Au+Au collisions at $\sqrt{s_{NN}}=$19.6, 130, and 200 GeV for various centralities [7]. The particle densities peak near midrapidity and increase with both $\sqrt{s_{NN}}$ and centrality.

An approximation of the total energy density in the system created at midrapidity in Au+Au collisions at $\sqrt{s_{NN}}=$200 GeV can be calculated from the charged particle pseudorapidity density, the average energy per particle and the volume from where the system was originated. Figure 3 shows the transverse momentum distributions of identified particles emitted at mid rapidity in central Au+Au collisions at $\sqrt{s_{NN}}=$200 GeV. The yield at low transverse momentum (p_T) was measured by PHOBOS [8] and the high p_T data was measured by PHENIX [9]. Fits to the yield make it possible to estimate the average transverse mass for all charged particles, $\langle m_T \rangle \simeq 600$ MeV/c, which is equal to the transverse energy at midrapidity. Under the assumption of a spherically symmetric distribution in momentum space, which would have equal average transverse and longitudinal momenta.

As for the volume where the system originated, elliptic flow results discussed below suggest that an upper limit of the time for the system to reach approximate equilibrium is of the order of 1–2 fm/c. Using the upper range of this estimate and assuming that the system expands during this time in both longitudinal and transverse directions one obtains a conservative estimate for the energy density produced at RHIC when the system reaches approximate equilibrium of $\epsilon \geq 3$ GeV/fm^3. This estimate is about 6 times the energy density inside the nucleons. Thus, a very high energy system is created whose description in terms of hadronic degrees of freedom is inappropriate.

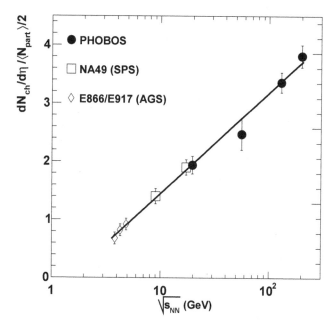

Fig. 1. Evolution of the midrapidity charged particle density $dN_{ch}/d\eta]_{|\eta|\leq1}$, per particating nucleon pair, $N_{part}/2$, as a function of collision energy. Solid line is a logarithmic extrapolation of the data from lower energies drawn to guide the eye.

3. Baryon Chemical Potential at RHIC Energies: Approach to a Baryon Free Environment

One of the most interesting results from heavy-ion collisions at lower energies was the observation that the production ratios for different particles with cross-section varying over several orders of magnitude can be described in terms of a statistical picture of particle production assuming chemical equilibrium [10]. One of the key components in the particle production mechanism is the baryon chemical potential μ_B which can be extracted from the kaon to antikaon and proton to antiproton ratios. The measurement of the ratios of charged antiparticles to particles produced at RHIC is shown in Fig. 4. This figure compares the antiparticle to particle ratios for both kaons and protons to the corresponding data at lower energies [11].

The system formed at RHIC is closer to having equal number of particles and antiparticles than that found at lower energies. The measured ratio for antiproton to proton production as function of the energy indicates that

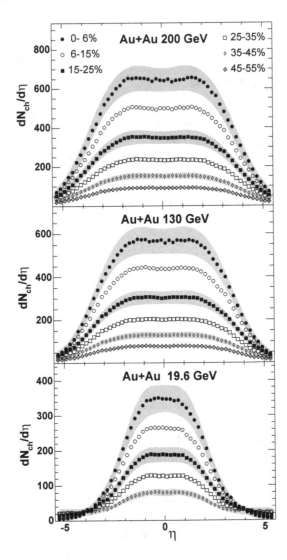

Fig. 2.   Pseudorapidity density of charged particles emitted in Au+Au collisions at three different values of the nucleon-nucleon center-of-mass energy. Data are shown for a range of centralities, labeled by the fraction of the total inelastic cross section in each bin, with smaller numbers being more central. Grey bands shown for selected centrality bins indicate the typical systematic uncertainties (90% C.L.). Statistical errors are smaller than the symbols.

the system approaches smaller values of μ_B at higher energies. Assuming a hadronization temperature of 160–170 MeV, a value of $\mu_B = 27$ MeV was

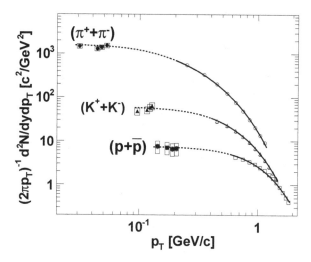

Fig. 3. Transverse momentum distributions of identified charged particles emitted near midrapidity in central Au+Au collisions at $\sqrt{s_{NN}}$=200 GeV. Invariant yield data shown are from PHENIX at higher momenta and PHOBOS at lower momenta. Boxes around the PHOBOS data indicate systematic uncertainties. Fits to PHENIX measurements are shown by solid curves ($\propto 1/[e^{(m_T/T)} + \epsilon]$, where $\epsilon = -1$ and $+1$ for mesons and baryons, respectively, m_T is the transverse mass, and T is the fit parameter). Note that the extrapolations (dashed curves) of the fit to the data at higher momenta are consistent with the low momentum yields.

found for central Au+Au collisions at $\sqrt{s_{NN}}$=200 GeV, showing that the system created at RHIC is close to a baryon-free medium.

4. Interaction Strength in High Energy Density Medium

The medium produced in heavy-ion collisions at RHIC energies was initially thought to consist of a weakly interacting plasma of quarks and gluons [12]. From experimental evidence it has been found however, that the nature of the system formed is not weakly interacting. As shown in Fig. 3, the production of particles with low p_T is consistent with the extrapolations from a fit to the distributions in the range from around 200 MeV to few GeV. If at RHIC, a medium of weakly interacting particles was formed, one would expect an enhancement of the production of particles with wavelengths roughly equal to the size of the collision volume (coherent pion production) [13]. The absence of an excess of particles at low transverse momentum is a manifestation of the strongly interacting nature of the medium produced at RHIC, which also gives rise to a large radial and elliptic flow signal.

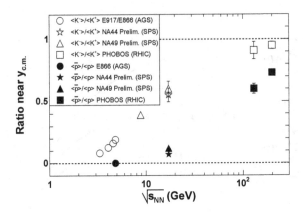

Fig. 4.   Ratios of identified antiparticles over particles measured near midrapidity in central collisions of Au+Au and PHOBOS at RHIC and Pb+Pb as a function of nucleon-nucleon center-of-mass energy. Error bars are statistical only.

Figure 5 shows the magnitude of elliptic flow (v_2) measured by PHO-BOS around midrapidity ($|\eta| \leq 1$) in Au+Au collisions at $\sqrt{s_{NN}}$=130 GeV and 200 GeV as a function of the number of participants (N_{part}) [14,15]. The elliptic flow signal is strong over a wide range in centrality and close to the value predicted by a relativistic hydrodynamics calculation [18]. The observation of an azimuthal asymmetry in the out-going particles is evidence of early interactions. Furthermore, from the strength of the observed elliptic flow and the known dimensions of the overlap region it can be estimated that the initially formed medium equilibrates in a time less than about 2 fm/c [16], the value used earlier in the calculation of the energy density.

Figure 6 shows the yield of charged particles per participant pair divided by a fit to the invariant cross section for proton-antiproton collisions (200 GeV UA1 [17]) plotted as a function of p_T, for the most peripheral and most central interactions. The dashed and solid lines show the expectation of the scaling of yields with the number of collisions ($\langle N_{coll} \rangle$) and the number of participants $\langle N_{part} \rangle$, respectively. The brackets show the systematic uncertainty. We can see that the general shape of the curves is only weakly dependent on the centrality and for p_T values up to 2 GeV/c there is an increase of the yield in Au+Au collisions compared to p$\bar{\text{p}}$. Above 2 GeV/c, the relative yield decreases for all centrality bins.

Two possible explanations have been suggested for this yield suppression as function transverse momentum relative to N_{coll} scaling for central collisions. The first ("initial state") model suggests that the effect is due to a

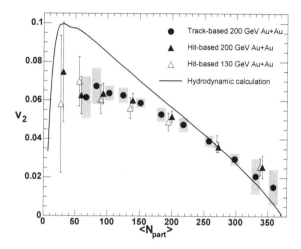

Fig. 5. Elliptic flow of charged particles near midrapidity ($|\eta| < 1$) as a function of centrality in Au+Au collisions at $\sqrt{s_{NN}}$=200 GeV (closed circles and triangles, are measurement using two different analysis methods) and at $\sqrt{s_{NN}}$=130 GeV (open triangles). Grey boxes show the systematic errors (90% C.L.) for the 200 GeV data. The curve shows the prediction from a relativistic hydrodynamics calculation.

modification of the wave functions of the nucleons in the colliding ions. This produces an effective increase of the interaction length in such a way that the nucleons in a nucleus will interact coherently with all the nucleons in the other nucleus in the longitudinal dimension [19]. This model predicts not only the yield quenching at high p_T, but also the scaling of the yield with the number of participants. The second ("final state") model suggests that the suppression comes from an increase of the energy loss of the partons traveling through the hot dense medium formed in the Au+Au collisions [20]. One way to discriminate between these models is to reduce the size of the hot and dense medium, by colliding d+Au instead of Au+Au. RHIC's 2003 run was dedicated mainly to d+Au collisions at $\sqrt{s_{NN}} = 200$ GeV. Details of the analysis and a broader discussion of p_T spectra are given in Ref. [21].

In Fig. 7 we present the nuclear modification factor R_{dAu} as function of p_T for four centrality bins defined as the ratio of the d+Au invariant cross section divided by the p$\bar{\text{p}}$ UA1 yield, scaled by the proton-antiproton invariant cross section (41 mb) and $\langle N_{coll} \rangle$. For the centrality bins there is a rapid rise of R_{dAu} reaching a maximum at around p_T= 2 GeV/c. For comparison, the R-factor is also plotted for the most central bin for Au+Au

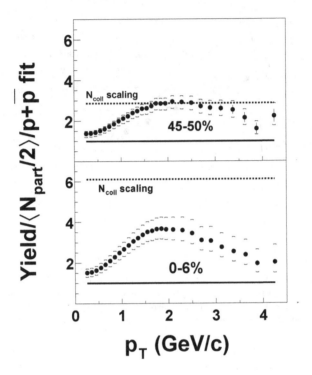

Fig. 6.   Ratio of the yield of charged hadrons in Au+Au as a function of p_T for the most peripheral and the most central bin to a fit of proton antiproton data scaled by $\langle N_{part}/2 \rangle$. The dashed(solid) line shows the expectation of $N_{coll}(N_{part})$ scaling relative to p$\bar{\text{p}}$ collisions. The brackets show the systematic uncertainty.

collisions at $\sqrt{s_{NN}} = 200$ GeV. In striking contrast to the behavior of R_{dAu}, R_{AuAu} also increases initially as function of p_T but above 2 GeV/c it decreases sharply.

The suppression of the inclusive yield observed in central Au+Au collisions is consistent with final-state interactions with in a dense medium crated in such collisions. Thus we find that the matter created at RHIC interacts strongly with high-p_T partons as expected for a partonic medium.

5. Final Remarks

In central A+Au collisions at the highest RHIC energies, a very high energy density medium is formed. Conservative estimates of its density are around 3 GeV/fm^3. This is greater than hadronic densities and it is inappropriate to

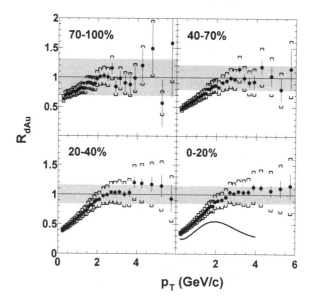

Fig. 7. Nuclear modification factor R_{dAu} as a function of p_T for four centrality bins. For the most central bin, the spectral shape for central Au+Au data relative to $p + \bar{p}$ is shown for comparison. The shaded area shows the uncertainty in R_{dAu} due to the systematic uncertainty in $\langle N_{coll} \rangle$ and the UA1 scale error (90% C.L.). The brackets show the systematic uncertainty of the d+Au spectra measurement (90% C.L.).

describe this medium in terms of hadronic degrees of freedom. The medium has been found to have a low baryon chemical potential. The transition to this state does not create abrupt changes in any observable including charged particle multiplicity, elliptic flow, interferometric radii, and other derived quantities which have been studied so far. This lack of strong signals will make it difficult to delineate the boundaries between high density highly interactive matter and hadronic states.

Acknowledgments

This work was partially supported by U.S. DOE grants DE-AC02-98CH10886, DE-FG02-93ER40802, DE-FC02-94ER40818, DE-FG02-94ER40865, DE-FG02-99ER41099, and W-31-109-ENG-38, by U.S. NSF grants 9603486, 0072204, and 0245011, by Polish KBN grant 1-P03B-062-27 (2004–2007), and by NSC of Taiwan Contract NSC 89-2112-M-008-024.

References

1. D. J. Gross, R. D. Pisarski, and L. G. Yaffe, *Rev. Mod. Phys.* **53**, 43 (1981).
2. R. D. Pisarski and F. Wilczek, *Phys. Rev.* **D29**, 338 (1984).
3. F. Karsch, *Nucl. Phys.* **A698**, 199 (2002).
4. B. B. Back *et. al.*, (PHOBOS), *Nucl. Inst. Meth.* **A499**, 603 (2003).
5. B. B. Back *et. al.*, (PHOBOS), *Phys. Rev.* **C70**, 021902(R) (2004).
6. B. B. Back *et. al.*, (PHOBOS), submitted to *Nucl. Phys A*; arXiv:nucl-ex/0410022 (2004).
7. B. B. Back *et. al.*, (PHOBOS), *Phys. Rev. Lett.* **91**, 052303 (2003).
8. B. B. Back *et. al.*, (PHOBOS), submitted to *Phys. Rev. Lett.*; arXiv:nucl-ex/0401006 (2004).
9. S. S. Adler *et. al.*, (PHENIX), *Phys. Rev.* **C69**, 034909 (2004).
10. P. Braun-Munzinger *et. al.*, *Phys. Lett. B* **465**, 15 (1999).
11. B. B. Back *et. al.*, (PHOBOS), *Phys. Rev. Lett.* **87** 102301 (2001).
12. E. V. Shuryak, arXiv:hep-ph/0405066.
13. W. Busza, Proceedings of NATO Advanced Study Institute on Particle Production in Highly Excited Matter, 149 (1993).
14. B. B. Back *et. al.*, (PHOBOS), *Phys. Rev. Lett.* **89** 222031 (2002).
15. B. B. Back *et. al.*, (PHOBOS), submitted to *Phys. Rev. C (RC)*; arXiv:nucl-ex/0407012 (2004).
16. P. F. Kolb and U. Heinz, Hydrodynamical description of ultrarelativistic heavy ion collisions. *World Scientific* (2004); arXiv:nucl-th/0305084 (2003).
17. G. Arnison et al., (UA1), Phys. Lett. B 118 (1982) 167.
18. U. W. Heinz, and H. Heiselberg, *Phys. Lett.* **500**, 232 (2001).
19. D. Kharzeev, E. Levin and L. McLerran, arXiv:hep-ex/0210332 (2002).
20. M. Gyulassy, I. Vitev, X. Wang and B. Zhang arXiv:hep-ex/0302077 (2003).
21. B. B. Back *et. al.*, (PHOBOS), *Phys. Rev. Lett.* **91** 072302 (2003).

CONTRIBUTIONS

SO(10) AS THE MINIMAL SUPERSYMETRIC GUT

ALEJANDRA MELFO

Centro de Física Fundamental, Universidad de Los Andes, Mérida, Venezuela

It is argued that the minimal renormalizable supersymmetric SO(10) GUT with the usual three generations of spinors has a Higgs sector consisting only of a light 10-dimensional and heavy 126 + $\overline{126}$ and 210 supermultiplets. It accounts correctly for fermion masses and mixings and has the MSSM with exact R-parity as the low-energy limits, yet it has only 26 real parameters, which makes it candidate for the minimal consistent supersymmetric grand unified theory.

Introduction   In the last few years, the accumulating evidence for neutrino masses has given new fuel to the search for a supersymmetric Grand Unified Theory. The situation seems to point away from the SU(5) scenario, which requires a lot of parameters in the Yukawa sector in order to accommodate massive neutrinos. SO(10), on the other hand, is custom fit to explain small neutrino masses in a simple and fairly predictive manner. The crucial question is the choice of the Higgs superfields that can achieve the breaking down to the MSSM.

One possibility is to use small representations and consider higher dimensional operators, suppressed by M_{Pl}, it naturally has a large number of free parameters; the other is to stick to the renormalizable theory and use large representations, at the price of becoming strong between M_{GUT} and the Planck scale. The second one is based on pure grand unification, with the hope that the Planck scale physics plays a negligible role, and it is the subject of this talk.

Towards Unification: Supersymmetric Pati-Salam   Quark-Lepton unification can be considered a first step towards the complete SO(10) unification of a family of fermions in a single representation.

To simplify the discussion, imagine a two-step breaking of the PS symmetry down to the MSSM. The first steps breaks G_{PS} down to its maximal subgroup, the LR [1] (Left-Right) group $SU(2)_L \times SU(2)_R \times U(1)_{B-L} \times SU(3)_c$, and it is simply achieved through the VEV of and adjoint representation ($(15,1,1)$ in the G_{PS} language) at a scale M_c. In turn, the breaking of the LR group can be achieved by having $SU(2)_R$ triplets fields, with $B - L = \pm 2$, acquiring a VEV at a lower scale M_R. Triplets will couple to fermions and give a mass to right-handed neutrino, providing the see-saw mechanism [2] at the renormalizable level. Here, they are contained in the $(10, 1, 3), (\overline{10}, 1, 3)$ representations.

There is a more profound reason for preferring the triplets. They have an even $B - L$ number, and thus preserve matter parity [3], defined as $M = (-1)^{3(B-L)}$, equivalent to the R-parity $R = M(-1)^{2S}$. This in turn means R-parity is not broken at a high scale. But then it can be easily shown that it cannot be broken afterwards, at the low energy supersymmetry breaking or electroweak scale: a spontaneous breakdown of R-parity through the sneutrino VEV (the only candidate) would result in the existence of a pseudo-Majoron with its mass inversely proportional to the right-handed neutrino mass. This is ruled out by the Z decay width [4].

It is straightforward to show that the standard model singlets take VEVs in the required directions to achieve the symmetry breaking. The interesting point here is that the breaking in the minimal model leaves a number of fields potentially light [5], in particular doubly-charged fields in the right-handed triplets get a mass only at the supersymmetry-breaking scale. These states are common in supersymmetric theories that include the Left-Right group, and have been subject of experimental search. In a similar way, a color octet in $(15, 1, 1)$ has a mass of order M_R^2/M_c, and could in principle be light. The unification constraints give the interesting possibility

$$10^3 GeV \leq M_R \leq 10^7 GeV \quad 10^{12} GeV \leq M_c \leq 10^{14} GeV$$

which gives a chance for the LHC discovering them at the TeV scale.

Surely the most interesting feature of a low scale of PS symmetry breaking is the possibility of having $U(1)_{B-L}$ monopoles, with mass $m_M \simeq 10M_c$. If produced in a phase transition via the Kibble mechanism, the requirement that their density be less than the critical density then implies $M_c \leq 10^{12} GeV$. Together with the potentially light states mentioned above, these intermediate mass monopoles could be the signature of PS unification.

Minimal SO(10)   There are a number of features that favor the choice of SO(10):

(1) A family of fermions is unified in a 16-dimensional spinorial representation, predicting the existence of right-handed neutrinos.

(**2**) LR symmetry is a finite gauge transformation in the form of charge conjugation.

(**3**) In the supersymmetric version, matter parity, equivalent to R-parity, is a gauge transformation a part of the center Z_4 of SO(10). By the same argument as before, R-parity remains exact at all energies [6]. The lightest supersymmetric partner is then stable and is a natural candidate for the dark matter of the universe.

(**4**) Its other maximal subgroup, besides $SU(5) \times U(1)$, is the PS group discussed above. It explains immediately the somewhat mysterious relations $m_d = m_e$ (or $m_d = 1/3 m_e$) of SU(5).

(**5**) The unification of gauge couplings can be achieved with or without supersymmetry.

(**6**) The minimal renormalizable version offers a connection between $b - \tau$ unification and a large atmospheric mixing angle in the context of the type II see-saw [7].

Such a theory is easily built with large representations in the Higgs sector [9,10], 210_H, $126_H + \overline{126}_H$, 10_H, as can be seen from its decompositions under G_{PS}:

$$
\begin{aligned}
\mathbf{10} &= (1,2,2) + (6,1,1) \\
\overline{\mathbf{126}} &= (\overline{10},3,1) + (10,1,3) + (15,2,2) + (6,1,1) \\
\mathbf{126} &= (10,3,1) + (\overline{10},1,3) + (15,2,2) + (6,1,1) \\
\mathbf{210} &= (1,1,1) + (15,1,1) + (15,1,3) + (15,3,1) + (6,2,2) + (10,2,2) + (\overline{10},2,2)
\end{aligned}
$$

We recognize here the representations containing the right-handed triplets (in 126_H and $\overline{126}_H$) and the $SU(4)_c$ adjoint (in 210_H) that we required in the previous section. The $(1,1,1)$ in 210_H achieves the first breaking $SO(10) \rightarrow G_{PS}$.

With this content the theory is not asymptotically free any more above M_{GUT}, and the SO(10) gauge couplings becomes strong at the scale $\Lambda_F \lesssim 10 M_{GUT}$. The Higgs superpotential is surprisingly simple

$$
\begin{aligned}
W_H = {} & m_{210}(210_H)^2 + m_{126}\overline{126}_H 126_H + m_{10}(10_H)^2 + \lambda(210_H)^3 \\
& + \eta 126_H \overline{126}_H 210_H + \alpha 10_H 126_H 210_H + \overline{\alpha} 10_H \overline{126}_H 210_H \quad (1)
\end{aligned}
$$

and it has a remarkably small number of couplings, seven complex parameters in total, which reduce to 10 real, after phase redefinition. The Yukawa sector is just

$$
W_Y = 16_F (y_{10} 10_H + y_{126} \overline{126}_H) 16_F \quad (2)
$$

with only 3 real (say) y_{10} couplings after diagonalization, and 6 x 2 = 12 symmetric y_{126} couplings, 15 in total. Adding the gauge sector, we end up with a theory with only 26 real parameters, the same as in the MSSM.

It can be shown that the required symmetry breaking is possible [10,13], and the usual minimal fine-tuning leaves the MSSM doublets light. Now, the success of gauge coupling unification in the MSSM favors a single step breaking of SO(10), so that $\langle (1,1,15)_{210} \rangle \simeq M_{GUT}$. As is well known [8], this induces a VEV for the doublets in $\overline{126}$

$$\langle (2,2,15)_{\overline{126}} \rangle \simeq \langle (2,2,1)_{10} \rangle \tag{3}$$

and the light Higgs is a mixture of (at least) $(2,2,1)_{10}$ and $(2,2,15)_{\overline{126}}$. Since $(2,2,15)_{\overline{126}}$ is an adjoint of $SU(4)_c$, being traceless it gives $m_\ell = -3m_d$, unlike $\langle (2,2,1)_{10} \rangle$, which implies $m_\ell = m_d$. In other words, the $\langle 10_H \rangle$ must be responsible for the $m_b \simeq m_\tau$ relation at M_{GUT}, and the $\langle \overline{126}_H \rangle$ for the $m_\mu \simeq 3m_s$ relation at M_{GUT}. In this theory, the Georgi-Jarlskog [14] program becomes automatic.

Another important feature is the connection between $b - \tau$ unification and a large atmospheric mixing angle through the BSV mechanism [7]. Since $\langle 10_H \rangle = \langle (2,2,1) \rangle$ is a Pati-Salam singlet, the difference between down quark and charged lepton mass matrices must come purely from $\langle \overline{126}_H \rangle$

$$M_d - M_e \propto y_{126} \tag{4}$$

Now, suppose the see-saw mechanism is dominated by the so-called type II; that is, neutrino masses come directly from the VEV of the left-handed triplet in $\overline{126}$, and thus through the same Yukawa coupling y_{126}. We have a very simple relation

$$M_\nu \simeq y_{126} \langle (3,1,\overline{10})_{126} \rangle \propto M_d - M_e \tag{5}$$

Let us now look at the 2nd and 3rd generations first. In the basis of diagonal M_e, and for the small mixing ϵ_{de}

$$M_\nu \propto \begin{pmatrix} m_\mu - m_s & \epsilon_{de} \\ \epsilon_{de} & m_\tau - m_b \end{pmatrix} \tag{6}$$

obviously, large atmospheric mixing can only be obtained for $m_b \simeq m_\tau$. The three generation numerical studies support a type II see-saw with the prediction of a large θ_{13} and a hierarchical neutrino mass spectrum [12].

Outlook   We have argued that the minimal supersymmetric GUT is the one based on SO(10) and the Higgs sector 210_H, $126_H + \overline{126}_H$, 10_H. Recently, a full analysis has being done of the symmetry breaking and the particle spectrum [13,15]. This should enable us to address in detail such issues as proton decay, neutrino masses and leptogenesis in this scenario. Work in this direction is already in progress by several groups, and hopefully will soon provide testable predictions.

References

1. J. C. Pati and A. Salam, Phys. Rev. D **10** (1974) 275. R. N. Mohapatra and J. C. Pati, Phys. Rev. D **11** (1975) 2558. G. Senjanović and R. N. Mohapatra, Phys. Rev. D **12** (1975) 1502. G. Senjanović, Nucl. Phys. B **153** (1979) 334.
2. P. Minkowski, Phys. Lett. B **67** (1977) 421. T. Yanagida, proceedings of the *Workshop on Unified Theories and Baryon Number in the Universe*, Tsukuba, 1979, eds. A. Sawada, A. Sugamoto, KEK Report No. 79-18, Tsukuba. S. Glashow, in *Quarks and Leptons, Cargèse 1979*, eds. M. Lévy. et al., (Plenum, 1980, New York). M. Gell-Mann, P. Ramond, R. Slansky, proceedings of the *Supergravity Stony Brook Workshop*, New York, 1979, eds. P. Van Niewenhuizen, D. Freeman (North-Holland, Amsterdam). R. Mohapatra, G. Senjanović, Phys.Rev.Lett. **44** (1980) 912
3. R. N. Mohapatra, *Phys. Rev.* **D 34**, 3457 (1986). A. Font, L. E. Ibáñez and F. Quevedo, *Phys. Lett.* **B228**, 79 (1989). S. P. Martin, *Phys. Rev.* **D46**, 2769 (1992).
4. C.S. Aulakh, K. Benakli, G. Senjanović, Phys. Rev. Lett. **79** (1997) 2188. C. S. Aulakh, A. Melfo, A. Rašin and G. Senjanović, Phys. Lett. B **459** (1999) 557.
5. C. S. Aulakh, A. Melfo and G. Senjanović, Phys. Rev. D **57**, 4174 (1998). Z. Chacko and R. N. Mohapatra, Phys. Rev. D **58** (1998) 015003 C. S. Aulakh, B. Bajc, A. Melfo, A. Rasin and G. Senjanovic, Phys. Lett. B **460** (1999) 325
6. C. S. Aulakh, B. Bajc, A. Melfo, A. Rašin and G. Senjanović, Nucl. Phys. B **597** (2001) 89.
7. B. Bajc, G. Senjanović and F. Vissani, Phys. Rev. Lett. **90** (2003) 051802.
8. K. S. Babu and R. N. Mohapatra, Phys. Rev. Lett. **70**, 2845 (1993).
9. T. E. Clark, T. K. Kuo and N. Nakagawa, Phys. Lett. B **115** (1982) 26. D. Chang, R. N. Mohapatra and M. K. Parida, Phys. Rev. D **30** (1984) 1052. X. G. He and S. Meljanac, Phys. Rev. D **41** (1990) 1620. D. G. Lee, Phys. Rev. D **49** (1994) 1417. D. G. Lee and R. N. Mohapatra, Phys. Rev. D **51** (1995) 1353.
10. C. S. Aulakh, B. Bajc, A. Melfo, G. Senjanović and F. Vissani, Phys. Lett. B **588**, 196 (2004).
11. B. Bajc, G. Senjanović and F. Vissani, arXiv:hep-ph/0402140.
12. H. S. Goh, R. N. Mohapatra and S. P. Ng, Phys. Lett. B **570** (2003) 215. H. S. Goh, R. N. Mohapatra and S. P. Ng, Phys. Rev. D **68** (2003) 115008.
13. B. Bajc, A. Melfo, G. Senjanović and F. Vissani, Phys. Rev. D **70** (2004) 035007.
14. H. Georgi and C. Jarlskog, Phys. Lett. B **86**, 297 (1979).
15. T. Fukuyama, A. Ilakovac, T. Kikuchi, S. Meljanac and N. Okada, arXiv:hep-ph/0401213. C. S. Aulakh and A. Girdhar, arXiv:hep-ph/0204097. C. S. Aulakh and A. Girdhar, arXiv:hep-ph/0405074.

A SUPERSYMMETRIC THREE-FAMILY MODEL
WITHOUT HIGGSINOS*

WILLIAM A. PONCE

Instituto de Física, Universidad de Antioquia,
A.A. 1226, Medellín, Colombia
wponce@naima.udea.edu.co

LUIS A. SÁNCHEZ

Escuela de Física, Universidad Nacional de Colombia,
A.A. 3840, Medellín, Colombia
lasanche@unalmed.edu.co

The supersymmetric extension of a three-family model based on the local gauge group $SU(3)_c \otimes SU(3)_L \otimes U(1)_X$ is constructed. Anomalies are canceled among the three families in a non trivial fashion. In the model the slepton multiplets play the role of the Higgs scalars and no higgsinos are needed.

1. Introduction

Two of the most intriguing puzzles in modern particle physics concerns the number of fermion families in nature and the pattern of fermion masses and mixing angles. One interesting attempt to answer to the question of family replication is provided by the 3-3-1 extension of the Standard Model (SM) of the strong and electroweak interactions. This extension is based on the local gauge group $SU(3)_c \otimes SU(3)_L \otimes U(1)_X$ which has among its best features that several models can be constructed so that anomaly cancellation is achieved by an interply between the families, all of them under he condition that the number of families equals the number of colors of $SU(3)_c$ (three-family models)[1].

*This work is supported by CODI, Universidad de Antioquia, and Universidad Nacional.

2. The Non-Supersymmetric Model

The model we are going to supersymmetrize is based on the local gauge group $SU(3)_c \otimes SU(3)_L \otimes U(1)_X$. It has 17 gauge bosons: one gauge field B^μ associated with $U(1)_X$, the 8 gluon fields G^μ associated with $SU(3)_c$ which remain massless after breaking the symmetry, and other 8 gauge fields associated with $SU(3)_L$ and that we write as[2]

$$\frac{1}{2}\lambda_\alpha A^\mu_\alpha = \frac{1}{\sqrt{2}} \begin{pmatrix} D^\mu_1 & W^{+\mu} & K^{+\mu} \\ W^{-\mu} & D^\mu_2 & K^{0\mu} \\ K^{-\mu} & \bar{K}^{0\mu} & D^\mu_3 \end{pmatrix},$$

where $D^\mu_1 = A^\mu_3/\sqrt{2} + A^\mu_8/\sqrt{6}$, $D^\mu_2 = -A^\mu_3/\sqrt{2} + A^\mu_8/\sqrt{6}$, and $D^\mu_3 = -2A^\mu_8/\sqrt{6}$. λ_i, $i = 1, 2, ..., 8$, are the eight Gell-Mann matrices normalized as $Tr(\lambda_i\lambda_j) = 2\delta_{ij}$. The electric charge operator Q is given by

$$Q = \frac{\lambda_{3L}}{2} + \frac{\lambda_{8L}}{2\sqrt{3}} + XI_3, \tag{1}$$

where $I_3 = Dg.(1, 1, 1)$ is the diagonal 3×3 unit matrix, and the X values are related to the $U(1)_X$ hypercharge and are fixed by anomaly cancellation. The sine of the electroweak mixing angle is given by $S^2_W = 3g^2_1/(3g^2_3 + 4g^2_1)$ where g_1 and g_3 are the coupling constants of $U(1)_X$ and $SU(3)_L$ respectively, and the photon field is given by

$$A^\mu_0 = S_W A^\mu_3 + C_W \left[\frac{T_W}{\sqrt{3}} A^\mu_8 + \sqrt{(1 - T^2_W/3)} B^\mu \right], \tag{2}$$

where C_W and T_W are the cosine and tangent of the electroweak mixing angle. There are two neutral currents in the model which we define as

$$Z^\mu_0 = C_W A^\mu_3 - S_W \left[\frac{T_W}{\sqrt{3}} A^\mu_8 + \sqrt{(1 - T^2_W/3)} B^\mu \right],$$

$$Z'^\mu_0 = -\sqrt{(1 - T^2_W/3)} A^\mu_8 + \frac{T_W}{\sqrt{3}} B^\mu, \tag{3}$$

where Z^μ_0 coincides with the weak neutral current of the SM.

The quark content for the three families is[2]: $Q_{iL} = (u_i, d_i, D_i)^T_L \sim (3, 3, 0)$, $i = 2, 3$, for two families, where D_{iL} are two exotic quarks of electric charge $-1/3$ (the numbers inside the parentheses stand for the $[SU(3)_c, SU(3)_L, U(1)_X]$ quantum numbers in that order); $Q_{1L} = (d_1, u_1, U)^T_L \sim (3, 3^*, 1/3)$, where U_L is an exotic quark of electric charge $2/3$. The right-handed quarks are: $u^c_{aL} \sim (3^*, 1, -2/3)$, $d^c_{aL} \sim (3^*, 1, 1/3)$, with $a = 1, 2, 3$, a family index, $D^c_{iL} \sim (3^*, 1, 1/3)$, and $U^c_L \sim (3^*, 1, -2/3)$.

The lepton content is given by the three anti-triplets $L_{\alpha L} = (\alpha^-, \nu^0_\alpha, N^0_\alpha)^T_L \sim (1, 3^*, -1/3)$, the three singlets $\alpha^+_L \sim (1, 1, 1)$, $\alpha = e, \mu, \tau$, and the vectorlike structure $L_{4L} = (N_4^0, E_4^+, E_5^+)_L^T \sim (1, 3^*, 2/3)$, and $L_{5L} = (N_5^0, E_4^-, E_5^-)_L^T \sim (1, 3, -2/3)$; where N_s^0, $s = e, \mu, \tau, 4, 5$, are five neutral Weyl states, and $E_\eta^-, \eta = 4, 5$ are two exotic electrons.

With the former quantum numbers it is just a matter of counting to check that the model is free of the following chiral anomalies: $[SU(3)_c]^3$ ($SU(3)_c$ is vectorlike); $[SU(3)_L]^3$ (seven triplets and seven anti-triplets), $[SU(3)_c]^2 U(1)_X$; $[SU(3)_L]^2 U(1)_X$; $[grav]^2 U(1)_X$ (the so called gravitational anomaly) and $[U(1)_X]^3$.

The minimal scalar sector, able to break the symmetry and to provide, at the same time, with masses for the fermion fields is given by [2]: $\chi_1^T = (\chi_1^-, \chi_1^0, \chi_1'^0) \sim (1, 3^*, -1/3)$, and $\chi_2^T = (\chi_2^0, \chi_2^+, \chi_2'^+) \sim (1, 3^*, 2/3)$, with vacuum expectation values (VEV) given by $\langle\chi_1\rangle^T = (0, v_1, V)$ and $\langle\chi_2\rangle^T = (v_2, 0, 0)$, with the hierarchy $V \gg v_1, v_2$. These VEV break the symmetry $SU(3)_c \otimes SU(3)_L \otimes U(1)_Y \longrightarrow SU(3)_c \otimes U(1)_Q$ in one single step.

The consistency of the model requires the existence of eight would-be Goldstone bosons in the scalar spectrum, out of which four are charged and four are neutral (one CP-even state and three CP-odd) in order to provide with masses for[3] $W^\pm$, $K^\pm$, K^0, $\bar{K}^0$, Z^0 and Z'^0.

This model sketched by Ponce[2,4] *et al*, has not been studied in the literature as far as we know. (A related model without the vector-like structure $L_{4L} \oplus L_{5L}$ and with a scalar sector of three triplets instead of two, has been partially analyzed by Foot, Long and Tran[5].)

3. The Supersymmetric Extension

When we introduce supersymmetry in the SM, the entire spectrum of particles is doubled as we must introduce the superpartners of the known fields, besides two scalar doublets ϕ_u and ϕ_d must be used in order to cancel the triangle anomalies; then the superfields $\hat{\phi}_u$ and $\hat{\phi}_d$, related to the two scalars, may couple via a term of the form $\mu\hat{\phi}_u\hat{\phi}_d$ which is gauge and supersymmetric invariant, and thus the natural value for μ is expected to be much larger than the electroweak and supersymmetry breaking scales. This is the so-called μ problem.

In a non supersymmetric model as the one presented in the former section, in which the Higgs fields transform as some of the lepton fields under the symmetry group, we can construct the SUSY extension with the scalar and the lepton fields acting as superpartners of each other, ending up with a SUSY model without Higgsinos[3], and free of chiral anomalies.

For three families we thus have the following chiral superfields: $\hat{Q}_a$, $\hat{u}_a$,

$\hat{d}_a$, $\hat{D}_i$, $\hat{U}$, $\hat{L}_a$, $\hat{e}_a$, and $\hat{L}_\eta$, plus gauge bosons and gauginos, where $a = 1, 2, 3$ is a family index, $i = 1, 2$, and $\eta = 4, 5$. The identification of the gauge bosons eigenstates in the SUSY extension follows the non-SUSY version.

The most general $SU(3)_c \otimes SU(3)_L \otimes U(1)_X$ invariant superpotential is

$$W = h_{ia}^u \hat{Q}_i \hat{u}_a \hat{L}_4 + h_i^U \hat{Q}_i \hat{U} \hat{L}_4 + h_{iab}^d \hat{Q}_i \hat{d}_a \hat{L}_b + h_{ija}^D \hat{Q}_i \hat{D}_j \hat{L}_a$$

$$+ h_a^{d\prime} \hat{Q}_1 \hat{d}_a \hat{L}_5 + h_i^{\prime D} \hat{Q}_1 \hat{D}_i \hat{L}_5 + h_{ab}^e \hat{L}_a \hat{e}_b \hat{L}_5 + \frac{1}{2} \lambda_{ab} \hat{L}_a \hat{L}_b \hat{L}_4$$

$$+ \mu \hat{L}_4 \hat{L}_5 + \lambda_{abi}^{(1)} \hat{u}_a \hat{d}_b \hat{D}_i + \lambda_{ai}^{(2)} \hat{U} \hat{d}_a \hat{D}_i + \lambda_{ijk}^{(3)} \hat{Q}_i \hat{Q}_j \hat{Q}_k$$

$$+ \lambda_{abc}^{(4)} \hat{u}_a \hat{d}_b \hat{d}_c + \lambda_{aij}^{(5)} \hat{u}_a \hat{D}_i \hat{D}_j + \lambda_{ab}^{(6)} \hat{U} \hat{d}_a \hat{d}_b + \lambda_{ij}^{(7)} \hat{U} \hat{D}_i \hat{D}_j, \qquad (4)$$

where summation over repeated indexes is understood, and the chirality, color and isospin indexes have been omitted.

The $\hat{u} \hat{d} \hat{D}$, $\hat{U} \hat{d} \hat{D}$, $\hat{U} \hat{d} \hat{d}$, $\hat{U} \hat{D} \hat{D}$, $\hat{u} \hat{d} \hat{d}$, $\hat{u} \hat{D} \hat{D}$, and $\hat{Q} \hat{Q} \hat{Q}$ terms violate baryon number and can possibly lead to rapid proton decay. We may forbid these interactions by introducing an anomaly free discrete Z_2 symmetry with the following assignments of Z_2 charge q

$$q(\hat{Q}_a, \hat{u}_a, \hat{U}, \hat{D}_i, \hat{d}_a, \hat{L}_\mu, \hat{\mu}) = 1; \quad q(\hat{L}_e, \hat{L}_\tau, \hat{e}, \hat{\tau}, \hat{L}_4, \hat{L}_5) = 0, \qquad (5)$$

where we have used $\hat{e}_1 \equiv \hat{e}$, $\hat{e}_2 \equiv \hat{\mu}$, $\hat{e}_3 \equiv \hat{\tau}$, $\hat{L}_1 \equiv \hat{L}_e$, $\hat{L}_2 \equiv \hat{L}_\mu$, and $\hat{L}_3 \equiv \hat{L}_\tau$. This is just one among several anomaly-free discrete symmetries available. This symmetry protects the model from a too fast proton decay, but the superpotential still contains operators inducing lepton number violation, which is desirable if we want to describe Majorana masses for neutrinos.

The Z_2 symmetry also forbids some undesirable mass terms for the spin $1/2$ fermions which complicate unnecessarily several mass matrices. But notice the presence of a μ term in the superpotential. So, contrary to the model by Mira *et al*[3], this model has a μ-term coming from the existence of the vector like structure $\hat{L}_{4L} \oplus \hat{L}_{5L}$.

References

1. F.Pizano and V.Pleitez, *Phys. Rev.* **D46**, 410, (1992), P.H.Frampton, Phys. Rev. Lett. **69**, 2887 (1992).
2. W.A.Ponce, Y.Giraldo and Luis A. Sánchez, *Phys. Rev.* **D67**, 075001 (2003).
3. J.M.Mira, W.A.Ponce, D.A.Restrepo and L.A.Sánchez, *Phys. Rev.* **D67**, 075002 (2003).
4. W.A.Ponce, J.B.Flórez and L.A.Sánchez, *Int. J. Mod. Phys.* **A17**, 643 (2002).
5. R.Foot, H.N.Long and T.A.Tran, *Phys. Rev.* **D53**, 437 (1996).

AREA-PRESERVING DIFFEOMORPHISMS GROUPS, THE MAJORANA REPRESENTATION OF SPINS, AND $SU(N)$*

JOHN D. SWAIN

Dept. of Physics,
Northeastern University,
Boston, MA 02115, USA
john.swain@cern.ch

It is by now common practice to regard large N limits of classical groups, especially $SU(N)$ for large N, as being connected in some way with groups $SDiff(M)$ of area-preserving diffeomorphisms of 2-dimensional manifolds. The limit-taking is problematic since one can rigorously demonstrate that $SDiff(M)$ never has the same de Rham cohomology nor homotopy as any $SU(N)$ or other large N classical group, or products thereof, even with $N = \infty$. Nevertheless, a connection can be made between $SU(n)$ for any n and a group action on a finite set of points on a sphere using the Majorana representation of spin-$\frac{n}{2}$ quantum states by sets of n points on a sphere. This allows for a realization of $SU(n)$ acting on such states, and thus a natural action on the two-dimensional sphere S^2. There is no need to work with a special basis of the Lie algebra of $SU(n)$, and there is a clear geometrical interpretation of the connection between $SU(n)$ and $SDiff(S^2)$.

1. Introduction

This talk is based on references [1,2,3] where the reader can find more complete references to the literature. For reasons of space, and with apologies to the many relevant authors, almost no other references are given here.

The basic point is that while there is a large literature claiming that area-preserving diffeomorphisms of various 2-surfaces can be in some sense approximated by the large N limits of classical groups, the approximation is very poor when it comes to the topology of these groups. A summary of these results can be found in [4].

---

*Work partially supported by the National Science Foundation.

Nevertheless, there is an interesting relationship between a certain $SU(N)$ action on unordered sets of points on the sphere which could have some bearing on the problem of connecting $SU(N)$ for large N with area-preserving diffeomorphisms of surface. We refer the reader interested in the proofs of the non-isomorphism of any $SDiff$ group with any large classical group to the references cited above.

2. The Majorana Representation of Spin

Note that a system with classical angular momentum $\vec{J}$ can be described by a single point on S^2 corresponding to the direction in which $\vec{J}$ points. The case in quantum mechanics is more subtle. A general state of spin j (in units of $\hbar$) must be represented by a collection of $2j$ points on the surface of a sphere, as first shown by Majorana [5] and later by Bacry [6] (for a nice summary of both papers, see reference [7]).

We reproduce the argument here for completeness, as it provides a connection between the action of $SU(2j)$ on the complex projective Hilbert space $\mathbb{C}^{2j}$ representing states of spin j and diffeomorphisms of S^2.

Let $\mathbb{CP}^{2j}$ denote the projective space associated with $\mathbb{C}^{2j+1}$. This is the space of $2j + 1$-tuples in $\mathbb{C}^{2j+1}$ considered equivalent if they differ by a complex scale factor λ. That is, two points $(a_1, a_2, ..., a_{2j+1})$, and $(\lambda a_1, \lambda a_2, ..., \lambda a_{2j+1})$ of $\mathbb{C}^{2j+1}$ are considered the same point of $\mathbb{CP}^{2j}$.

Now consider the set P_{2j} of nonzero homogeneous polynomials of degree $2j$ in two complex variables, x, and y, which we associate to $(2j+1)$-tuples as follows:

$$(a_1, a_2, ..., a_{2j+1}) \rightarrow a_1 x^{2j} + a_2 x^{2j-1} y + \cdots + a_{2j+1} y^{2j} \tag{1}$$

Now (by the fundamental theorem of algebra) the polynomials can be factored into a product of $2j$ (not necessarily different) homogeneous terms linear in x and y. For each polynomial in P_{2j} we can associate an element of $\mathbb{C}^{2j+1}$ by writing it as a product of $2j$ factors:

$$a_1 x^{2j} + a_2 x^{2j-1} y + \cdots + a_{2j+1} y^{2j} = (\alpha_1 x_1 - \beta_1 y_1)(\alpha_2 x_2 - \beta_2 y_2) \cdots (\alpha_{2j} x_{2j} - \beta_{2j} y_{2j}) \tag{2}$$

for some complex α_i, β_i.

The coefficients of x and y in each term then also correspond to a point in $\mathbb{C}^{2j+1}$ which we had labelled by the a_i. This factorization is not unique, in that the terms can be permuted, and one can multiply any two factors by α and α^{-1} respectively. Thus we see that the whole space $\mathbb{CP}^{2j}$ in one-to-one correspondence with unordered sets of $2j$ points in $\mathbb{CP}^1$ corresponding to the $2j$ terms in equation (2) where we identify $(x_i, y_i) = \lambda(x_i, y_i)$ for any

λ. But $\mathbb{CP}^1$ is S^2 and so we have the required result : *States of angular momentum $2j$ can be represented by single points in $\mathbb{CP}^{2j}$ or unordered sets of $2j$ (not necessarily distinct) points on S^2.*

Another, though perhaps less explicit, way to understand the Majorana representation is to note that a state of spin-j can be written as a totally symmetric product of $2j$ spin-1/2 wavefunctions.

3. $SU(N)$ and Its Action on Spin States

Recall that $\mathbb{CP}^n$ is the set of lines in $\mathbb{C}^{n+1}$ and can be written as the coset space

$$\mathbb{CP}^n = SU(n+1)/U(n) \tag{3}$$

Now for any coset space $S = G/H$, G acts transitively on S (that is, for any two points p and q of S, there is an element g of G which takes p to q, so that $gp = q$). This means that $SU(2n+1)$ has a natural action on $\mathbb{CP}^{2n}$, and thus on sets of points of S^2 such that any unordered set of points is carried to any other unordered set of points by a suitable transformation in $SU(2n+1)$.

So far, what has been presented is valid for any finite n. For each finite n then, we have a realization of the action of $SU(n)$ on S^2.

Now with this action on S^2, $SU(n)$ will *always* (even for finite n) contain transformations which carry two distinct points into the same point on S^2, and thus which will not correspond to diffeomorphisms of S^2. (Note that $SDiff(S^2)$ is k-fold transitive for every positive integer k [8] where we recall that the action of a group G on a manifold M is said to be k-fold transitive if for any two arbitrary sets of k *distinct* points $(p_1, p_2, p_3, \ldots, p_k)$ and $(q_1, q_2, q_3, \ldots, q_k)$ of M there is an element of G which takes p_i to q_i for all $i = 1, \ldots, k$).

Thus in the limit of very large n, the permutations of sets of points on the sphere become much larger than the set of all diffeomorphisms of the sphere. In particular, any finite N approximation, $SU(N)$ of $SDiff(S^2)$ will contain mappings which do not correspond to elements of $SDiff(S^2)$. Thus we see, with the $SU(N)$ action defined here, that

$$\lim_{N \to \infty} SU(N) \not\simeq SDiff(S^2) \tag{4}$$

This can also be seen from another geometric view, where the Lie algebra of $SDiff(S^2)$ is that of divergenceless vector fields on S^2. Clearly, points pushed along the integral curves of these vector fields must not meet, and yet we see that there exist elements of $SU(N)$ acting on S^2 which will not satisfy this requirement, pushing two points together. Of course one can argue that perhaps in some sense "many" or "most" of the transformations will not carry two distinct points into one. In fact this argument is implicit in the assumptions about the limiting procedures which associate the Lie algebras of $SDiff(M)$ and of $SU(\infty)$ for various choices of 2-manifold M [9], but this does not evade the fact that $SU(\infty)$ contains transformations which are clearly not in $SDiff(M)$. This suggests that one is looking not so much as isomorphisms between $SDiff(M)$ and $SU(\infty)$, but rather some sort of unitary representation via embedding.

It is interesting to note that $SU(\infty)$ as realized here, is, in fact, so large, and capable of such dramatic topology-changing distortions of S^2 that it may in fact be a useful description not of area-preserving diffeomorphisms of S^2, but rather of a wider class of deformations of S^2 including those which result in punctured spheres, 2-manifolds of different topologies, and 2-manifolds which have degenerated into 1-manifolds, or even a single point. Similar ideas have been put forth in reference[9].

Acknowledgements

As always, I would like to acknowledge the support of the US National Science Foundation. I would also especially like to thank all my Latin American colleagues, and especially the organizing committee, for a wonderful week in Peru.

References

1. John Swain, http://arXiv.org/abs/hep-th/0405003.
2. John Swain, http://arXiv.org/abs/hep-th/0405004.
3. John Swain, http://arXiv.org/abs/hep-th/0405002.
4. John Swain, "The Topology of Area-Preserving Diffeomorphism Groups and Large N Limits of Classical Groups", Proceedings of PASCOS '04, to appear.
5. E. Majorana, Nuov. Cim. **9** (1932) 43.
6. H. Bacry, J. Math. Phys. **15** (1974) 1686.
7. L. C. Biedenharn and J. D. Louck, Angular Momentum in Quantum Physics, (Addison-Wesley, London, 1981)
8. W. M. Boothby, S. Kobayashi, and H. C. Wang, Ann. Math. **77** (1963) 329.
9. B. de Wit and H. Nicolai, Proceedings, 3rd Hellenic School on Elementary Particle Physics (Corfu, 1989) eds. E. N. Argyres et al. (World Scientific, Singapore) p. 777.

ON THE MAGNETIZED KERR-NEWMAN BLACK HOLE ELECTRODYNAMICS

E. P. ESTEBAN

Physics Department, University of Puerto Rico-Humacao, Humacao, PR 00791
KITP Scholar, Kavli Institute, University of Santa Barbara, California, CA 93106, USA

Because of its intrinsic theoretical interest and its potential astrophysical applications, we have carried out an electrodynamical study of the magnetized Kerr-Newman metric. This electrovac metric is an exact solution of the combined Maxwell-Einstein equations and can represent the exterior space-time of a rotating and charged black hole immersed in an external magnetic field. Although this metric was found by Ernst-Wild three decades ago, it is not until recently that renewed interest has arisen. There are several reasons to justify this enterprise: (1) Nowadays faster computers and improved computational algebra packages make possible to tackle the study of this otherwise mathematically complex metric; (2) It was recently proved that the magnetized Kerr-Newman black hole's charge is astrophysically significant ; (3) A physical understanding of its parameters have been just recently achieved; (4) There are not references of this metric in the literature. In this short contribution we carried out an analytical electrodynamical study of the magnetized Kerr-Newman metric. This study is limited to a spacetime region in which the dimensional parameter β ($= MB \ll 1$), where M and B, are the black hole mass, and the asymptotical external magnetic field, respectively. We present analytical expressions for the electromagnetic fields, electrodynamical invariants, and the magnetic flux, associated to rotating and charged black holes immersed in external magnetic fields. With regard to astrophysical applications, we propose to use instead of the standard Schwarzschild and Kerr vacuum black holes, a magnetized Kerr-Newman black hole as a central engine to model AGN's, and galactic hard $\gamma-$ ray (EGRET) sources.

1. The Magnetized Kerr-Newman Metric

In the seventies, an exact electrovac solution to the Einstein-Maxwell equations was found by Ernst-Wild [1]. The Ernst-Wild [1] electrovac solution is a generalization of the Kerr-Newman metric and can be interpreted as providing a model for the exterior space-time due to a charged and rotating black hole which is placed in an external magnetic field. From now on, this solution will be called the magnetized Kerr-Newman metric. It possesses a non-singular event horizon, and could therefore be used to study the effect of an external magnetic field on the properties of a black hole. A peculiar feature of the magnetized Kerr-Newman electrovac solution is that due to the presence of the magnetic field is not asymptotically flat. This feature in so far as possible astrophysical applications has interest in its own right.

The magnetized Kerr-Newman metric is generated by a Harrison type transformation [2] applied to the familiar Kerr-Newman solution. In a Boyer-Lindquist coordinate system the magnetized Kerr-Newman solution correspond to the following line element (in units such that c=G=1)

$$ds^2 = f_{mkn}^{-1}(-2P^{-2} \, d\zeta \, d\zeta^* + \rho^2 dt^2) - f_{mkn} \, (| \, \Lambda_0 \, |^2 \, d\varphi - \omega_{mkn} \, dt)^2, \quad (1)$$

where

$$d\zeta = \frac{1}{\sqrt{2}}(\frac{dr}{\sqrt{\Delta}} + i \, d\theta), \tag{2}$$

$$\rho = \sqrt{\Delta} \, \sin\theta, \tag{3}$$

$$P = \frac{1}{\sqrt{A}\sin\theta}, \tag{4}$$

with

$$\Delta = r^2 - 2\,M\,r + a^2 + e^2, \tag{5}$$

$$A = (r^2 + a^2)^2 - \Delta \, a^2 \sin^2\theta, \tag{6}$$

$$\Sigma = r^2 + a^2 \cos^2\theta. \tag{7}$$

In Eq. (1) ω_{mkn} and f_{mkn} can be obtained from the following

$$\nabla\omega_{mkn} = | \, \Lambda \, |^2 \, \nabla\omega + \frac{\rho}{f}(\Lambda^*\nabla\Lambda - \Lambda\nabla\Lambda^*), \tag{8}$$

$$f_{mkn} = \frac{f}{| \, \Lambda \, |^2} = -\frac{A \, \sin^2\theta}{| \, \Lambda \, |^2\Sigma}, \tag{9}$$

where $\nabla = \sqrt{\Delta}\frac{\partial}{\partial r} + i\frac{\partial}{\partial \theta}$, $\omega = \frac{a \, (2\,M\,r - e^2)}{A}$, and the asterisk (*) means complex conjugate.

Also, the complex functions $(\Lambda, \Phi, \varepsilon)$ can be explicitly written as

$\Lambda = \Lambda_r + i \, \Lambda_i, \, \Phi = \Phi_r + i \, \Phi_i$, and $\varepsilon = \varepsilon_r + i \, \varepsilon_i$, with

$$\varepsilon_r = -\left(r^2 + a^2\right) \sin^2 \theta - e^2 \cos^2 \theta - \frac{2 \, a^2 \, \sin^2 \theta}{\Sigma} \, (M \, r \, \sin^2 \theta + e^2 \cos^2 \theta), \tag{10}$$

$$\varepsilon_i = 2 \, M \, a \, (3 - \cos^2 \theta) \, \cos\theta - \frac{2 \, a \, \cos\theta \, \sin^2 \theta}{\Sigma} \, (\, r \, e^2 - a^2 \, M \, \sin^2 \theta), \tag{11}$$

$$\Lambda_r = 1 + B \, (\Phi_r - \frac{1}{4} \, B \, \varepsilon_r), \tag{12}$$

$$\Lambda_i = B \, (\Phi_i - \frac{1}{4} \, B \, \varepsilon_i \,), \tag{13}$$

$$\Phi_r = \frac{e \, a \, r}{\Sigma} \, \sin^2 \theta, \tag{14}$$

$$\Phi_i = - \, e \, \cos\theta \, (1 + \frac{a^2 \sin^2 \theta}{\Sigma}). \tag{15}$$

Notice in Eq. (1) that the term Λ_0, which is Λ calculated at $\theta = 0$ or π, is not present in Ernst-Wild original derivation [1]. It was introduced later by Aliev and Gal'tsov [3], to eliminate the existence of a conical singularity. This anomaly was pointed out first by Hiscock [4].

The magnetized Kerr-Newman metric has four parameters. Two of them, B and M are associated with the external magnetic field and the black hole's mass respectively. The remaining two parameters, a and e, are related but not equal to the black hole's angular momentum and charge, respectively.

Notice that the concept of total mass or energy in the magnetized Kerr-Newman metric can not be defined as in the standard vacuum metrics because the magnetic field diverges at spatial infinity. To obtain a finite electromagnetic energy, Li [5] truncated the electromagnetic field with a spherical surface, and, thus, proved that a Kerr black hole immersed in a uniform magnetic field acquires a net charge, although different than the one predicted by Wald [6]. Also, Karas and Vokrouhlicky [7] found that the mass contained within the magnetized Kerr-Newman black hole satisfy the well-known Smarr formula (which they proved that is also valid in non-asymptotically flat space-times).

With regards to whether a black hole could be charged or not, as previously mentioned, Li [5], Karas [7], and also Aliev and Gal'tsov [3] have proved rigorously that the presence of a magnetic field implies the existence of a black hole with a nonzero net charge. Moreover, Li [5] has also proven that the black hole's charge is of astrophysical importance when the black hole is rapidly rotating.

In addition, Kim et al [10], have showed that a slightly charged and magnetized Kerr black hole has a non-vanishing magnetic flux at the black hole's event horizon, and therefore the viability of the Blandorf-Znakek mechanism to extract black hole's rotational energy still holds, which is not the case when AGN's central engine is a vacuum Kerr black hole.

These results strongly suggest that the magnetized Kerr-Newman metric may play a more important role as compared to the standard Kerr-Newman metric, where the lack of an external magnetic field brings as a consequence the neutralization of the black hole's charge.

2. The Procedure

In a locally non-rotating reference frame the orthogonal components of the electromagnetic field for a magnetized Kerr-Newman black hole $H_{rmkn}, H_{\theta mkn}, E_{\theta mkn}, E_{rmkn}$ can be derived from a complex potential (Φ')

$$H_{rmkn} + i\, E_{rmkn} = \frac{1}{\sqrt{A}\,\sin\theta}\,\frac{\partial\,\Phi'}{\partial\,\theta}, \tag{16}$$

$$H_{\theta mkn} + i\, E_{\theta mkn} = -\sqrt{\frac{\Delta}{A}}\frac{1}{\sin\theta}\frac{\partial\,\Phi'}{\partial\,r}, \tag{17}$$

where $\Phi' = \frac{1}{\Lambda}(\Phi - \frac{1}{2}B\,\varepsilon)$. After a lengthy but straightforward procedure we obtain from Eqs. (16)-(17) the following

$$
\begin{aligned}
E_{rmkn} = & \frac{Q_{mkn}\,(a^2+r^2)\,(r^2-a^2\,\cos^2\theta)}{\sqrt{A}\,\Sigma^2} \\
& + \frac{B\,a\,M}{\sqrt{A}\Sigma^2}(a^2-r^2)\,(-r^2+(a^2+3\,r^2)\,\cos^2\theta+a^2\,\cos 4\theta) \\
& + \frac{B\,a\,Q_{mkn}^2\,r}{\sqrt{A}\,\Sigma^3}(-2a^2r^2-3r^4+(6\,a^4+12\,a^2r^2+9\,r^4-a^4\,\cos^2\theta \\
& + 2\,a^2\,r^2\,\cos^2\theta+a^4\,\cos^4\theta)\cos^2\theta),
\end{aligned}
\tag{18}
$$

$$
\begin{aligned}
H_{rmkn} = & \frac{2\,a\,Q_{mkn}\,r\,(a^2+r^2)\,\cos\theta}{\sqrt{A}\,\Sigma^2} + \frac{B\,\cos\theta}{\sqrt{A}} \\
& \times (\Delta_k + \frac{2\,M\,r\,(r^4-a^4)}{\Sigma^2}\,) + \frac{Q_{mkn}^2\,B\,\cos\theta}{\sqrt{A}} \\
& \times (1 + \frac{2\,(a^2+r^2)\,(a^2(a^2+2r^2)\,\cos^2\theta-2\,r^4-3\,a^2\,r^2)}{\Sigma^3}),
\end{aligned}
\tag{19}
$$

$$H_{\theta mkn} = \frac{a\, Q_{mkn}\, \Delta_{QB}^{1/2}\, (r^2 - a^2 \, \cos^2\theta)\, \sin\theta}{\sqrt{A}\, \Sigma^2}$$

$$+ \frac{B\Delta_{QB}^{1/2}\, \sin\theta}{\sqrt{A}\, \Sigma^2} \tag{20}$$

$$(-r^5 - a^2 M r^2 + a^2 (a^2 M - M r^2 - 2r^3)\, \cos^2\theta + a^4\, (M-r)\, \cos^4\theta)$$

$$+ \frac{2B\Delta_{QB}^{1/2}\, a^2\, Q_{mkn}^2\, r\, \sin\theta}{\sqrt{A}\, \Sigma^3}\, ((3\, a^2 + 4\, r^2)\cos^2\theta - r^2),$$

$$E_{\theta mkn} = -\frac{a^2\, Q_{mkn}\, r\, \Delta_{QB}^{1/2}\, \sin 2\theta}{\sqrt{A}\, \Sigma^2} + \frac{a^3\, B\, M\, r\, \Delta_{QB}^{1/2}}{\sqrt{A}\Sigma^2}(1 + \cos^2\theta)\sin 2\theta$$

$$- \frac{a\, B\, \Delta_{QB}^{1/2}\, Q_{mkn}^2}{2\sqrt{A}\Sigma^3} \tag{21}$$

$$(-3r^2\, (2a^2 + r^2) + 2\, a^2\, (a^2 + 3\, r^2)\, \cos^2\theta + a^4\, \cos^4\theta)\, \sin 2\theta,$$

where $\Delta_k = r^2 - 2Mr + a^2$, $\Delta_Q = \Delta_k + Q_{mkn}^2 - 4aQ_{mkn}BM$, and Q_{mkn} is the total black hole's charge which was calculated as follows

$$Q_{mkn} = -|\Lambda_0|^2\, \Phi_i'\, (r_+, 0) = -\frac{e^3}{4}B^2 + e + 2\, a\, M\, B. \tag{22}$$

An inspection of Eq. (22) suggests that a charged black hole when embedded in a ionized and magnetized interstellar medium, will selectively accrete charge. This build up of a black hole's charge is in part due to the coupling between the black hole's angular momentum and the magnetic field. In fact, a charged black hole in a magnetic field will build up positive or a negative net charge.

Eqs. (18)–(21) are new in the literature and represent the electromagnetic fields associated with the magnetized Kerr-Newman metric. They are nicely decoupled. For instance, the first term of each of these equations corresponds to the electromagnetic fields associated to the Kerr-Newman metric, while the second term are those corresponding to the magnetized Kerr metric [9]. The third term in Eqs.(18)-(21) is induced by a coupling between Q_{mkn}^2 and B.

Also of interest, it is the magnetic flux (F_{mkn}) an invariant quantity given by

$$F_{mkn} = 2\pi\, |\Lambda_0|^2\, (\Phi_r'\, (r_+, \theta) - \Phi_r'\, (r_+, 0)), \tag{23}$$

where in the above equation, r_+ is the magnetized Kerr-Newman's event horizon (which is not located at the same place as in the standard Kerr-Newman black hole). We have calculated from Eq. (23), the magnetic flux **for any value** of Q_{mnk} across an axisymmetric cap with $\theta = \pi/2$, as a physical boundary. Thus

$$F_{mkn} = \frac{2\pi}{1 + \frac{2aQ_{mkn}B}{r_+}} \left(\frac{aQ_{mkn}}{r_+} + \frac{B}{2}r_+(1 - \frac{a^4}{r_+^4}) + \frac{2a^3BMQ_{mkn}}{r_+^2} \right) - 3\pi Q_{mkn}^2 B \,.$$

$$(24)$$

As is very well known, the Blanford-Znajek process to extract rotational energy from a black hole requires a magnetic flux different of zero at the black hole's event horizon. This fact holds in Eq. (24) even when $a- > r_+$. This result extend Kim's conclusions [10], and it keeps the viability of the Blandford-Znakek process in black hole physics.

On the other hand, from Eqs. (18)–(21) we can construct the electromagnetic tensor, and calculate from it, the invariant $G = \varepsilon^{\mu\nu\lambda\rho}F_{\mu\nu}F_{\lambda\rho} = \vec{E}\cdot\vec{B}$. It turned out that for the magnetized Kerr-Newman metric the invariant G is non-zero. This implies that a component of the electric field will be parallel to a magnetic line of force. In other words, most particle charge will be confined by the black hole's magnetic lines geometry which, thus may in fact explain at least in first approximation the observed jet's collimation. The another invariant

$F = F^{\mu\nu}F_{\mu\nu} = 2\left(B^2 - E^2\right)$, can be used to estimate the production of electron-positron pairs by gamma-ray sources following closely a procedure given by Heyl [8] for a magnetized Kerr black hole.

References

1. Ernst F.J. and Wild, W. J., J. Math. Phys. 17, 182 (1976).
2. Harrison B. K., J. Math. Phys., 9, 1744 (1968).
3. Aliev A.N. and Gal'tsov, D. V., Sov. Phys. Usp., 32, 75 (1989).
4. Hiscock W. A., J Math. Phys. 17, 182 (1976).
5. Li L, Phys. Rev. D, 61, 084033 (2000).
6. Wald R. M., Phys Rev. D. 10, 6, 1680, (1974).
7. Karas W. and Vokrouhlicky, D., Class. Quantum Grav. 7, 391 (1990).
8. Heyl J. S., Phys. Rev D, 63, 064028, (2001).
9. King A. R., Lasota J.P., and Kundt, Phys D, 12, 3037 (1975).
10. Kim H., Lee C. H., Lee H. K., Phys. Rev. D., 63, 064037-1 (2001).

SUPERNOVA NEUTRINOS AND THE ABSOLUTE SCALE OF NEUTRINO MASSES — A BAYESIAN APPROACH*

ENRICO NARDI

INFN, Laboratori Nazionali di Frascati, C.P. 13, I00044 Frascati, Italy
Instituto de Física, Universidad de Antioquia, A.A.1226, Medellín, Colombia

We apply Bayesian methods to the study of the sensitivity to neutrino masses of a Galactic supernova neutrino signal. Our procedure makes use of the full statistics of events and is remarkably independent of astrophysical assumptions. Present detectors can reach a sensitivity down to $m_\nu \sim 1$ eV. Future megaton detectors can yield up to a factor of two improvement; however, they will not be competitive with the next generation of tritium β-decay and neutrinoless 2β-decay experiments.

1. Introduction

It was realized long time ago that neutrinos from a Supernova (SN) can provide valuable informations on the neutrino masses. The basic idea relies on the time-of-flight (tof) delay Δt that a neutrino of mass m_ν and energy E_ν traveling a distance L would suffer with respect to a massless particle:

$$\Delta t = \frac{L}{v} - L \approx 5.1\,\text{ms} \left(\frac{L}{10\,\text{kpc}}\right)\left(\frac{10\,\text{MeV}}{E_\nu}\right)^2\left(\frac{m_\nu}{1\,\text{eV}}\right)^2. \tag{1}$$

Indeed the detection of about two dozens of neutrinos from SN1987 allowed to set model independent upper limits at the level of $m_{\bar\nu_e} < 30\,\text{eV}$.[1] Since SN1987A, several proposal have been put forth to identify the best ways to measure the neutrino tof delays. Often, these approaches rely on the identification of "timing" events that are used as benchmarks for measuring the time lags, as for example the simultaneous emission of gravitational waves[2] or the abrupt interruption of the neutrino flux due to a further

---

*Work done in collaboration with J. Zuluaga. Supported in part by COLCIENCIAS.

collapse into a black hole.[3] The less model dependent limits achievable with these methods are at the level of $m_\nu \lesssim 3\,\mathrm{eV}$, and tighter limits are obtained only under specific assumptions. A different method to extract informations on the neutrino masses that uses the full statistics of the signal and does not rely on any benchmark event was proposed in Ref. 4 and developed in Ref. 5. Here we resume the results obtained by applying this method to determine the sensitivity of the SuperKamiokande (SK) water Čerenkov and KAMLAND scintillator detectors, and of the planned experimental facilities Hyper-Kamiokande and LENA.

2. The Method

In real time detectors, supernova $\bar{\nu}_e$ are revealed through to the positrons they produce via charged current interactions, that provides good energy informations as well. Each $\bar{\nu}_e$ event corresponds to a pair of energy and time measurements (E_i, t_i). In order to extract the maximum of information from a high statistics SN neutrino signal, all the neutrino events have to be used in constructing a suitable statistical distribution, as for example the Likelihood, that can be schematically written as:

$$\mathcal{L} \equiv \prod_i \mathcal{L}_i = \prod_i \left\{ \phi(t_i) \times F(E_i; t_i) \times \sigma(E_i) \right\}. \tag{2}$$

$\mathcal{L}_i$ represents the contribution to the Likelihood of a single event with the index i running over the entire set of events, $\sigma(E)$ is the $\bar{\nu}_e$ detection cross-section and $F(E; t)$ is the energy spectrum of the neutrinos, whose time profile can be reconstructed rather accurately directly from the data.[4,5] The main problem in constructing the Likelihood (2) is represented by the (unknown) time profile of the neutrino flux $\phi(t)$. We construct a flux model by requiring that it satisfies some physical requirements and a criterium of simplicity: *i)* the analytical flux function must go to zero at the origin and at infinity; *ii)* it must contain at least two time scales for the neutrino emission corresponding to the fast rising initial phase of shock-wave breakout and accretion, and the later Kelvin-Helmholtz cooling phase; *iii)* it must contain the minimum possible number of free parameters. The following model for the flux has all the required behaviors:

$$\phi(t; \lambda) = \frac{e^{-(t_a/t)^{n_a}}}{[1 + (t/t_c)^{n_p}]^{n_c/n_p}} \begin{cases} \sim e^{-(t_a/t)^{n_a}} & (t \to 0) \\ \sim (t_c/t)^{n_c} & (t \to \infty). \end{cases} \tag{3}$$

The five parameters that on the l.h.s of (3) have been collectively denoted with λ are: two time scales t_a for the initial exponentially fast rising phase and t_c for the cooling phase, two exponents n_a and n_c that control their specific rates and one additional exponent n_p that mainly determines the width of the "plateau" between the two phases. Since we are interested only in the neutrino mass squared m_ν^2, irrespectively of the particular values of the *nuisance parameters* λ, starting from the Likelihood (2) we will need to evaluate the *marginal posterior probability* $p(m_\nu^2|D)$, that is the probability distribution for m_ν^2 given the data D. This is done by *marginalizing* the posterior probability with respect to the nuisance parameters:

$$p(m_\nu^2|D, I) = N^{-1} \int d\lambda \, \mathcal{L}(D; m_\nu^2, \lambda) \, p(m_\nu^2, \lambda|I) \,. \qquad (4)$$

$p(m_\nu^2, \lambda|I)$, that in Bayesian language is called the *prior probability of the model*, allows us to take into account any available prior information on the parameters m_ν^2 and λ. We will use flat priors for all the λ's and, to exclude unphysical values of m_ν^2, a step function $\Theta(m_\nu^2) = 1$, (0) for $m_\nu^2 \geq 0$, (< 0).

The dependence on m_ν^2 could be directly included in the flux (2) by redefining the time variable according to (1). However, it is more convenient to proceed as follows: given a test value for the neutrino mass, first the arrival time of each neutrino is shifted according to its time delay, and then the Likelihood is computed for the whole time-shifted sample. Subtilities in the evaluation of the Likelihood contribution $\mathcal{L}_i(t_i, E_i)$ arising from the uncertainty ΔE_i in the energy measurement are discussed in Refs. 4, 5.

We have tested the method applying it to a large set of synthetic Monte Carlo (MC) neutrino signals, generated according to two different SN models: *SN model I* corresponds to the simulation of the core collapse of a 20 $M_\odot$ star[6] carried out by using the Livermore Group code.[8] In this simulation $\nu_{\mu,\tau}$ opacities were treated in a simplified way, and this resulted in quite large (and probably unrealistic) differences in their average energies with respect to $\bar{\nu}_e$. *SN model II* corresponds to a recent hydrodynamic simulation of a 15 $M_\odot$ progenitor star[7] carried out with the Garching group code.[9] This simulation includes a more complete treatment of neutrino opacities and results in a quite different picture, since the antineutrino spectra do not differ for more than about 20%. In both cases the effects of neutrino oscillations in the SN mantle have been properly included in the simulations. Note that the two types of neutrino spectra of SN model I and II fall close to the two extremes of the allowed range of possibilities. This gives us confidence that the results of the method are robust with respect to variations in the spectral characteristics.

Table 1. Results for the fits to m_ν^2: a)-c) SK for different SN distances, d) SK plus KamLAND, e) Hyper-Kamiokande, f) LENA. All masses are in eV.

Detector	MODEL 1		MODEL 2	
	$\overline{m}_{up} \pm \Delta m_{up}$	$\sqrt{m_{min}^2}$	$\overline{m}_{up} \pm \Delta m_{up}$	$\sqrt{m_{min}^2}$
a) SK (10 kpc)	1.0 ± 0.2	1.0	1.1 ± 0.3	1.2
b) SK (5 kpc)	–	–	1.1 ± 0.3	1.0
c) SK (15 kpc)	–	–	1.6 ± 0.6	1.4
d) SK+KL (10 kpc)	1.0 ± 0.2	0.9	1.1 ± 0.3	1.0
e) HK (10 kpc)	0.4 ± 0.1	0.4	0.5 ± 0.1	0.5
f) LENA (10 kpc)	0.9 ± 0.2	0.9	0.9 ± 0.3	0.9

3. Results

To test the sensitivity of our method, we have analyzed a large number of neutrino samples, grouped into different ensembles of about 40 samples each. For each ensemble we vary in turn the SN model (model I and II), the SN-earth distance (5, 10, and 15 kpc) and the detection parameters specific for two operative (SK and KamLAND) and two proposed (Hyper-Kamiokande and LENA) detectors. The results of our recent detailed analysis[5] are summarized in Table 1. In columns 2 and 4 we give the 90% c.l. upper limits that could be put on m_ν in case its value is too small to produce any observable delay. In columns 3 and 5 we estimate for which value of m_ν the massless neutrino case can be rejected at least in 50% of the cases. These results confirm the claim[4] that detectors presently in operation can reach a sensitivity of about 1 eV, that is seizable better than present results from tritium β-decay experiments, competitive with the most conservative limits from neutrinoless double β-decay, less precise but remarkably less dependent from prior assumptions than cosmological measurements. However, in spite of a sizeable improvement, future detectors will not be competitive with the next generation of tritium β-decay and neutrinoless double β decay experiments.

References

1. D. N. Schramm, Comments Nucl. Part. Phys. **17**, 239 (1987).
2. D. Fargion, Lett. Nuovo Cim. **31**, 499 (1981); N. Arnaud *et. al.*, Phys. Rev. D **65**, 033010 (2002).
3. J. F. Beacom R. N. Boyd and A. Mezzacappa, Phys. Rev. D **63**, 073011 (2001); Phys. Rev. Lett. **85**, 3568 (2000).
4. E. Nardi and J. I. Zuluaga, Phys. Rev. D **69**, 103002 (2004); E. Nardi, arXiv:astro-ph/0401624.

5. E. Nardi and J. I. Zuluaga, Submitted to Nucl. Phys. B.
6. S. E. Woosley *et. al.*, Astrophys. J. **433**, 229 (1994).
7. G. Raffelt *et. al.*, arXiv:astro-ph/0303226; R. Buras, Private Communication.
8. J. R. Wilson and R. W. Mayle, Phys. Rept. **227**, 97 (1993).
9. M. Rampp and H. T. Janka, Astron. Astrophys. **396**, 361 (2002).

LOOP QUANTUM GRAVITY AND ULTRA HIGH ENERGY COSMIC RAYS

J. ALFARO*

Facultad de Física, Pontificia Universidad Católica de Chile
Casilla 306, Santiago 22, Chile.
jalfaro@puc.cl

G. A. PALMA

Department of Applied Mathematics and Theoretical Physics,
Center for Mathematical Sciences, University of Cambridge
Wilberforce Road, Cambridge CB3 0WA, UK
G.A.Palma@damtp.cam.ac.uk

For more than a decade we have been observing the arrival of extragalactic cosmic rays with energies greater than 1×10^{20} eV, which correspond to the highest energy events ever registered. The measure of such energies violates the theoretical limit known as the Greisen-Zatsepin-Kuzmin cutoff, which predicts a strong suppression of cosmic rays with energies above $\sim 4 \times 10^{19}$ eV. Here we show how the effects of loop quantum gravity corrections can affect the predicted spectrum of ultra high energy cosmic rays (those cosmic rays with energies greater than $\sim 4 \times 10^{18}$ eV). We shall also analyze new strong constraints on the parameters of loop quantum gravity.

A detailed comprehension of the origin and nature of ultra high energy cosmic rays(UHECR) is far from being achieved; the way in which the cosmic ray spectrum reveals to us still is a mystery. In this sense, of particular interest is the detection of protons with energies above 1×10^{20} eV [1,2,3], which, according to theory, should not be observed. A first difficulty comes from the lack of reasonable mechanisms for the acceleration of particles to such energies. A second difficulty consists in the existence of the cosmic microwave background radiation (CMBR), whose presence necessarily

*Work partially supported by Fondecyt 1010967.

produce a friction over the propagation of UHECR making them release energy and, therefore, affecting their possibility to reach great distances. A first estimation of the characteristic distance that UHECR can reach before losing most of their energy was simultaneously made in 1966 by K. Greisen [4] and G. T. Zatsepin & V. A. Kuzmin [5], showing that cosmic rays with energies greater than 4×10^{19} eV should not travel more than ~ 100 Mpc. Since there are no known active objects in our neighborhood capable to act as sources of such cosmic rays and since their arrival is mostly isotropic (without privileging any local source), we are forced to conclude that these cosmic rays come from distances larger than 100 Mpc. This is known as the Greisen-Zatsepin-Kuzmin (GZK) anomaly.

More detailed approaches to the GZK-cutoff feature have been made since the first estimation. For example Berezinsky et al. [6] and S. T. Scully & F. W. Stecker [7] have made progress in the study of the spectrum $J(E)$ which UHECR should present (that is to say, the flux of arriving particles as a function of the observed energy E). In these approaches it is assumed that the sources are uniformly distributed in the Universe and an emission flux $F(E_g)$ with a power law $F(E_g) \propto E_g^{-\gamma_g}$, where E_g is the energy of the emitted particle and γ_g is the generation index. The main quantity for the calculation of the UHECR spectrum is the energy loss $-E^{-1}dE/dt$, which takes into consideration two chief contributions: the energy loss due to redshift attenuation and the energy loss due to collisions with the CMBR. This last contribution depends, at the same time, on the cross section σ and the inelasticity K of the interactions produced during the propagation of protons in the extragalactic medium, as well as on the CMBR spectrum (the inelasticity is the fractional difference $K = \Delta E/E$, where $\Delta E = E - E_f$ is the difference between the initial energy E of the proton and its final energy E_f as a result of an interaction). The most important reactions taking place in the description of proton's propagation (and which produce the release of energy in the form of particles) are the pair creation $p + \gamma \to p + e^- + e^+$, and the photo-pion production $p + \gamma \to p + \pi$. This last reaction happens through several channels (for example the baryonic Δ and N, and mesonic ρ and ω resonance channels, just to mention some of them) and is the main reason for the appearance of the GZK cutoff.

Using the methods of V. Berezinsky et al. [6] we can reproduce the UHECR spectrum for uniformly distributed sources. Fig. 1 shows the obtained spectrum $J(E)$ of UHECR and the Akeno Giant Air Shower Array (AGASA) observed data [3]. To reconcile the data of the low energy region ($E < 4 \times 10^{19}$ eV), where the pair creation dominates the energy loss, it is necessary to have a generation index $\gamma_g = 2.7$ (with the additional supposition that sources does not evolve). It can be seen that for events with energies $E > 4 \times 10^{19}$ eV, where the energy loss is dominated by the photo-pion production, the predicted spectrum does not fit well the data.

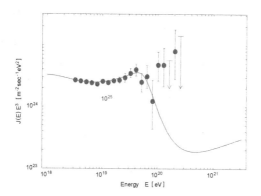

Fig. 1. UHECR spectrum and AGASA observations. The figure shows the UHECR spectrum $J(E)$ multiplied by E^3, for uniform distributed sources, without evolution, and with a maximum generation energy $E_{\mathrm{max}} = \infty$. Also shown are the AGASA observed events. The best fit for the low energy sector ($E < 4 \times 10^{19}$ eV) corresponds to $\gamma_g = 2.7$.

The fact that the GZK anomaly involves the highest energy events registered up to now has motivated the notion that the necessary framework to explain this phenomena could be of a quantum-gravitational nature [8,9,10]. To deepen this idea we have adopted the loop quantum gravity (LQG) theory [11], one of the proposed alternatives for the yet nonexistent theory of quantum gravity. It is possible to study LQG through effective theories that take into consideration matter-gravity couplings. In this line, in the works of J. Alfaro *et al.* [12,13], the effects of the loop structure of space at the Planck level are treated semiclassically through a coarse-grained approximation. An interesting feature of this kind of methods is the explicit appearance of the Planck scale l_p and the appearance of a new length scale $\mathcal{L} \gg l_p$ (called the "weave" scale), such that for distances $d \ll \mathcal{L}$ the quantum loop structure of space is manifest, while for distances $d \geq \mathcal{L}$ the continuous flat geometry is regained. The presence of these two scales in the effective theories has the consequence of introducing Lorentz invariance violations (LIV's) to the dispersion relations $E = E(p)$ for particles with energy E and the momentum p. Take for instance the modified dispersion relation for fermions [12]; holding the first order contributions in both scales we can write [10] (using natural units $c = \hbar = 1$)

$$E_{\pm}^2 = p^2 + 2\alpha p^2 + \eta p^4 \pm 2\lambda p + m^2, \tag{1}$$

where the sign $\pm$ depends on the helicity state of the particle, m is the particle mass, and the corrections α, η and λ depend on the scales $\mathcal{L}$ and l_p as follows: $\alpha = \kappa_\alpha (l_p/\mathcal{L})^2$, $\eta = \kappa_\eta l_p^2$ and $\lambda = \kappa_\lambda l_p/2\mathcal{L}^2$, with κ_α, κ_η and κ_λ being adimensional parameters of order 1. Since the Lorentz symmetry is present in the full LQG theory [14] and the κ coefficients appears because of the quantum-gravity vacuum expectation values of matter fields coupled to gravity, we shall assume that LID's introduced in (1) come from a spontaneous Lorentz symmetry breaking. The above corrections can be analyzed through several methods (see, for example, [15,17]). Here we shall explore them through their effects on the propagation of ultra high energy cosmic rays.

Casting the modified dispersion relation (1) in the form $E^2 = p^2 + f(p)$ (where $f(p)$ corresponds to the addition of the LQG corrections and the mass square m^2) and using the fact that the Lorentz symmetry is spontaneously broken, we can develop an algebraic equation for the inelasticity K for the photo-pion production (or any other meson production). This equation is

$$(1 - K)\sqrt{s} = F + \left(F^2 - f_p[(1 - K)E]\right)^{1/2} \cos\theta, \tag{2}$$

where θ is the angle between the directions of propagation of the initial and final protons in the center of mass reference system, and F is a quantity that can be written in terms of K and E as follows:

$$F = \frac{1}{2\sqrt{s}} \left(s + f_p[(1 - K)E] - f_\pi[KE]\right). \tag{3}$$

Solving equation (2) and averaging in the angle θ, we can obtain a modified version of the inelasticity for the photo-pion production.

An interesting thing of having a spontaneous Lorentz symmetry breaking is that the quantity $f(p)$ can be interpreted as an effective mass square for particles carrying momentum p. This fact can be used to modify any quantity involved in the calculation of the energy loss: it will be enough to use the prescription $m^2 \rightarrow f(p)$. Since we can calculate the inelasticity K of a given process, we will be able to estimate the energy carried by any final state and, at the same time, to estimate the value of the effective mass for any final particle.

Introducing these modifications to the different quantities involved in the propagation of protons (like the cross section σ and inelasticity K), we are able to find a modified version for the UHECR energy loss due to collisions. Since the only relevant correction for the GZK anomaly is α, we concentrated our analysis in the particular case $f(p) = 2\alpha p^2 + m^2$. To simplify our model we restricted our treatment to the case $\alpha > 0$ and taken only $\alpha_m \neq 0$, where α_m is assumed to have the same value for mesons π, ρ and ω.

Using again the methods of Berezinsky *et al.* [6] and introducing the discussed modifications, we can find the modified version of the UHECR spectrum for $\alpha_m \neq 0$. Fig. 2 shows the AGASA observations and the predicted UHECR spectrum in the case $\alpha_m = 1.5 \times 10^{-22}$ ($\mathcal{L} \simeq 6.7 \times 10^{-18}$ eV^{-1}), for three different maximum generation energies $E_{\max}$. These are, curve 1: 5×10^{20} eV; curve 2: 1×10^{21} eV; and curve 3: 3×10^{21} eV. The Poisson

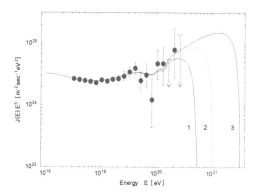

Fig. 2. Modified UHECR spectrum and AGASA observations. The figure shows the modified spectrum $J(E)$ multiplied by E^3, for uniform distributed sources and without evolution, for the case $\alpha_m = 1.5 \times 10^{-22}$ ($\mathcal{L} \simeq 6.7 \times 10^{-18}$ eV^{-1}). Three different maximum generation energies $E_{\max}$ are shown; these are, curve 1: 5×10^{20} eV; curve 2: 1×10^{21} eV; and curve 3: 3×10^{21} eV.

probabilities of an excess in the five highest energy bins for the three curves are $P_1 = 3.6 \times 10^{-4}$, $P_2 = 2.6 \times 10^{-4}$ and $P_3 = 2.3 \times 10^{-4}$. Meanwhile, the Poisson χ^2 for the eight highest energy bins are $\chi_1^2 = 10$, $\chi_2^2 = 10.9$ and $\chi_3^2 = 11.2$ respectively. It is noteworthy the possibility of reconciling the data with finite maximum generation energies (which is impossible for conventional models, where infinite maximum generation energies $E_{\max}$ are needed for the best fit).

Finally we want to mention a recent development [13,18], where the dispersion relations for fermions and bosons are generalized to include an extra factor Υ, which measure a possible non-canonical scaling of the gravitational expectation values in the semiclassical state. It should be noticed that this new factor always appear in the dispersion relations in the form $\left(\frac{\ell_P}{\mathcal{L}}\right)^{\Upsilon}$, so for instance the parameter α in our equation (1) gets an extra factor $\left(\frac{\ell_P}{\mathcal{L}}\right)^{\Upsilon-1}$. This freedom can be used to move the scale $\mathcal{L}$ down if needed, so that the cosmic ray momentum p always satisfied the bound $p\mathcal{L} \leq 1$, without changing our prediction of the spectrum of UHECR.

Bounds on LIV Parameters. We have seen that the conventionally obtained theoretical spectrum provides a very good description of the phenomena up to an energy $\sim 4 \times 10^{19}$ eV. The main reaction taking place in this well described region is the pair creation $\gamma + p \rightarrow p + e^+ + e^-$. So, we will require that the threshold condition for pair creation not be substantially altered by the new corrective terms.

Example: The Cubic Correction. A commonly studied correction which has appeared in several recent works, and which deserves our attention, is the case of a cubic correction of the form

$$E^2 = p^2 + m^2 + \xi\, p^3, \tag{4}$$

(where ξ is an arbitrary scale). It is interesting to note that strong bounds can be placed over deformation $f(p) = \xi\, p^3$. We will assume in this section, that ξ is an universal parameter (an assumption followed by most of the works in this field). The threshold condition for the pair production is

$$\omega E + \xi \frac{m_e m_p}{(m_p + 2m_e)^2} E^3 \geq m_e(m_p + m_e). \tag{5}$$

Since we cannot infer modifications in the description of pair production in the cosmic ray spectra up to energies $E \sim 4 \times 10^{19}$ eV, we must at least impute the following inequality

$$
\begin{aligned}
|\xi| &< \frac{(m_p + m_e)(m_p + 2m_e)^2}{m_p\, E_{\mathrm{ref}}^3} \\
&= 3.26 \times 10^{-41} \text{ eV}^{-1} \\
&(= 3.98 \times 10^{-13}\, l_p),
\end{aligned} \tag{6}
$$

where we have used $E_{\mathrm{ref}} = 3 \times 10^{19}$ eV. This last result shows the strong suppression over ξ. As a consequence, the particular case $|\xi| = l_p = 8 \times 10^{-28}$ eV^{-1} should be discarded. A bound like (6) seems to have been omitted up to now for the GZK anomaly analysis.

Acknowledgements

The work of JA is partially supported by Fondecyt 1010967. He acknowledges the hospitality of the organizers of V Silafae and PUC(Perú); and financial support from an Ecos(France)-Conicyt(Chile) project.The work of GAP is partially supported by MIDEPLAN and DAMTP.

References

1. D.J. Bird *et al.*, Phys. Rev. Lett. **71**, 3401 (1993).
2. D.J. Bird *et al.*, Astrophys. J. **424**, 491 (1994); **441**, 144 (1995).
3. M. Takeda *et al.*, Phys. Rev. Lett. **81**, 1163 (1998). For an update see M. Takeda *et al.*, Astrophys. J. **522**, 225 (1999).
4. K. Greisen, Phys. Rev. Lett. **16**, 748 (1966).
5. G.T. Zatsepin and V.A. Kuzmin, Zh. Eksp. Teor. Fiz., Pisma Red. **4**, 114 (1966).
6. V. Berezinsky, A.Z. Gazizov and S.I. Grigorieva, hep-ph/0107306; hep-ph/0204357.
7. S.T. Scully and F.W. Stecker, Astropart. Phys. **16**, 271 (2002).
8. J. Ellis, N.E. Mavromatos and D.V. Nanopoulos, Phys. Rev. D **63**, 124025 (2001).
9. G. Amelino-Camelia, Phys. Lett. B **528**, 181 (2002).
10. J. Alfaro and G. Palma, Phys. Rev. D **65**, 103516(2002);Phys.Rev.**D67**:083003(2003).
11. M. Gaul and C. Rovelli, Lect. Notes Phys. **541**, 277 (2000).
12. J. Alfaro, H.A. Morales-Técotl and L.F. Urrutia, Phys. Rev. Lett. **84**, 2318 (2000).
13. J. Alfaro, H.A. Morales-Técotl and L.F. Urrutia, Phys. Rev. D **65**, 103509 (2002).
14. T. Thiemann, Class. Quant. Grav. **15**, 1281 (1998).
15. D. Sudarsky, L. Urrutia and H. Vucetich,Phys.Rev.Lett.**89**:231301(2002).
16. S. Coleman & S.L. Glashow, Phys. Rev. D **59**, 116008 (1999).
17. T.J. Konopka and S.A. Major, New J. Phys. **4**, 57 (2002).
18. J. Alfaro, H.A. Morales-Técotl and L.F. Urrutia, Quantum gravity and spin 1/2 particles effective dynamics, Phys.Rev.**D66**:124006(2002).

$Q\bar{Q}$ BOUND STATES IN AN EXTENDED QCD_2 MODEL

PEDRO LABRAÑA* and JORGE ALFARO[†,*]

Facultad de Física, Pontificia Universidad Católica de Chile,
Vicuña Mackena 4860, Macul, Santiago de Chile, Casilla 306, Chile.
*plabrana@puc.cl
†jalfaro@puc.cl

ALEXANDER A. ANDRIANOV [†]

V. A. Fock Department of Theoretical Physics, St. Petersburg State University, 1, ul.
Ulianovskaya, St. Petersburg 198504, Russia and Istituto Nazionale di Fisica Nucleare,
Sezione di Bologna via Irnerio 46, Bologna 40126, Italy.
andrianov@bo.infn.it

We study an extended QCD model in $2D$ obtained from QCD in $4D$ by compactifying two spatial dimensions and projecting onto the zero-mode subspace.
This system is found to induce a dynamical mass for transverse gluons – adjoint scalars in QCD_2, and to undergo a chiral symmetry breaking with the
full quark propagators yielding non-tachyonic, dynamical quark masses, even
in the chiral limit. We construct the hadronic color singlet bound-state scattering amplitudes and study quark-antiquark bound states which can be classified
in this model by their properties under Lorentz transformations inherited from
$4D$.

We study a QCD reduced model in $2D$ which can be formally obtained
from QCD in $4D$ by means of a classical dimensional reduction from $4D$ to
$2D$ and neglecting heavy K-K (Kaluza-Klein) states. Thus only zero-modes
in the harmonic expansion in compactified dimensions are retained. As a
consequence, we obtain a two dimensional model with some resemblances of

---

*The work of P.L. is supported by a Conicyt Ph.D. Fellowship (Beca Apoyo Tesis Doc-
toral). The work of J.A. is partially supported by Fondecyt # 1010967.
†The work of A.A. is partially supported by RFBR grant and the Program "Universities
of Russia: Basic Research".

the real theory in higher dimension, that is, in a natural way adding boson matter in the adjoint representation to QCD_2. The latter fields being scalars in $2D$ reproduce transverse gluon effects[2,3]. Furthermore this model has a richer spinor structure than just QCD_2 giving a better resolution of scalar and vector states which can be classified by their properties inherited from $4D$ Lorentz transformations. The model is analyzed in the light cone gauge and using large N_c limit. The contributions of the extra dimensions are controlled by the radiatively induced masses of the scalar gluons as they carry a piece of information of transverse degrees of freedom. We consider their masses as large parameters in our approximations yet being much less than the first massive K-K excitation. This model might give more insights into the chiral symmetry breaking regime of QCD_4. Namely, we are going to show that the inclusion of solely lightest K-K boson modes catalyze the generation of quark dynamical mass and allows us to overcome the problem of tachyonic quarks present in QCD_2.

We start with the QCD action in $(3+1)$ dimensions:

$$S_{QCD} = \int d^4 x \left[-\frac{1}{2\tilde{g}^2} tr(G^2_{\mu\nu}) + \bar{\Psi} \left(i\gamma^\mu D_\mu - m \right) \Psi \right]. \tag{1}$$

Follow the scheme of[4] we proceed to make a dimensional reduction of QCD, at the classical level, from $4D$ to $2D$. For this we consider the coordinates $x_{2,3}$ being compactified in a 2-Torus, respectively the fields being periodic on the intervals $(0 \le x_{2,3} \le L = 2\pi R)$. Next we assume L to be small enough in order to get an effective model in $2D$ dimensions. Then by keeping only the zero K-K modes, we get the following effective action in $2D$, after a suitable rescaling of the fields:

$$S_2 = \int d^2 x \; tr \left[-\frac{1}{2} F^2_{\mu\nu} + (D_\mu \phi_1)^2 + (D_\mu \phi_2)^2 \right] + \bar{\psi}_1 \left(i\gamma^\mu D_\mu - m \right) \psi_1$$

$$+ \; \bar{\psi}_2 \left(i\gamma^\mu D_\mu - m \right) \psi_2 - i \frac{g}{\sqrt{N_c}} \left(\bar{\psi}_1 \gamma^5 \phi_1 \psi_2 + \bar{\psi}_2 \gamma^5 \phi_1 \psi_1 \right) \tag{2}$$

$$- \; i \frac{g}{\sqrt{N_c}} \left(\bar{\psi}_1 \gamma^5 \phi_2 \psi_1 - \bar{\psi}_2 \gamma^5 \phi_2 \psi_2 \right) + \frac{g^2}{N_c} tr[\phi_1, \phi_2]^2 \;.$$

Where we have defined the coupling constant of the model $g^2 = N_c \tilde{g}^2 / L^2$. We expect[5] the infrared mass generation for the two-dimensional scalar gluons ϕ_i. To estimate the masses of scalar gluons ϕ_i we use the Schwinger-Dyson equations as self-consistency conditions, we get:

$$M^2 = \frac{2N_c \tilde{g}^2}{L^2} \int^\Lambda \frac{d^2 p}{(2\pi)^2} \frac{1}{p^2 + M^2} = \frac{N_c \tilde{g}^2 \Lambda^2}{8\pi^3} \log \frac{\Lambda^2 + M^2}{M^2} \;, \tag{3}$$

thus M^2 brings an infrared cutoff as expected. We notice that the gluon mass remains finite in the large-N_c limit if the QCD coupling constant de-

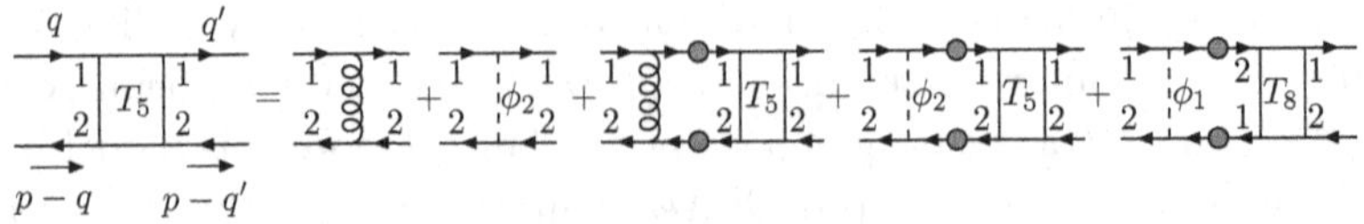

Fig. 1.   Inhomogeneous Bethe-Salpeter equation for quark-antiquark scattering amplitude T_5.

creases as $1/N_c$ in line with the perturbative law of $4D$ QCD. We adopt the approximation $M \ll \Lambda \simeq 1/R$ to protect the low-energy sector of the model and consider the momenta $|p_{0,1}| \sim M$. Thereby we retain only leading terms in the expansion in p^2/Λ^2 and M^2/Λ^2, and also neglect the effects of the heavy K-K modes in the low-energy Wilson action. We observe that the limit $M \ll \Lambda$ supports consistently both the fast decoupling of the heavy K-K modes and moderate decoupling of scalar gluons[4], the latter giving an effective four-fermion interaction different from[6]. Also allow us to define the "heavy-scalar" expansion parameter $A = g^2/(2\pi M^2) = 1/log\frac{\Lambda^2}{M^2} \ll 1$. Now we proceed to the study of bound states of quark-antiquark. In our reduction we have four possible combinations of quark bilinears to describe these states with valence quarks: $(\psi_1 \bar{\psi}_1), (\psi_1 \bar{\psi}_2), (\psi_2 \bar{\psi}_1), (\psi_2 \bar{\psi}_2)$. We need to compute the full quark-antiquark scattering amplitude T in the different channels. As an example we are going to show the computing of T_5 which correspond to the scattering $(q_1 + \bar{q}_2 \longrightarrow q_1 + \bar{q}_2)$. It satisfies, the equation given graphically in Fig.1, in the large N_c limit and in ladder exchange approximation (non-ladder contribution are estimated to be of higher order in the A expansion). Notice that in the equation for T_5 the amplitude T_8 appears, which correspond to the process $q_2 + \bar{q}_1 \longrightarrow q_1 + \bar{q}_2$. This means that the equations for T_5 and T_8 are coupled. In Fig.1 all internal fermion lines correspond to dressed quark propagators which were determined in[4] by using the large N_c limit and the one-boson exchange approximation. There are two kind of solutions for the dressed quark propagators. In one solution, the perturbative one, we have tachyonic quarks in the chiral limit as in QCD_2 and our model could be interpreted as a perturbation from the result of chiral QCD in $2D$. But we are not allowed to consider that possibility because the spectrum for the lowest $q\bar{q}$ bound states becomes imaginary if one takes into account the scalar field exchange. The second solution, the non-perturbative one, supports non-tachyonic quarks with masses going to zero, in the chiral limit and also yield real masses for the $q\bar{q}$ bound states. Having found the full quark propagator we proceed to solve the

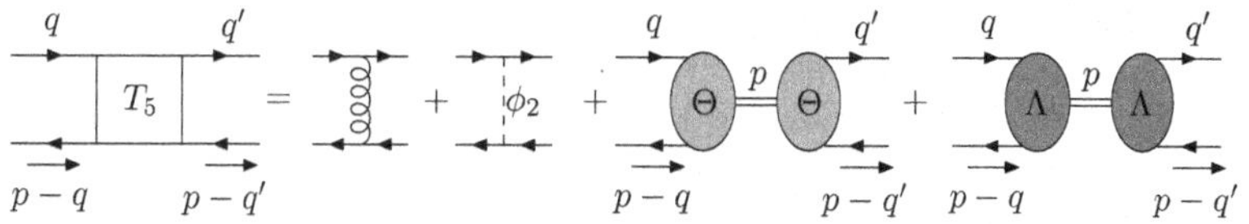

Fig. 2. $q\bar{q}$ scattering T_5 in the color-singlet channel.

inhomogeneous Bethe-Salpeter equation for T_5 and T_8. We obtain for T_5 (see Figure 2):

$$T_5^{\alpha\beta,\gamma\delta}(q,q';p) = -i\frac{g^2}{N}\frac{\gamma_-^{\alpha\delta}\gamma_-^{\beta\gamma}}{(q-q')_-^2} - i\frac{g^2}{N}\frac{\sigma_3^{\alpha\gamma}\sigma_3^{\beta\delta}}{[(q-q')^2 - M^2 + i\epsilon]} + \qquad (4)$$

$$+\sum_j \Theta_j^{\alpha\gamma}(q;p)\frac{i}{[p^2 - \bar{m}_j^2 + i\epsilon]}\Theta_j^{\beta\delta}(q';p) + \sum_k \Lambda_k^{\alpha\gamma}(q;p)\frac{i}{[p^2 - \tilde{m}_k^2 + i\epsilon]}\Lambda_k^{\beta\delta}(q';p).$$

There are no continuum states in the quark-antiquark amplitude– only bound states at $p^2 = \bar{m}^2$ and $p^2 = \tilde{m}^2$, whose residue yield the bound state wave functions $\Theta(p,r)$ and $\Lambda(p,r)$. These bound states have a direct interpretation in term of Dirac Bilinears of the theory in $4D^4$. In particular, the pseudoscalar $4D$ states are related with $\Lambda_k(r;p)$ which could be interpreted as the vertex: (quark)-(antiquark)-($4D$ pseudoscalar meson). The masses of the bound states and the vertex functions are fixed by the solution of the homogeneous Bethe-Salpeter equation which, in our model, yield to a eigenvalue problem, generalization of the integral equation found by 't Hooft[1] in QCD_2. For example the following eigenvalue integral equation determine the mass spectrum of the pseudoscalar bound states (in the chiral limit):

$$\tilde{m}^2\,\phi(x) = -\frac{g^2}{\pi}\int_0^1 dy\,\frac{\mathbf{P}}{(x-y)^2}\,\phi(y) - A\frac{\Sigma_0^2}{x^{1-\beta}}\int_0^1 dy\,\frac{\phi(y)}{(1-y)^{1-\beta}} \qquad (5)$$

$$-A\frac{\Sigma_0^2}{(1-x)^{1-\beta}}\int_0^1 dy\,\frac{\phi(y)}{y^{1-\beta}} + A\frac{\Sigma_0^2}{[(1-x)x]^{1-\beta}}\int_0^1 dy\,\phi(y) + A\Sigma_0^2\int_0^1 dy\,\frac{\phi(y)}{[(1-y)y]^{1-\beta}},$$

where $\beta = A/2$. This equation was analyzed for small A and a massless analytic solution for the ground state was found. We interpreted this solution as a "pion" of the model. The massive states were estimated working with the regular perturbation theory, starting from 't Hooft[1] solutions. The meson sector of the theory was also examined revealing a massive spectrum.

References

1. G. 't Hooft, Nucl. Phys. **B75** (1974) 461.
2. S. Dalley and I.R. Klebanov, Rhys. Rev. **D47** (1993) 2517;

3. H.-C. Pauli and S.J. Brodsky, Rhys. Rev. **D32** (1985) 1993 and 2001;
   F. Antonuccio and S. Dalley, Phys. Lett. **B376** (1996) 154.
4. J. Alfaro, A. A. Andrianov and P. Labraña, JHEP 0704: 067 (2004).
5. S. Coleman, Comm. Math. Phys. **31** (1973) 461.
6. M. Burkardt, Phys. Rev. **D56** (1997) 7105.

OBSERVATIONAL CONSTRAINTS ON LORENTZ SYMMETRY DEFORMATION

J. ALFARO* and M. CAMBIASO†

Facultad de Física, Pontificia Universidad Católica de Chile
Casilla 306, Santiago 22, Chile
*jalfaro@puc.cl
†mcambias@puc.cl

Lorentz symmetry is nowadays being subject to severe scrutiny, both theoretically and experimentally. In this work we focus on the bounds imposed by observations of ultra-high energy astronomy to the parameters of Lorentz symmetry deformation when this deformation is expressed as modifications to the dispersion relations of free particles.

1. Motivations

1.1. *Formal issues*

To begin with it is good to remind that being the Lorentz group a non-compact group it will never be possible to test Lorentz symmetry up to arbitrary high energies. However one still wishes to answer what happens at energies well above those accesible in particle accelerators. Equivalently, the structure of spacetime at scales near the Planck length, $\ell_P = \sqrt{\frac{G_N \hbar}{c^3}} \approx 10^{-33}$ cm. needs to be understood. On merely theoretical grounds, departures from Lorentz symmetry are present in many effective models coming from theories that claim to clarify some issues regarding the quantum nature of spacetime, such as in String Theory[1] and Loop Quantum Gravity[2]. Particularly we will be interested in those theories in which Lorentz symmetry deformations are manifested as modified dispersion relations, *aka* MDRs.

1.2. On the detectability of these effects

The common belief is that the modifications to the kinematics of free particles will only be seen at high energies. Also and according to some of the effective models mentioned above, the effects would be proportional to the distance travelled by the particles, therefore Cosmic Rays, whose energies range from the TeV up to 10^{20} eV, seem a promising laboratory to probe such theories. Furthermore, by allowing tiny modifications to the particles' kinematics one may solve long standing problems related with the so called threshold anomalies of Cosmic Rays near $E \sim 10^{20}$ eV and TeV photons, the GZK[3] and TeV-γ paradoxes, respectively.

2. A Parameterization of MDRs

A plausible parameterization for the modified dispersion relations is the following:

$$E_a^2 = \vec{p}_a^2 + m_a^2 + \sum_n C_a^{(n)} |\vec{p}_a|^n. \tag{1}$$

where the coefficients $C_a^{(n)}$ are dimensional parameters and a labels the particle species. In the high-energy regime where the new effects should arise, $|\vec{p}| \gg m$ therefore $|\vec{p}| \sim E$, rendering the above expansion meaningful. Besides, since the corrections encoded in the $C_a^{(n)}$ are expected to be present only when E/E_P is non-negligible, for $E_P \equiv \sqrt{\frac{\hbar c^5}{G_N}} \approx 1.2 \times 10^{28}$ eV, then it is expected that the corrections be suppressed by powers of such factor. This is a characteristic feature of most of the models know so far and has the obvious consequence that if measuring evidence in favour of $C_a^{(3)} \neq 0$ is hard, harder will it be to measure a similar thing for $C_a^{(4)} \neq 0$. Therefore we will focus on the bounds encountered for lowest order terms, i.e., corrections to the "speed of light" of particles and on cubic corrections such as $\frac{1}{\Lambda} p^3$.

3. Parameter Constraining Rationale and Some Models

Threshold analysis[4] readily allows for constraining the $C_a^{(n)}$, because for certain values of these coefficient an observed reaction may be forbidden, or vice-versa, an argument used to explain why do we actually see cosmic rays with energies above the GZK cutoff[5]. Another technique is based on astronomical interferometry[6], where the above MDRs imply different group and phase velocities which ultimately result in randomly varying phase on the two arms of the interferometer. If so, a total smearing of any interference pattern would occur, therefore the presence of interference pattern for some specific distant objects may rule out a particular MDR. A third prediction is the difference between time of flights of two particles emitted with slightly different energies from a source such as a GRB[7]. Modern instrumentation would allow to measure such an effect. Finally we comment the Loop Quantum Gravity framework in which the MDR found could allow for an explanation of the GZK-paradox[5] and also yield birefringence effects[8], which can be bounded from data on the polarization of some radio galaxies[9].

4. Observational Bounds

4.1. *Quadratic corrections*

Since the quadratic corrections amounts for modifications to the "speed of light" of particles, the bound thus far obtained are in the form of differences between the maximal attainable velocities for different particles, and basically come from threshold analysis for different reactions. Some of the bounds are for protons and the Δ resonance[10] or protons and photons[4] or for protons and electrons[5].

(i) $C_p^{(2)} < C_\Delta^{(2)} - 1 \times 10^{-25}$,

(ii) $C_p^{(2)} < C_\gamma^{(2)} - 1 \times 10^{-23}$,

(iii) $|C_p^{(2)} - C_e^{(2)}| < 9.8 \times 10^{-22}$.

4.2. *Cubic corrections*

The bounds for $C_a^{(3)}$ come basically from threshold analysis of TeV photons[11] coming from the Blazars Mkn 501 and 421, from the interferometric observations[6], from polarization effects and the observation of the polarization in UV light[9] from the radio galaxy 3C256, from the synchrotron radiation by electrons cycling around the magnetic fields in the Crab Nebula[12] and other observations. However we want to stress the most stringent bound so far, one coming from allowing the MDRs to explain the GZK-paradox but also by demanding that such modifications do not alter the region of the spectrum which has the highest accuracy, with the assumption that this correction be universal[5].

(i) $C^{(3)} < 2.5 \times 10^{-28}$ eV^{-1} (for details see[11]),

(ii) $C^{(3)} < 7.4 \times 10^{-31}$ eV^{-1} (for details see[6]),

(iii) $C^{(3)} < 8.3 \times 10^{-33}$ eV^{-1} (for details see[9]),

(iv) $C^{(3)} < 7 \times 10^{-35}$ eV^{-1} (for details see[12]),

(v) $C^{(3)} < 2 \times 10^{-37}$ eV^{-1} (for details see[13]),

(vi) $C^{(3)} < 3.26 \times 10^{-41}$ eV^{-1} (for details see[5]).

5. Conclusions

Before we conclude we mention that there are other kinds of bounds coming from low-energy physics as well as precision clock tests to probe the limits of validity of Special Relativity. About the bounds here presented we emphasize that regardless the model from which a particular $C_a^{(n)}$ was motivated, the observational bounds apply in general. The validity of the observations being for the astronomers to decide. On the experimental side, we mention that the future observations of the Pierre Auger Collaboration will start taking data at an estimated rate of 3000 events with $E > 10^{19}$ eV per year and 30 events per year for energies $E \geq 10^{20}$ eV, therefore their observations are being eagerly waited for since they may help in providing clear-cut evidence for any eventual deviations from Lorentz symmetry.

Acknowledgements

M. C. wishes to thank the organizing committee of V-Silafae and also, acknowledges partial support from DIPUC (Dirección de Investigación y Postgrado de la Pontificia Universidad Católica de Chile).

References

1. J. R. Ellis, N. E. Mavromatos and D. V. Nanopoulos, Phys. Rev. D **63**, 124025 (2001).
2. J. Alfaro, H. A. Morales-Tecotl and L. F. Urrutia, Phys. Rev. Lett. **84**, 2318 (2000).
3. K. Greisen, Phys. Rev. Lett. **16**, 748 (1966).
   G. T. Zatsepin and V. A. Kuzmin, JETP Lett. **4**, 78 (1966) [Pisma Zh. Eksp. Teor. Fiz. **4**, 114 (1966)].
4. S. R. Coleman and S. L. Glashow, Phys. Rev. D **59**, 116008 (1999)
5. J. Alfaro and G. Palma, Phys. Rev. D **67**, 083003 (2003).
6. R. Lieu and L. W. Hillman, Astrophys. J. **585**, L77 (2003)
7. G. Amelino-Camelia, J. R. Ellis, N. E. Mavromatos, D. V. Nanopoulos and S. Sarkar, Nature **393**, 763 (1998).
8. R. Gambini and J. Pullin, Phys. Rev. D **59**, 124021 (1999).
9. R. J. Gleiser and C. N. Kozameh, Phys. Rev. D **64**, 083007 (2001).
10. O. Bertolami and C. S. Carvalho, Phys. Rev. D **61**, 103002 (2000)
11. F. W. Stecker, Astropart. Phys. **20**, 85 (2003).
12. T. Jacobson, S. Liberati and D. Mattingly, Nature **424**, 1019 (2003).
13. J. R. Ellis, N. E. Mavromatos, D. V. Nanopoulos and A. S. Sakharov, arXiv:astro-ph/0309144.

VARIABLE-MASS DARK MATTER AND THE AGE OF THE UNIVERSE[*]

U. FRANCA

SISSA / ISAS
Via Beirut 4
34014 Trieste, Italy
urbano@sissa.it

R. ROSENFELD

Instituto de Física Teórica-UNESP
Rua Pamplona, 145
01405-900, São Paulo, SP, Brazil
rosenfel@ift.unesp.br

Models with variable-mass particles, called VAMPs, have been proposed as a solution to the cosmic coincidence problem that plagues dark energy models. In this contribution we make a short description of this class of models and explore some of its observational consequences. In particular, we show that fine tuning is still required in this scenario and that the age of the Universe is considerably larger than in the standard Λ-CDM model.

1. Introduction

The Wilkinson Microwave Anisotropy Probe (WMAP) satellite [1], along with other experiments, has confirmed that the universe is very nearly flat, and that there is some form of dark energy (DE) that is the current dominant energy component, accounting for approximately 70% of the critical density. Non-baryonic cold dark matter (DM) contributes around 25% to

---

[*]Presented by R. Rosenfeld at the V SILAFAE, July 12-16, Lima, Peru.

the total energy density of the universe and the remaining 5% is the stuff we are made of, baryonic matter.

DE is smoothly distributed throughout the universe and its equation of state with negative pressure is causing its present acceleration. It is generally modelled using a scalar field, the so-called quintessence models, either slowly rolling towards the minimum of the potential or already trapped in this minimum [2].

An intriguing possibility is that DM particles could interact with the DE field, resulting in a time-dependent mass. In this scenario, dubbed VAMPs (VAriable-Mass Particles)[3], the mass of the DM particles evolves according to some function of the dark energy field ϕ [3,4,5,6,7,8,9,10,11]. In this case, the DM component can have an effective equation of state with negative pressure that could present the same behaviour as DE.

We studied a model with exponential coupling between DM and DE, since it presents a tracker solution where the effective equation of state of DM mimics the effective equation of state of DE and the ratio between DE and DM energy density remains constant after this attractor is reached. This behavior could solve the "cosmic coincidence problem", that is, why are the DE and DM energy densities similar today.

This type of solution also appears when the DE field with a exponential potential is not coupled to the other fluids. In fact, Liddle and Scherrer [12] showed that for a non-coupled DE, the exponential potential is the only one that presents stable tracker solutions. In this case, however, it is not able to explain the current acceleration of the universe and other observational constraints, unless we assume that the field has not yet reached the fixed point regime [13].

In this contribution, we review our results presented in [14].

2. A Simple Exponential VAMP Model

In the exponential VAMP model, the potential of the DE scalar field ϕ is given by

$$V(\phi) = V_0 \, e^{\beta\phi/m_p}, \tag{1}$$

where V_0 and β are positive constants and $m_p = M_p/\sqrt{8\pi} = 2.436 \times 10^{18}$ GeV is the reduced Planck mass in natural units, $\hbar = c = 1$. Dark matter is modelled by a scalar particle χ of mass

$$M_\chi = M_{\chi 0} \, e^{-\lambda(\phi-\phi_0)/m_p} \,, \tag{2}$$

where $M_{\chi 0}$ is the current mass of the dark matter particle (hereafter the index 0 denotes the present epoch, except for the potential constant V_0) and λ is a positive constant.

In this case we can show that

$$\dot{\rho}_\chi + 3H\rho_\chi(1 + \omega_\chi^{(e)}) = 0 \ , \tag{3}$$

$$\dot{\rho}_\phi + 3H\rho_\phi(1 + \omega_\phi^{(e)}) = 0 \ , \tag{4}$$

where

$$\omega_\chi^{(e)} = \frac{\lambda \dot{\phi}}{3Hm_p} = \frac{\lambda \phi'}{3m_p} \ , \tag{5}$$

$$\omega_\phi^{(e)} = \omega_\phi - \frac{\lambda \dot{\phi}}{3Hm_p} \frac{\rho_\chi}{\rho_\phi} = \omega_\phi - \frac{\lambda \phi'}{3m_p} \frac{\rho_\chi}{\rho_\phi} \ , \tag{6}$$

are the effective equation of state parameters for dark matter and dark energy, respectively. Primes denote derivatives with respect to $u = \ln(a) = -\ln(1 + z)$, where z is the redshift, and $a_0 = 1$.

At the present epoch the energy density of the universe is divided essentially between dark energy and dark matter. In this limit, there is a fixed point solution for the equations of motion of the scalar field such that

$$\Omega_\phi = 1 - \Omega_\chi = \frac{3}{(\lambda + \beta)^2} + \frac{\lambda}{\lambda + \beta} \tag{7}$$

$$\omega_\chi^{(e)} = \omega_\phi^{(e)} = -\frac{\lambda}{\lambda + \beta}. \tag{8}$$

The equality between ω_χ and ω_ϕ in the attractor regime comes from the tracker behavior of the exponential potential [7,11] in this regime.

However, this is only valid in the fixed point regime. If we want to know what happens before, the equations must be solved numerically. The density parameters for the components of the universe and the effective equations of state for the DE and DM for a typical solution are shown in figure 1. Notice that the transition to the tracker behavior in this example is occurring presently.

3. Cosmological Constraints and the Age of the Universe in the Exponential VAMP Model

We have calculated the age of the universe for the models that satisfy the Hubble parameter and the dark energy density observational constraints

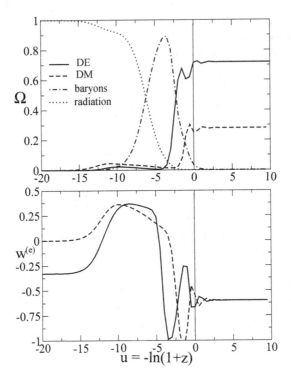

Fig. 1. *Top panel*: Density parameters of the components of the universe as a function of $u = -\ln(1+z)$ for $\lambda = 3$, $\beta = 2$ and $V_0 = 0.1\tilde{\rho}_c$. After a transient period of baryonic matter domination (dot-dashed line), DE comes to dominate and the ratio between the DE (solid line) and DM (dashed line) energy densities remains constant. *Bottom panel*: Effective equations of state for DE (solid line) and DM (dashed line) for the same parameters used in top panel. In the tracker regime both equations of state are negative.

$(h = 0.72 \pm 0.08, 0.6 \leq \Omega_{\phi 0} \leq 0.8)$. We have used stepsizes $\Delta\lambda = \Delta\beta = 0.2$ and $\Delta V_0 = 0.05\tilde{\rho}_c$ for the region $\lambda = [0.01, 20]$, $\beta = [0.01, 20]$, $V_0 = [0.1\tilde{\rho}_c, 0.8\tilde{\rho}_c]$, generating 1.4×10^5 models.

Fitting the distribution of models as function of ages, the age of the universe in this VAMP scenario was found to be

$$t_0 = 15.3^{+1.3}_{-0.7} \text{ Gyr} \quad 68\% \text{ C.L.} , \qquad (9)$$

which is considerably higher than the age of models of non-coupled dark energy [1,15]. This seems natural, since in these models the CDM also has an effective negative equation of state and accelerates the universe. This result is very conservative, since it relies only on the well established limits on the Hubble constant and the dark energy density today.

4. Conclusion

VAMP scenario is attractive since it could solve the problems of exponential dark energy, giving rise to a solution to the cosmic coincidence problem. However, we found that in order to obtain solutions that can provide realistic cosmological parameters, the constant V_0 has to be fine tuned in the range $V_0 = [0.25\tilde{\rho}_c, 0.45\tilde{\rho}_c]$ at 68% C. L. This implies that the attractor is being reached around the present epoch. In this sense, the model is not able to solve the coincidence problem.

A generic feature of this class of models is that the universe is older than non-coupled dark energy models. Better model independent determination of the age of the universe could help to distinguish among different contenders to explain the origin of the dark energy.

Acknowledgments

R. R. would like to thank the organizers of the V SILAFAE for their efforts in providing a nice atmosphere for the participants and to CNPq for financial support.

References

1. D. N. Spergel *et al.*, Astrophys. J. Supp. **148**, 175 (2003).
2. See, *e.g*, P. J. E. Peebles, B. Ratra, Rev. Mod. Phys. **75**, 559 (2003) and references therein.
3. G. W. Anderson, S. M. Carrol, astro-ph/9711288.
4. J. A. Casas, J. García-Bellido, M. Quirós, Class. Quantum Grav. **9**, 1371 (1992).
5. G. R. Farrar, P. J. E. Peebles, Astrophys. J. **604**, 1 (2004).
6. M. B. Hoffman, astro-ph/0307350.
7. L. Amendola, Phys. Rev. D **62**, 043511 (2000).
8. L. Amendola, D. Tocchini-Valentini, Phys. Rev. D **64**, 043509 (2001).
9. L. Amendola, Mon. Not. Roy. Astron. Soc. **342**, 221 (2003).
10. M. Pietroni, Phys. Rev. D **67**, 103523 (2003).
11. D. Comelli, M. Pietroni, A. Riotto, Phys. Lett. B **571**, 115 (2003).
12. A. R. Liddle, R. J. Scherrer, Phys. Rev. D **59**, 023509 (1998).
13. U. Franca, R. Rosenfeld, JHEP **10**, 015 (2002).
14. U. Franca, R. Rosenfeld, Phys. Rev. D **69**, 063517 (2004).
15. M. Kunz, P. S. Corasaniti, D. Parkinson, E. J. Copeland, Phys. Rev. D **70**, 041301 (2004).

DYNAMICAL STUDY OF SPINODAL DECOMPOSITION IN HEAVY ION COLLISIONS

A. BARRAÑÓN[1] and J. A. LÓPEZ[2]

[1]Universidad Autónoma Metropolitana - Azcapotzalco, México D.F., México
[2]University of Texas at El Paso, El Paso, Texas 79968, U.S.A.

Nuclei undergo a phase transition in nuclear reactions according to a caloric curve determined by the amount of entropy. Here, the generation of entropy is studied in relation to the size of the nuclear system.

1. Introduction

Collisions at hundreds of $MeVs$ break the nuclei into several fragments[1]. This fragmentation led to the determination[2,3] of the relation between the system's temperature and its excitation energy, *i.e.* the caloric curve, CC. A striking feature of the CC is the dependence of the transition temperature, *i.e.* the "plateau" temperature T_p, with the mass of the breaking system. Recent observations[4] show that lighter systems, *i.e.* with masses between 30 to 60 nucleons have a $T_p \approx 9\ Mev$, while systems with masses between 180 to 240 yield a $T_p \approx 6\ MeV$. This variation of T_p has been linked to the entropy generated during the reaction[5,6]. Here these collisions are used to find the relationship between the entropy generated in the reaction and the size of the fragmenting system.

2. Caloric Curve and Entropy

Ions fuse in collisions and reach some maximum density and temperature, to then expand cooling and reducing the density[7,8]. In the density-temperature plane[9], this corresponds to a displacement from normal saturation density and zero temperature, ($n_o \approx 0.15\ fm^{-3}$, $T = 0$), to some higher values of (n, T), to then approach $n \to 0$ and $T \to 0$ asymptotically.

These reactions include phase changes and non-equilibrium dynamics, and are best studied with molecular dynamics. The model we use is "$LATINO$"[10], combined with the fragment-recognition code $ECRA$[11]; these, without adjustable parameters, have described reactions[12], phase transitions[13], criticality[14], and the CC[5,6].

The right panel of figure 1 shows trajectories in the $n - T$ plane of collisions simulated with "$Latino$" and corresponding to central collisions of $Ni + Ni$ at energies between 800 to 1600 MeV. As during the reaction, nucleons evaporate leaving a smaller system, the trajectories follow different paths on the plane according to the residual size of the nuclear system. The plot shows the cases of systems with 49, 75, and 91 nucleons.

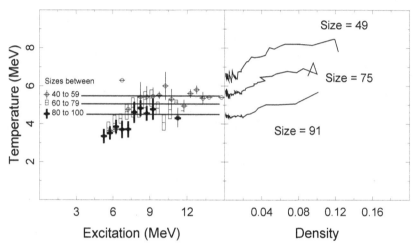

Fig. 1. Caloric curves (left panel) and density-temperature trajectories (right panel) of $Ni + Ni$ central collisions at varying energies. Density values are in fm^{-3}.

As noted before[5,6], these trajectories are determined by the entropy generated in the beginning of the reaction. These reactions can help to relate the caloric curves to system size, and to the amount of entropy generated.

2.1. *Mass dependence of the caloric curve*

Central collisions were performed for $Ni + Ag$ at different beam energies. During the collision, n and T of the 50% most central particles were determined as a function of time. At the same time, $ECRA$ was used to identify the fragment source size, fragmentation times, fragment stability, and other quantities of interest. This information was then used to obtain the caloric curve as a function of the source size.

The left panel of figure 1 shows caloric curves obtained by *Latino* in central collisions of $Ni + Ag$ at energies between 1000 to 3800 MeV. The three groups correspond to different masses: circles for collisions with residual sources of 40 to 59 nucleons, rectangles for 60 to 79, and crosses for 80 to 100 nucleons. The symbol sizes denote the standard deviations of T and excitation energy of each energy bin. The lines, average values of T of the last few points of each mass range, clearly show the inverse relationship between the transition temperature, T_p, and source size, as observed experimentally[4].

2.2. *Entropy dependence of the caloric curve*

The collision takes a $T = 0$ quantum-system to a hot and dense "quasi-classical" state through a non-equilibrium path, which then expands and decomposes spinodaly. In terms of entropy, the system goes from an initial $S = 0$ state to a high-entropy state, to then coast into an isentropic expansion until a violent spinodal decomposition increases entropy again.

Latino + ECRA allow the evaluation of the entropy during the reaction. Neglecting quantum effects, the entropy per particle S can be quantified in units of Boltzmann's constant as for a classical gas[16] through $S = \log \left[\frac{1}{n} \left(\frac{3T}{2} \right)^{3/2} \right] + S_o$, where S_o depends on the nucleon mass[5].

Figure 2 shows the entropy of central collisions of $Ni + Ni$ at energies between 600 to 2000 MeV plotted against the size of the residual fragmenting source. The bottom curve corresponds to the entropy achieved at maximum heating and compression, *i.e.* the value at which the system expands until it decomposes; the one responsible for determining T_p. The top curve, on the other hand, is the asymptotic value of S, and it includes all other increases of entropy that took during the phase change and expansion. Both of these curves show that lighter systems are more apt to generate entropy, and that the phase change and final expansion succeed in increasing the initial entropy.

3. Conclusions

These results confirm previous results[5,6], the initial stage of the reaction reaches a value of S which defines the trajectory of the compound nucleus into a spinodal decomposition. The transition temperature T_p is thus defined by the intersection of the isentrope and the spinodal. The present study further confirms that the amount of entropy generated initially in the

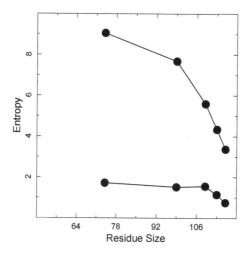

Fig. 2.   Entropy of collisions of $Ni + Ni$ at varying energies *versus* residual size.

reaction varies inversely with the size of the fragmenting system. This inverse relationship is also maintained after the phase change. Unfortunately, as concluded in previous studies, the observed relationship between S and source size was obtained with $LATINO$, and does not explain what causes lighter systems to generate more entropy than heavier ones in heavy ion collisions. We propose to continue with this study in the near future.

References

1. P. F. Mastinu *et al.*, *Phys. Rev. Lett.* **B76**, 2646 (1996).
2. J. Pochodzalla *et al.*, *Prog. Part. Nucl. Phys* **39**, 43 (1997).
3. H. Feldmeier *et al.*, *Gross properties of nuclei and nuclear excitations*, GSI, Darmstadt, (1999).
4. J. B. Natowitz *et al.*, *Phys. Rev.* **C65**, 034618 (2002).
5. A. Barrañón, J. Escamilla, and J. A. López, *Phys. Rev.* **C69**, 014601 (2004).
6. A. Barrañón, J. Escamilla and J. A. López, *Bras. J. Phys.*, **34**, 904 (2004).
7. G. F. Bertsch, and P. Siemens, *Phys. Lett.* **B126**, 9 (1983).
8. J. A. López and P. Siemens, *Nucl. Phys.* **A314**, 465 (1984).
9. J. A. López and C. O. Dorso, *Lecture Notes on Phase Transitions in Nuclear Matter*, World Scientific, Singapore, (2000).
10. A. Barrañón *et al.*, *Rev. Mex. Fís.* **45-Sup.2**, 110 (1999).
11. A. Strachan and C. O. Dorso, *Phys. Rev.* **C56**, 995 (1997).
12. A. Chernomoretz *et al.*, *Phys. Rev.* **C65**, 054613 (2002).
13. A. Barrañón, C. O. Dorso and J. A. López, *R. Mex. Fís.* **47-Sup.2**, 93 (2001).
14. A. Barrañón *et al.*, *Heavy Ion Phys.* **17-1**, 59 (2003).
15. A. Barrañón, C. O. Dorso and J. A. López, *Infor. Tecnológica* **14**, 31 (2003).
16. K. Huang, *Stat. Mech., 2nd edition*, John Wiley & Sons, New York, (1987).

PREDICTIONS FOR SINGLE SPIN ASYMMETRIES IN INCLUSIVE REACTIONS INVOLVING PHOTONS*

V. GUPTA

CINVESTAV Unidad Merida, Mexico
virendra@mda.cinvestav.mx

C. J. SOLANO SALINAS

Universidad Nacional de Ingenieria, Peru
and
CINVESTAV Unidad Merida, Mexico
jsolano@uni.edu.pe

H. S. MANI

The Institute of Mathematical Sciences, India
hsmani@imsc.res.in

A phenomenological model used in explaining polarization phenomena and left-right asymmetry in inclusive p-p scattering is considered for reactions involving photons. In particular, the reactions (a) $\gamma + p(\uparrow) \to \pi^{\pm} + X$ and (b) $p(\uparrow) + p \to \gamma + X$ are considered where γ = resolved photon . Predictions for hyperon asymmetry (in (a) and (b)) provide further tests of this particular model.

1. Introduction

Polarization phenomena in high energy proton-proton scattering has been studied experimentally [1] and theoretically [2,3] for over two decades. Left-right asymmetry (A_N) has been measured [4] in the scattering of a trans-

---

*The work is also supported in part by Conacyt-Mexico, Project No. 28265E.

versely polarized proton $(p(\uparrow))$ with an unpolarized proton (p) in the inclusive reactions of the type

$$p(\uparrow) + p \rightarrow h + X \qquad (1)$$

where h is a hadron and typically is $\pi^{\pm}$, Λ etc.

Of the various models, which reproduce some features of the data for A_N and $P(H)$, the most succesful one is perhaps the "orbiting valence quark model" [3], is very interesting in that it relates A_N and $P(H)$ as being due to the same underlying production mechanism in terms of the constituent quarks of the protons and the produced hadron.

The basic idea is that a transversely polarized quark $q^P(\uparrow)$ or $q^P(\downarrow)$ in the projectile (P) proton combines with the appropriate quarks or antiquarks from the 'sea' of the target proton to form the observed hadron h in reaction Eq. (1). Further, the produced hadron h moves preferentially to the left of the beam direction, in the upper side of the production plane . That is if $\vec{S} \cdot \vec{n} > 0$, where $\vec{S}$ represents the transverse polarization of $q^P(\uparrow)$, which is polarized upwards w.r.t. the production plane. While, h formed from $q^P(\downarrow)$, with $\vec{S} \cdot \vec{n} < 0$, will move preferentially to the right. This gives rise to a left-right asymmetry. The same reasoning gives rise to a net $P(H)$ in the sub-sample of hyperons going left since H is assumed to retain the polarization of the q^P which forms it [3].

Based on these ideas, we consider reactions (other than in Eqs. (1) and (2)) to provide further tests of this model. In particular, in S2 we consider the left-right asymmetry in the reaction

$$e^- + p(\uparrow) \rightarrow e^- + \pi^{\pm} + X \qquad (2)$$

The photon from the $ee\gamma$ vertex is considered to be 'real' with 'resolved' hadronic components.

The left-right asymmetry A_N^γ of the photon in

$$p(\uparrow) + p \rightarrow \gamma + X \qquad (3)$$

is considered in S3. This reaction has a much larger cross-section than the reaction

$$p(\uparrow) + p \rightarrow e^+ + e^- + X \qquad (4)$$

considered earlier [3].

2. Resolved Photon-Proton Reactions

Left-right asymmetry in $e^- + p(\uparrow) \rightarrow e^- + \pi^{\pm} + X$.

We consider the photon from the $ee\gamma$ - vertex with $M^2 = (mass)^2 \leq (100 MeV)^2$. At such small values of M^2 the 'resolved' photon $-\gamma$, containing hadronic components dominates in contrast to the 'direct' photon which has a point-like interaction. The study of jet production in $e^- + e^+ \rightarrow e^- + e^+ + X$ and $e^- + p \rightarrow e^- + X$ has led to the determination of the photon structure functions : $q^{\gamma(M^2)}(x, Q^2)$ for q = u,d,s quarks and $g^{\gamma(M^2)}(x, Q^2)$ for the gluons [5,6]. Here, Q^2 is the momentum transfer scale between the γ and proton in the effective reaction

$$\gamma + p(\uparrow) \rightarrow \pi^{\pm} + X \tag{5}$$

We now consider the left-right asymmetry of the $\pi^{\pm}$ in this reaction.

In the picture under discussion, the u_v valence quark of the proton, with structure function $u_v^p(x^p, Q^2)$, combines with the $\bar{d}$ in the resolved γ, with structure function $\bar{d}^{\gamma}(x^{\gamma}, Q^2)$, to form a π^+ meson. Similarly, d_v^p will combine with the $\bar{u}$ in the photon to form π^-. We consider the reaction in Eq. (7) in the photon-parton center of mass frame with total energy $\sqrt{s}$. In this frame [7] the Feynman parameter x_F and the longitudinal momentum $p_{||}$ for the emitted pion are given by $x_F = x^p - x^{\gamma}$ and $p_{||} = \frac{x_F \sqrt{s}}{2}$, for the proton fragmentation region $x_F > 0$. The normalized number density of the observed pion in a given kinematic region D is

$$N(x_F, Q|s, i) = \frac{1}{\sigma_m} \int_D d^2 p_T \frac{d^3 \sigma(x_F, \vec{p}_T, Q|s, i)}{dx_F d^2 p_T} \tag{6}$$

where $i = \uparrow$ or $\downarrow$ refers to the transverse spin of the proton, and σ_{in} is the total inelastic cross-section. Then,

$$\Delta N(x_F, Q|s) = N(x_F, Q|s \uparrow) - N(x_F, Q|s \downarrow) = C_{\gamma} \Delta D(x_F, Q|s) \tag{7}$$

where C_{γ} is a flavor independent constant which must be determined fitting experimental data. It is expected to lie between 0 and 1, like the corresponding constant $C \simeq 0.6$ [3] for reaction in Eq. (1). In Eq. (7), ΔN has been taken to be proportional to $\Delta D(x_F, Q|s) = D(x_F, Q, +|s) - D(x_F, Q, -|s)$ where $D(x_F, Q, \pm|s)$ is the normalized number density eg. of the $(u_v \bar{d})$ which give the π^+.

The $\pm$ in $D(x_F, Q, \pm|s)$ refers to the polarization of the valence quark with respect to proton spin direction. We must point out that here, and in the rest of this work, that we are using parton distribution functions which go to zero at $x_F = 1$ and have a finite value for $x_F = 0$. In the model [3], for

π^+ one has $D^{\pi^+}(x_F, Q, \pm|s) = K_\pi u_v^p(x^p, Q^2, \pm)\bar{d}^\gamma(x^\gamma, Q^2)$, where K_π is a constant. Thus, the asymmetry [8]

$$A_{\gamma_p}^{\pi^+}(x_F, Q|s) = \frac{C_\gamma K_\pi [\Delta u_v^p(x^p, Q^2)]\bar{d}^\gamma(x^\gamma, Q^2)}{N_0(x_F|s) + K_\pi u_v^p(x^p, Q^2)\bar{d}^\gamma(x^\gamma, Q^2)} \qquad (8)$$

where the numerator is $\Delta N(x_F, Q|s)$ and Eqs. (9-10) have been used, so that $\Delta u_v^p(x^p, Q^2) = u_v^p(x^p, Q^2, +) - u_v^p(x^p, Q^2, -)$. In the denominator, $N_0^\gamma(x_F|s)$ stands for the non-direct part of π^+ production and the second term is $[D^{\pi^+}(x_F, Q, +|s) + D^{\pi^+}(x_F, Q, -|s)]$. In general, the non-direct part N_0^γ for reaction in Eq. (5) and the non-direct part N_0 for reaction in Eq. (1) will be different. However, one expects that their behaviour with respect to x_F will be similar. From fits to experimental data in 9, for $h = \pi^\pm$ in Eq. (1), $N_0(x_F|s)$ is known. Assuming $N_0^\gamma(x_F|s) \simeq N_0(x_F|s)$, the asymmetry in Eq. (8) is plotted in Fig. 1 using the experimentally known $\bar{d}^\gamma$ and u_v^p [6,10]. The non-direct part $N_0(x_F|s)$ is significant for small x_F and negligible for $x_F > 0.5$. One expects $N_0^\gamma(x_F|s)$ also to be negligible for large x_F.

In Fig. 1 we can see that results for photon–proton reaction are qualitatively similar to those for pion production in proton–proton reactions. the latter are recalculated using the new proton distribution functions CTEQ5 [10] and show asymmetry for lower x_F values than before [8,9]. Note that the asymmetry in both cases is qualitatively the same for $x_F > 0.8$.

3. Left-Right Asymmetry in $p(\uparrow) + p \rightarrow \gamma + X$

In this reaction, if the projectile proton is polarized then one would expect to observe A_N^γ the left-right asymmetry of the emitted photon in the projectile fragmentation region. The process at the quark level is $q_v + \bar{q}_s \rightarrow \gamma + gluon$, where $q_v = u$ or d and $\bar{q}_s$ is from the 'sea' of the other proton. Formation of γ through $u\bar{u}$ is larger than through $d\bar{d}$ by a factor 4. For a polarized projectile proton, the model gives the asymmetry

$$A_N^\gamma(x_F^\gamma) = \frac{\tilde{C}_\gamma K_\gamma [4 \triangle u_v^P(x^P)\bar{u}_s^T(x^T) + \triangle d_v^P(x^P)\bar{d}_s^T(x^T)]}{N_0(x_F^\gamma) + K_\gamma [4u_v^P(x^P)\bar{u}_s^T(x^T) + d_v^P(x^P)\bar{d}_s^T(x^T)]} \qquad (9)$$

where the superscripts P and T refer to the projectile and the target and $\tilde{C}_\gamma$ and K_γ are constants to be determined. The expected asymmetry for this case is plotted in Fig. 2 where we can see that in comparison to pion production (Fig. 1) the asymmetry becomes important even for small x_F values.

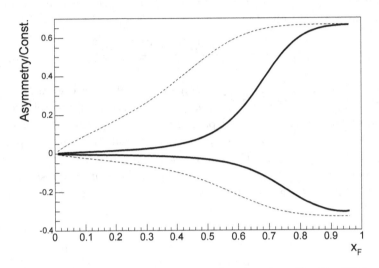

Fig. 1. Left-right asymmetry for pions as a function of x_F (See Eq. (8)) for γp reactions in Eq. (5) is plotted as solid lines. The parton distribution GRS99 for the photon and CTEQ5 for the proton were used. For comparison, the corresponding quantity for reaction in Eq. (1) is plotted using the same parton distributions (dashed lines). In both cases the positive values of asymmetry are for π^+ and the negative values for π^-. In the ordinate the constant is C_γ (C) for reactions in Eq. (5) (Eq. (1)).

4. Concluding Remarks

In this paper, tests of a particular phenomenological model are given for some new processes in sections 2-4. In particular, relations like $A_{\gamma p}^{\pi^+} \sim \frac{2}{3} C_\gamma$ (for $x_F > 0.8$) in $S2$, provide new and simple tests of the model. It is interesting to note that $A_N^{\pi^+} \sim \frac{2}{3} C$ (Fig. 1) and $A_N^{\pi^+} \sim \frac{2}{3} \tilde{C}_\gamma$ (Fig. 2) have similar limits. We expect that with the new generations of experiments in HERA, and maybe at Jefferson Lab, will be able to measure the effects discussed above.

References

1. A. Lesnik *et al.*, Phys. Rev. Lett. **35**, 770 (1975); G. Bunce *et al.* Phys. Rev. Lett. **36**, 1113 (1976); K. Heller *et al.*, Phys. Lett. **68B**, 480 (1977); K. Heller *et al.*, Phys. Rev. Lett. **51**, 2025 (1983); J. Felix *et al.*, Phys. Rev. Lett. **76**, 22 (1996).
2. B. Anderson, G. Gustafson and G. Ingelman, Phys. Lett. **85 B**, 417 (1979); T.A. DeGrand and H.I. Miettinen, Phys. Rev. D **24**, 2419 (1981); J. Szwed, Phys. Lett. **105 B**, 403 (1981); J. Soffer and N.A. Törnqvist, Phys. Rev. Lett.

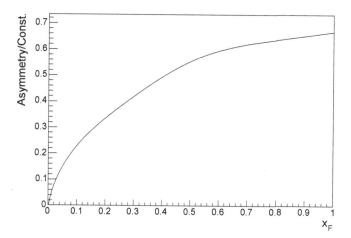

Fig. 2. Left-right asymmetry for photons as a function of x_F in proton–proton interactions Eq. (3). For the non-direct formation we used $N_0^{pp \to \pi}$ The ordinate is $A_N^\gamma(x_F^\gamma)/\tilde{C}_\gamma$ given in Eq. (9).

68, 907 (1992); Y. Hama and T. Kodama, Phys. Rev. D 48, 3116 (1993).

3. LiangZuo-tang, and MengTa-chung, Z. Phys. A344, 171 (1992); C. Boros, LiangZuo-tang, and MengTa-chung, Phys. Rev. Lett. 70, 1751 (1993); Phys. Rev. D51, 4867 (1995); Phys. Rev. D54, 4680 (1996).

4. S. Saroff *et al.*, Phys. Rev. Lett. 64, 995 (1990); E704 Collaboration, D.L. Adams *et al.*, Phys. Lett. B 345, 569 (1995); 276, 531 (1992); 261, 201 (1991).

5. M. Drees, and R.M. Godbole, Pramana 41, 83 (1993) and hep-ph/9508221.

6. M.Glück, E.Reya, and I.Schienbein, Phys. Rev. D60, 054019 (1999). These are used in Fig. 1.

7. Note $s = (p^p + p^\gamma)^2 \simeq 2p^p.p^\gamma$, $Q^2 = (x^p p^p + x^\gamma p^\gamma)^2 \simeq s x^p x^\gamma$ and $(x^p \vec{p}_{||}^{\,p} + x^\gamma \vec{p}_{||}^{\,p}) = \vec{p}_{||}^{\,\pi}$. Last equation gives $x^p - x^\gamma = x_F$ in the CM-frame. Superscripts p and γ refer to the proton and the photon while x^p and x^γ are the fraction of the momentum carried by their constituent involved in the reaction.

8. Since here $Q^2 = m_\pi^2$, one has $x^p x^\gamma \simeq m_\pi^2/s$ which means that small values of x^γ will be probed.

9. LiangZuo-tang, and MengTa-chung, Phys. Rev. D49, 3759 (1994).

10. CTEQ collaboration, Eur.Phys. J. C 12, 375-392 (2000). These are used in Figs. 1 and 2.

11. E791 collaboration, E.M. Aitala et al., Phys. Lett. B 496, 9-18 (2000).

12. J. Ashman et al. Nucl. Physics B328, 1 (1989).

13. M. Goshtasbpour, and G.P. Ramsey, Phys. Rev. D55, 1244 (1997).

BOSONIZATION AND THE GENERALIZED
MANDELSTAM-HALPERN OPERATORS*

HAROLD BLAS

ICET-Universidade Federal de Mato Grosso
Av. Fernando Correa, s/n, Coxipó,
78060-900, Cuiabá - MT - Brazil
blas@cpd.ufmt.com

The generalized massive Thirring model (GMT) with N_f [=number of positive roots of $su(n)$] fermion species is bosonized and shown to be equivalent to a generalized sine-Gordon model (GSG). Each fermion species is mapped to its corresponding soliton in the spirit of particle/soliton duality of Abelian bosonization.

1. Introduction

The transformation of Fermi fields into Bose fields, called *bosonization*, provided in the last years a powerful tool to obtain non-perturbative information in two-dimensional field theories. [1] The Abelian and non-Abelian bosonizations have been derived in Refs. 2 and 3, respectively. In the non-Abelian developments the appearance of solitons in the bosonized model, which generalizes the sine-Gordon solitons, to our knowledge has not been fully explored. The interacting multi-flavor massive fermions deserves a consideration in the spirit of the particle/soliton duality of the Abelian bosonization. In Refs. 4, 5 it has been shown that the generalized massive Thirring model (GMT) is equivalent to the generalized sine-Gordon model (GSG) at the classical level; in particular, the mappings between spinor bilinears of the GMT theory and exponentials of the GSG fields were established on shell and the various soliton/particle correspondences were uncovered.

---

*This work is supported by FAPEMAT-CNPq.

The bosonization of the GMT model has been performed following a hybrid of the operator and functional formalisms. [6,7] In this talk we summarize those results.

2. Functional Integral and Operator Approaches

The two-dimensional massive Thirrring model with current-current interactions of N_f (Dirac) fermion species is defined by the Lagrangian density

$$\frac{1}{k'} L_{GMT}[\psi^j, \overline{\psi}^j] = \sum_{j=1}^{N_f} \{i\overline{\psi}^j \gamma^\mu \partial_\mu \psi^j - m^j \, \overline{\psi}^j \psi^j\} - \frac{1}{4} \sum_{k,l=1}^{N_f} \left[\hat{G}_{kl} \, J_k^\mu J_{l\,\mu}\right] (1)$$

where the m^j's are the mass parameters, the overall coupling k' has been introduced for later purposes, the currents are defined by $J_j^\mu = \overline{\psi}^j \gamma^\mu \psi^j$, and the coupling constant parameters are represented by a non-degenerate $N_f \times N_f$ symmetric matrix $\hat{G} = \hat{g} G \hat{g}$, $\hat{g}_{ij} = g_i \delta_{ij}$, $G_{jk} = G_{kj}$.

The GMT model (1) is related to the weak coupling sector of the $su(n)$ affine Toda model coupled to matter (ATM) in the classical treatment of Refs. 4 and 5. We shall consider the special cases of $su(n)$ ($n = 3, 4$). In the $n = 3$ case the currents at the quantum level must satisfy

$$J_3^\mu = \hat{\delta}_1 J_1^\mu + \hat{\delta}_2 J_2^\mu, \tag{2}$$

where the $\hat{\delta}_{1,2}$ are some parameters related to the couplings $\hat{G}_{kl}$. Similarly, in the $n = 4$ case the currents at the quantum level satisfy

$$J_4^\mu = \hat{\sigma}_{41} J_1^\mu + \hat{\sigma}_{42} J_2^\mu, \quad J_5^\mu = \hat{\sigma}_{51} J_1^\mu + \hat{\sigma}_{53} J_3^\mu, \quad J_6^\mu = \hat{\sigma}_{62} J_2^\mu + \hat{\sigma}_{63} J_3^\mu, \tag{3}$$

where the $\hat{\sigma}_{ia}$'s (i=4,5,6; a=1,2,3) are related to the couplings $\hat{G}_{kl}$.

The cases under consideration $n = 3, 4$ correspond to $N_f = 3, 6$, respectively; however, most of the construction below is valid for $N_f > 6$. In the hybrid approach the Thirring fields are written in terms of the "generalized" Mandelstam "soliton" $\Psi^j(x)$ and σ_j fields

$$\psi^j(x) = \Psi^j(x)\sigma^j, \quad j = 1, 2, 3, ..., N_f; \tag{4}$$

where

$$\Psi^j(x) = (\frac{\mu}{2\pi})^{1/2} K_j \, e^{-i\pi\gamma_5/4} : e^{-i\left(\frac{\beta_j}{2}\gamma_5 \Phi^j(x) + \frac{2\pi}{\beta_j} \int_{x^1}^{+\infty} \dot{\Phi}^j(x^0, z^1) dz^1\right)} : \tag{5}$$

$$\sigma^j = e^{\frac{i}{2}\left(\eta_j - \frac{\sqrt{4\pi}}{\Delta_j}\tilde{\xi}_j\right)} \tag{6}$$

$$= e^{-\frac{i}{2}g_j \ell^j}. \tag{7}$$

In (5) the factor K_j makes the fields anti-commute for different flavors[8]. The Lagrangian in terms of purely bosonic fields becomes

$$\frac{1}{k'}L'_{eff} = \sum_{j,k=1}^{N_f} \frac{1}{2}\left[C_{jk}\, \partial_\mu \Phi_j \partial^\mu \Phi_k + E_{jk}\, \partial_\mu \xi_j \partial^\mu \xi_k + F_{jk}\, \partial_\mu \eta_j \partial^\mu \eta_k \right] +$$
$$\sum_{j=1}^{3} M^j \cos(\Phi_j), \tag{8}$$

where C_{jk}, D_{jk}, E_{jk} and M^j are some parameters.

Notice that each Ψ^j is written in terms of a non-local expression of the corresponding bosonic field Φ^j and the appearance of the couplings β_j in (5) in the same form as in the standard sine-Gordon construction of the Thirring fermions [2]; so, one can refer the fermions $\Psi^j(x)$ as generalized SG Mandelstam-Halpern soliton operators.

2.1. *The su(3) case*

The GMT model (1) with three fermion species, with the currents constraint (2), bosonizes to the GSG model (8) with three Φ bosons and the constraint $\beta_3 \Phi_3 = \beta_1 \Phi_1 \delta_1 + \delta_2 \beta_2 \Phi_2$, by means of the "generalized" bosonization rules

$$i\bar{\psi}^j \gamma^\mu \partial_\mu \psi^j = \frac{1}{2}(1-\rho_j)(\partial_\mu \Phi^j)^2, \quad j = 1,2,3; \tag{9}$$

$$m_j \bar{\psi}^j \psi^j = M_j \cos\left(\beta_j \Phi^j\right), \quad \beta_j^2 = \frac{4\pi}{1 + \frac{g_j^2}{\pi}\frac{1}{4G_{lm}^j}} \tag{10}$$

$$\bar{\psi}^j \gamma^\mu \psi^j = -\frac{\beta_j}{2\pi}\, \epsilon^{\mu\nu} \partial_\nu \Phi_j, \tag{11}$$

where ρ_j can be written in terms of g_j and the correlation functions on the right hand sides must be understood to be computed in a positive definite quotient Hilbert space of states $H \sim \frac{H'}{H_o}$ [6].

2.2. *The su(4) case*

In the $su(4)$ GSG case [7] one has the constraints $\Phi_4 = \sigma_{41}\Phi_1 + \sigma_{42}\,\Phi_2$, $\Phi_5 = \sigma_{51}\Phi_1 + \sigma_{53}\,\Phi_3$, and $\Phi_6 = \sigma_{62}\Phi_2 + \sigma_{63}\,\Phi_3$, with σ_{ij} being some parameters.

The bosonization rules become

$$\bar{\psi}^j \gamma^\mu \psi^j = -\frac{\hat{\beta}_i \, \epsilon^{\mu\nu} \partial_\nu \Phi_i}{\pi M_i^- M_i^+ - (\frac{z_i}{2}) (g_i)^2 \, M_i^+} \quad i = 1, 2, 3, ..., 6; \qquad (12)$$

$$i \bar{\psi}^j \gamma^\mu \partial_\mu \psi^j = \frac{1}{2} (1 - \hat{\rho}_j)(\partial_\mu \Phi^j)^2, \quad j = 1, 2, 3..., 6; \qquad (13)$$

$$m_j \bar{\psi}^j \psi^j = M_j \cos\left(\hat{\beta}_j \Phi^j\right). \qquad (14)$$

The parameters $M_i^{\pm}$, $\hat{\beta}_j$ and $\hat{\rho}_j$ can be expressed in terms of g_i, G_{ij}. The z_i's are regularization parameters.

3. Discussions

We have considered the bosonization of the multiflavour GMT model with N_f species. One must emphasize that the classical properties of the ATM model [4,5,9] motivated the various insights considered in the bosonization procedure of the GMT model. The form of the quantum GSG model (8) is similar to its classical counterparts in Refs. 4 and 5, except for the field renormalizations, the relevant quantum corrections to the coupling constants and the presence of the un-physical η and ξ fields . The bosonization results presented in this talk can be useful to study particle/soliton duality and confinement in multi-fermion two-dimensional models, extending the results of Refs. 10 and 11.

Acknowledgments

The author thanks Prof. M.C. Araújo and ICET-UFMT for hospitality.

References

1. E. Abdalla, M.C.B. Abdalla and K.D. Rothe, Non-perturvative methods in two-dimensional quantum field theory, 2nd edition (World Scientific, Singapore, 2001).
2. S. Coleman, *Phys. Rev.* **D11** (1975) 2088;
   S. Mandelstam, *Phys. Rev.* **D11** (1975) 3026.
3. E. Witten, *Commun. Math. Phys.* **92** (1984) 455.
4. J. Acosta, H. Blas, *J. Math. Phys.* **43** (2002) 1916;
   H. Blas, to appear in *Progress in Soliton Research* (Nova Science Publishers, 2004), hep-th/0407020.
5. H. Blas, *JHEP* **0311** (2003) 054.
6. H. Blas, *Eur. Phys. J.* **C37** (2004) 251.
7. H. Blas, invited paper for *Progress in Boson Research* (Nova Science Publishers, 2005), hep-th/0409269.

8. Y. Frishman and J. Sonnenschein, *Phys. Reports* **223** (1993) 309.
9. H. Blas, *Nucl. Phys.* **B596** (2001) 471;
   H. Blas and B.M. Pimentel, *Annals Phys.* **282** (2000) 67; see also hep-th/0005037.
10. H. Blas and L.A. Ferreira, *Nucl. Phys.* **B571** (2000) 607.
11. H. Blas, *Phys. Rev.* **D66** (2002) 127701; see also hep-th/0005130.

IS THERE REALLY AN ENTANGLED STATE FOR FAR AWAY TWIN PHOTONS?

HOLGER G. VALQUI

Facultad de Ciencias of Universidad Nacional de Ingeniería, Lima, Peru

A certain school of physicists based on the following assumptions: *(i) Photons acquire a polarization state only after they have passed through a polarizer (which some claim to be a stipulation of one of the postulates of the Copenhagen Interpretation of Quantum Mechanics), (ii) The states of twin photons are correctly represented by $\chi(1,2) = \sqrt{2}(u \otimes v + v \otimes u)$, where u is an arbitrary state (of polarization) and v is orthogonal to u, and (iii) Numerous experiments directly or indirectly confirm the theoretical predictions.*, concludes that there exists an extraordinary entangled state of both photons, which permits that each of the photons somehow be sensible to the interaction which affects the other one, notwithstanding the great distance which might exist between them. In the present exposition I show that: *(i) There are experimental reasons to maintain that photons have a definite polarization state before and after having passed through a polarizer, (ii) According to the principles of Quantum Mechanics the "official" representation $\chi(1,2) = \sqrt{2}(u \otimes v + v \otimes u)$ can not be right.*, and *(iii)Significant doubts exist concerning the accuracy of the critical measurements and their interpretation.* At the last I conclude that the alleged extraordinary entanglement of well separated photons might be a delusion. Finally, to settle the matter, I propose a crucial experiment.

1. Classical Polarization Experiments

01) Electromagnetic field polarization is usually characterized by the vector of electric field E. In the case of light there are several ways to obtain linear polarized light and circular polarized light. Among the devices which produce polarized light I want to mention the so called linear and circular polarizers: $L(\theta)$, DC and LC, respectively. The following schema shows the better known experimental results, where θ is the angle between the polarizer's orientation and a given direction, and J is the intensity of light:

(1) $Natural\ light(J) + [L(\theta),\ DC\ or\ LC] \longrightarrow J/2$

263

(2) *Natural light(J) + [L(θ), DC or LC]*
   +[L(θ), DC or LC, respectively] ⟶ J/2
(3) *Natural light(J) + L(θ) + [DC or LC] ⟶ J/4*
(4) *Natural light(J) + [DC or LC] + L(θ) ⟶ J/4*
(5) *Natural light(J) + L(θ) + L(α) ⟶ (J/2)cos²(θ − α)* (Malus Law)
(6) *Natural light(J) + [DC or LC] + [LC or DC, respectively] ⟶ 0*

In every one of the above experiments it is assumed that light impinges perpendicularly upon the plane of the polarizers, which are parallel to each other. Each one of the above mentioned experimental results are independent of the value of the intensity J, provided the apparatus are able to measure such intensity.

02) The classical experiments listed above are dependent on the reading of the light intensity, J. But they can also be interpreted via the quantum model: According to this light is composed of particles called photons,

(1) Photons are carriers of light energy , and such energy's value is not modified if a photon passes through a polarizer.
(2) When a photon impinges upon a polarizer it can be absorbed by the polarizer or it can pass through it. If a photon passes a polarizer it acquires a fixed polarization state induced by the polarizer (we say the photon has acquired the polarization of the polarizer).
(3) If a photon which has a known polarization state u_α (because it already has passed a L(α) polarizer) impinges upon a polarizer L(θ) the probability that it passes through the polarizer is equal to $cos^2(\theta - \alpha)$ and the probability that the polarizer absorbs the photon is $sen^2(\theta - \alpha)$. This is called the Malus' Probabilistic Law.
(4) The Hilbert polarization state is two dimensional, so that an adequate linear combination of each pair of orthogonal states can represent correctly any photon's polarization state.
(5) To each polarization state of the photon corresponds a well determined polarization direction in the plane that is perpendicular to the momentum of the photon.

03) Using the quantum model of photons one can explain the above experimental results as follows:

(1) If a photon pass through a polarizer it acquires the polarization of the polarizer.
(2) The polarization states induced by a CD polarizer and by a CL polarizer are orthogonal to each other; they form a basis for the state space.

(3) The classical Malus' Law is a consequence of the probabilistic law, applied to a large number of photons ($\approx 10^{20}$), which transform a probabilistic result in a deterministic one. Additionally the polarization states induced by polarizers $L(\theta)$ and $L(\theta + \pi/2)$ are orthogonal to each other, and form a basis for the Hilbert polarization space.

(4) The linear polarization state is equal to the addition, with coefficients which have the same absolute values, of CD and CL polarization states. This happens in the state space, not in physical space.

(5) The polarization state CD (or CL) is equal to the addition, with coefficients which have the same absolute values, of linear polarization states induced by $L(\theta)$ and $L(\theta + \pi/2)$ polarizers.

(6) Natural light is composed of a very large number of photons, which are randomly polarized without a preferred direction.

2. Is it Possible to Measure the Polarization State of an Isolated Photon?

01) We have seen above that a linear polarizer $L(\theta)$ can only give very little information about the polarization state of a photon (which has previously been polarized by a polarizer of unknown direction): If the impinging photon is absorbed by $L(\theta)$ we only can say that it had a polarization different to the polarization of the polarizer; on the other hand, if the photon passes through the polarizer, then we can only know that its polarization was not orthogonal to the polarizer.

So, a polarizer by itself cannot give information about the polarization of a photon impinging on it.

What then is the possible meaning of the measurement of the polarization of a photon by means of a linear polarizer $L(\theta)$?

In order to build an answer consider the following:

A beam of N photons, which have already been polarized in an unknown direction α, impinges on a linear polarizer $L(\theta)$ and N' photons pass through it (acquiring the polarization u_θ, is it then true that $N'/N = cos^2(\theta - \alpha)$?

The answer, as we know, is dependent on the value of N:

If N is a small number, say $N < 10$, then the fractions N'/N, for different values of N, are independent of each other. With increasing values of the number N the fraction N'/N becomes less oscillating, and only when N is a very large number does the value of the fraction become independent of the value of N. Thus only in the case that N1 and N2 are both large numbers can we practically write $N_1'/N_1 = N_2'/N_2$; this fraction (for a large N) is called the probability that the photon passes through the polarizer, and it is equal to $cos^2(\theta - \alpha)$. Additionally, being N big enough, N' should not be too small for it would cast some doubt on the true value of the very small fraction N'/N.

02) Could it be possible that a polarizer is unable to measure the polarization of a photon because before passing through the polarizer the photon has no polarization?

The answer is clearly negative as it is shown by experiments. Exp2 says that if a photon passes through a polarizer Π_1 , then it surely will pass through a second polarizer Π_2 , similar and parallel to the first one. That means that, in this case, the photon before impinging on Π_2 already had a well defined polarization.

But, on the other hand, could it be possible that there exist **virgin photons,** that is photons which because of not having passed through any polarizer have not acquired a well defined state of polarization?

First we must understand that *according to the quantum model* every state of a physical system is, by hypothesis, well determined - which does not mean that such state is cognizable.

In order to know the state of a macroscopic system we must undertake the measurement of the properties which characterize such a state. But we cannot avoid modifying the values of the same properties we want to measure. In the case of macroscopic systems we usually can suppose that the measurement's perturbations could be made arbitrarily insignificant. However such assumption cannot be valid anymore in the case of the microscopic systems. In regard to the existence of non cognizable states we might remember Feynman's point of view (Lectures on Physics, III, p. 2-8): It is not necessary (nor convenient) to discard ideas or concepts which we cannot measure, as is the case with the polarization of an isolated photon.

Second, if virgin photons exist then experiments to distinguish such photons from the usual ones should exist. Apparently no such experiments exist. Consequently there is no inconsistency in assuming that every photon has a well defined polarization state notwithstanding the fact that we might be unable to determine such polarization state by means of a polarizer.

Third, according to the experimental results, Exp1, when natural light impinges on a polarizer, only half of the photons are able to pass. This fact is easily explained under the assumption that the impinging photons were randomly polarized, which is justified because that light is not generated by a point source. How can such a fact be explained with help of the hypothesis of virgin photons?

3. Is the "Official" Representation of Twin Photons Correct?

(1) Let ϕ_1, ϕ_2 be two orthonormal polarization states in a Hilbert space, $< \phi_j, \phi_k >= \delta_{jk}$, and let u_1, u_2 be its corresponding spatial orientations, hence $u_i \bullet u_j = \delta_{ij}$.

(2) To describe simultaneously the polarization state of two non interacting photons, F_1 and F_2, we can use the tensor product of the two Hilbert spaces, $H^{(1)} \otimes H^{(2)}$. If $\{\phi_j^{(1)}\}$ and $\{\phi_k^{(2)}\}$ are bases in such spaces, then $\{\phi_j^{(1)} \otimes \phi_k^{(2)}\}$ is a basis in the product space, and a general form for the polarizations state of both photons is $\psi = \sum_j q_{jk} \phi_j^{(1)} \otimes \phi_k^{(2)}$.

(3) Photons (of the same energy) are identical particles, and if they have the same momentum we can identify their polarizations spaces, $H^{(1)} = H^{(2)}$. Consequently we can write the general polarization state as $\psi = \sum_{jk} q_{jk}(\phi_j^{(1)} \otimes \phi_k^{(2)} + \phi_k^{(1)} \otimes \phi_j^{(2)})$, where we can see that $q_{jk} = q_{kj}$.

(4) In the case of twin photons, F_1 y F_2, (which are produced 'almost' simultaneously by the same source) with unknown polarization states μ y η, the function of the composite state can be written as $\chi(1,2) = \sqrt{2}(\mu \otimes \eta + \eta \otimes \mu)$. But because of angular momentum conservation generally means that the individual states should be orthogonal, $< \mu, \eta >= 0$

(5) Suppose that the measurement of the polarization state of the distant photon F_1 has given the state u , corresponding to the u-orientation of the polarizer. Now, in order to analyze the unknown polarization states μ and η, use the known state u, and an orthogonal one, v, as a basis. Thus we can write $\mu = \alpha_1 u + \beta_1 v$, $\eta = \alpha_2 u + \beta_2 v$. But condition $< \mu, \eta >= 0$ implies $\alpha_1 * \alpha_2 + \beta_1 * \beta_2 = 0$. We can verify that $\alpha_1 = sin\theta$, $\beta_1 = cos\theta$, $\alpha_2 = cos\theta$, $\beta_2 = -sin\theta$, where θ is an arbitrary real number, satisfy such equation. That is, the states of the twin photons before measurement are correctly expressed (disregarding normalization) by the composite state $\chi(1,2) = (sin\theta\ u\ +\ cos\theta\ v) \otimes (cos\theta\ u\ -\ sin\theta\ v)\ +\ (cos\theta\ u\ -\ sin\theta\ v) \otimes (sin\theta\ u\ +\ cos\theta\ v)$ or $\chi(1,2) = u \otimes (sin2\theta\ u\ +\ cos2\theta\ v)\ -\ v \otimes (sin2\theta\ v\ -\ cos2\theta\ u)$

(6) Now if measurement of F_1's state gives, as presupposed, the state u, then the composite state should collapse as follows $\chi(1,2) \longrightarrow u \otimes (sin2\theta\ u\ +\ cos2\theta\ v)$ where it is shown that the state of photon F2 remains unknown.

(7) On the other hand, followers of Copenhagen Interpretation assume that in the case of twin photons the composite state should be given by the same former formula $\chi(1,2) = \sqrt{2}(\mu \otimes \eta + \eta \otimes \mu)$, but they don't relate the unknown states μ, η to a known basis. *This is the trick that ensures the mathematical presence of entanglement.* If the measured state of F_1 is u, they claim the collapse is of the form: $\chi(1,2) \longrightarrow u \otimes v$, which is equivalent to assuming $\theta = 0$ in the previous analysis.

The "official" explanation of measurement could be paraphrased as follows: the first photon, F1 , which impinges on the polarizer placed in its path (both photons do not arrive at their polarizer exactly at the same time), get polarized, and automatically the distant photon F_2 acquires an orthogonal polarization state which enables it to pass through the second polarizer, placed perpendicular to the first one. Or, if the first photon is absorbed, the second one will be absorbed too.

Note that according to this interpretation the second polarizer is not necessary for the entanglement; it is only necessary to make the act of entanglement evident.

4. Are the Published Experimental Results Reliable?

(1) Almost all reports on experiments for detecting entanglement assume the reliability of the numerical results. In the 1980s several critical experiments were carried out by A.Aspect and coworkers, confirming the existence of two-photon entanglement. Or so it was believed. But, for example, in an interview published in Le Monde Quantique, Aspect acknowledges that there were several experimental difficulties because of the instruments' efficiencies; he mentions that the best photomultipliers had no more than 40% efficiency. In general the statistical analysis of the numerical results is mentioned only briefly, shielding the fact that the appearance of entanglement might be product of the experimenter's belief confirmation. Additionally we must take into account that the detected photons are not produced by an isolated atom but by a large number of atoms, which being excited almost simultaneously, some of them could also be de-excited almost simultaneously.

(2) According to the interpretation presented in this paper, if the direction corresponding to photon F_1 were given by the angle θ , hence the one corresponding to F_2 would be $\theta + \phi/2$. Now, if F_1 and F_2 would impinge on polarizers L(α) and L(β), respectively, then we will have the following values for the probabilities p_{12} (both photons pass through), p_{00}

(both photons are absorbed), p_{10} (F_1 passes through, F_2 is absorbed) and p_{02} (F_1 is absorbed, F_2 passes through):

$p_{12} = cos^2(\alpha - \theta)sen^2(\beta - \theta)$, $p_{00} = sen^2(\alpha - \theta)cos^2(\beta - \theta)$,
$p_{10} = cos^2(\alpha - \theta)cos^2(\beta - \theta)$, $p_{02} = sen^2(\alpha - \theta)sen^2(\beta - \theta)$

But these isolated probabilities have no physical significance. To attain physical meaning from them a statistical counting has to be prepared If we believe that there is no special direction preferred by the source when producing the twin photons, we may assume that such polarization-directions are uniformly distributed in interval $(0, 2\pi)$. Hence, integrating within such interval we can get the mean values of such probabilities:

$P_{12} = [3 - 2cos^2(\alpha - \beta)]/8$, $P_{00} = [3 - 2cos^2(\alpha - \beta)]/8$,
$P_{10} = [1 + 2cos^2(\alpha - \beta)]/8$, $P_{02} = [1_2cos^2(\alpha - \beta)]/8$,

where the sum equals unity.

(3) If, as is planned in experiments, the polarizers are perpendicular to each other, $|\alpha - \beta| = \pi/2$, then the probabilities mean values become $P_{12} = 3/8$, $P_{00} = 3/8$, $P_{10} = 1/8$, $P_{02} = 1/8$ On the other hand, the existence of entanglement gives a very different numerical result: $P_{12} = P_{00} = 1/2, P_{10} = P_{02} = 0$.

But we must not forget that in order to verify the above numerical values one has to count: i) The total number of twin photons produced by the source in the time interval considered, which cannot be deduced only from the coincidence readings, ii) The total number of passed photons on each side, which can be considered one of a twin pair, and iii) The total number of absorbed photons on each side separately.

5. A Falsifying Experiment, According to Popper

As I have said above the phenomenon of entanglement does no require the presence of two polarizers, each to different sides of the source. The twin photons are not produced simultaneously (only almost so) and the distance of the polarizer to the source cannot be exactly the same, hence they cannot arrive simultaneously to the corresponding polarizer. The photon which arrives first gets a particular polarization state, the other photon complies taking a polarization state orthogonal to the first one, and so is able to pass through the polarizer which has been oriented perpendicular to the polarizer at the other side of the source. Here I show very schematically the usual experimental setup, where the coincidence detectors and connection are not shown.

In the proposed experimental setup I eliminate one of the entanglement-polarizers and add two polarizers, one on each side of the source, oriented

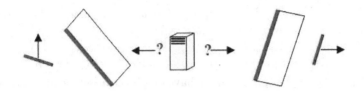

Fig. 1.   Usual experimental setup.

as indicated in the diagram.

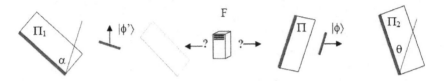

Fig. 2.   Experimental setup with two polarizers.

When the photon to the right, say photon F_1, acquires the state $|\Phi>$ induced by the polarizer II, photon F_2 automatically gets the orthogonal state $|\Phi'>$ without having passed through a polarizer. After that F_1 impinges on polarizer II_1, which has an orientation θ, and F_2 impinges on polarizer II_2 which has an orientation α. Hence according to the probabilistic Malu's law the probabilities that F_1 passes II_1 and F_2 passes II_2 are $p_1(\theta) = cos^2\theta$, $p_2(\alpha) = sen^2\alpha$, respectively.

In order to get more clear numerical result we may choose $\alpha = \theta = 0$, where F_1 will pass through but F_2 will be absorved, or $\alpha = \theta = \pi/2$, where F_1 will be absorved and F_2 will pass through.

But, what if photon F_2 is not affected by the measurement performed on F_1?

Insuch a case photon F_1 will have acquired the state $|\Phi>$ induced by polarizer II while F_2 will maintain its unknown polarization state, which will produce random readings when impinging on polarizer II_2

References

1. B. d'Espagnat, "The Quantum Theory of Reality", *Sci.Amer.*, Nov. (1979).
2. A. Aspect, P. Grangier, and G. Roger, *Phys.Rev.Lett.* *47*, 91 (1982).
3. D. Mermin, "Is the moon there when nobody looks?", *Physics Today.*, Apr. (1985).
4. Santos Marshall, *Physics Today*, Nov. (1985). "It arises because the vast

majority (more than 95%) of all signals emitted by the calcium atoms are not detected by the photomultipliers. Because one has to analize coincidences, well over 99% of all cascade signals are not analized. *So can one be sure that the coincidences actually analized constitute a fair sample from the cascade population?"*

5. An interview with A. Aspect, "En la criba de la experiencia", *El Mundo Cuántico, Alianza*, Alianza Editorial, (1990).

6. J.G. Cramer, "The transactional iterpretation of quantum mechanics", *Rev.Mod.Phy.*, Vol.58/3, July (1986).

7. R. Feynman et al, *Lecture on Physics III*.

8. K. Wodkiewicz, *Optical Tests of Quantum Mechanics in Quantum Optics*.

9. Greenberger and Zeilinger,, "Quantum Theory: still crazy after all these years", *Physics World*, Sep. (1995).

10. H.G. Valqui, "El error inicial", *Revciuni*, 7, Feb.1 (2003).

11. H.G. Valqui, "Corrección: Un experimento crucial para verificar el entanglement de los fotones", *Revciuni*, 8, Feb.1 (2004).

USE OF NEURAL NETWORKS TO CLASSIFY COSMIC-RAY SHOWERS ACCORDING TO THEIR MUON/EM RATIO*

H. SALAZAR and L. VILLASEÑOR†

Facultad de Ciencias Físico-Matemáticas, BUAP, Puebla Pue., 72570, México

The ratio of muon/EM (μ/EM) in extensive air showers (EAS) is an important parameter to measure the primary mass of cosmic rays. The existence of very strong correlations among amplitude, charge deposition, 10%-90% and 10%-50% rise-times for isolated electrons, isolated muons and extensive air showers in a water Cherenkov detector allows one to separate cleanly these components by using simple cuts in the bi-dimensional parameter space of amplitude vs rise-time. We used these signals to form artificial traces for EAS by adding pre-determined numbers of real electron and muon traces to form showers with known values of μ/EM. Real EAS data were best reproduced this way by using exponential arrival times for electrons and muons with decay values narrowly distributed around 15 ns for muons and 25 ns electrons. The artificial EAS were in turn used to train Kohonen neural networks with the aim to classify EAS events according to their μ/EM ratio. We found that neural networks are able to classify EAS in a better-than- random way, encouraging further work along this direction.

1. Introduction

The measurement of μ/EM, i.e., the muon contents relative to the electromagnetic (electrons, positrons and photons) contents of extensive air showers, at each of the ground stations of giant arrays of water Cherenkov detectors, such as the Pierre Auger Observatory [1], allows in principle a

*This work was partially supported by Coordinacin de la Investigación Científica-UMSNH, Vicerectoria de Investigación-BUAP and CONACYT projects G32739-E and G38706-E.
†On leave of absence from Instituto de Física y Matemáticas, Universidad Michoacana de San Nicolás de Hidalgo, Morelia Mich., 58040, México.

measurement of the composition of primary cosmic rays. Simulation of EAS of ultra high energy show that extensive air showers initiated by Fe nuclei have around 50% more muons at ground level than proton-initiated EAS of cosmic-ray primaries of the same energy; in turn gamma primaries produce less muons, for the same energy of the primary cosmic rays. In the case of a hybrid cosmic-ray detector, such as the Pierre Auger Observatory, μ/EM, along with the shower maximum of the EAS, as measured with fluorescence telescopes, are the main parameters for measuring composition of cosmic rays with energies around 1020 eV.

2. Results and Discussion

n previous papers we have reported on the charge deposition of electrons from muon decays in a water Cherenkov detector [2], and we have demonstrated that by doing precise measurements of the amplitude, charge and rise-time of the signals taken with the same detector in an inclusive sample of secondary cosmic rays, we can easily identify isolated electrons, isolated muons and extensive air showers [3], see Figure 1. Identification of the signals for isolated electrons and muons, see Fig. 2, permits one to form artificial signals for EAS by adding pre-determined numbers of real electron and muon traces to form shower traces with known values of μ/EM. These artificial signatures for EAS can in turn be used to train specific neural networks with the aim to use them to classify real EAS according to their μ/EM ratio. The obvious advantadge of using real electron and muon signals is that one does not have to rely on simulations of the detector response to muons, electrons and EAS. By summing the real signal traces of electrons and muons, an example of which is shown in Fig. 2, for varying numbers of electrons and muons we formed artificial EAS with known contents of muons and electrons. For the purpose of using a specific set of EAS, we selected a real EAS data sample with a mean charge of 8 VEM from our inclusive muon data, see Fig. 1. Then we formed artificial EAS in three categories: 8+0e, 4+33e and 0+66e by using 8, 4 or 0 muon and 0, 33 or 66 electron traces from a database formed with thousands of real traces of isolated electrons and isolated muons as shown in Fig. 1. The number of muons and electrons in the artificial EAS were chosen to make the mean values of these artificial EAS equal to the 8 VEM value of real EAS. The mean values for the amplitude, 10-50% and 10-90% rise-times were 1.16±8 V, 16.7± a 0.9 ns and 50.8±2.0 ns, respectively. We found out that the mean values of the amplitude, 10-50% and 10-90% rise-times for the signals of the artificial EAS formed were very sensitive to the arrival time distributions used to combine the electron and muon signals. The corresponding values of real data were reproduced best by using exponential arrival times for electrons and muons with decay values narrowly distributed around 15 ns for muons and 25 ns electrons.

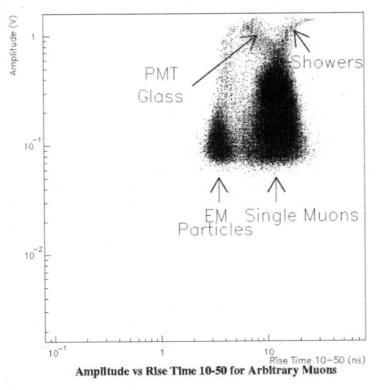

Amplitude vs Rise Time 10-50 for Arbitrary Muons

Fig. 1.　Amplitude vs rise-time from 10% to 50% for signals from a water Cherenkov detector taken with an inclusive trigger with an amplitude threshold of 30 mV. Simple cuts on this plot allow one to separate isolated electrons and isolatated muons whose traces are subsequently used to form artificial shower signals which, in turn, are used to train neural networks to classify EAS acording to their μ/EM ratio.

This first result is in agreement with the known fact that on average muons arrive first than electrons. Since the charge was matched in the artificial and real EAS, we used only the 10-50% rise time, 10-90% rise time, amplitude and AC/DC, as the primary features in the classification process of EAS. We used Kohonen neural networks with these four features

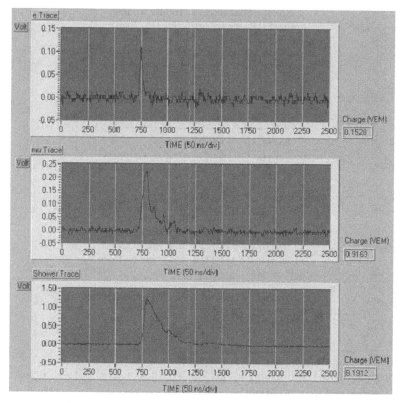

Fig. 2. Typical oscilloscope traces for an isolated electron, with a typical charge of around 0.12 VEM and rise time 10-50% of around 5 ns; an isolated muon with typical charge of around 1 VEM and rise time 10-50% of around 12 ns; and an extensive air shower with a charge of 8.2 VEM and a rise time 10-50% of 15 ns.

as input parameters and different numbers of dimensions and of neurons per dimension. We found that the network that gives the highest rate of correct classification has two dimensions with 8 neurons in the first dimension and 4 neurons in the second dimension. We trained each of such neural networks with sets of 100 artificial traces for each category and then measured the rate of correct identification on different sets of 100 traces.

Table 1 shows the results for the classification into two distinct categories labeled 8+0e and 4+33e. Likewise, table 2 shows the results for the classification into two distinct categories labeled 8+0e and 0+66e. As seen from these tables, neural networks classify in a way which is better than random, i.e., the diagonal numbers are higher than 50% and the off-diagonal

Table 1. Classification results of EAS trained by using artificial EAS with $8\mu+0e$ and $4\mu+33e$. The numbers correspond to the percentage of times an EAS of the type given by the first raw is identified as an EAS of the type given by the left colum.

EASType	$8\mu+0e$	$4\mu+33e$	Data
$8\mu+0e$	68%	36%	75%
$4\mu+33e$	32%	64%	25%

Table 2. Classification results of EAS trained by using artificial EAS with $8\mu+0e$ and $0\mu+66e$. The numbers correspond to the percentage of times an EAS of the type given by the first raw is identified as an EAS of the type given by the left colum.

EASType	$8\mu+0e$	$0\mu+66e$	Data
$8\mu+0e$	84%	17%	84%
$0\mu+66e$	16%	83%	16%

lower than 50%. In addition, we get the result that real data samples run into the trained neural networks are classified as containing mostly muons.

Table 3. Classification results of EAS trained by using artificial EAS with $8\mu+0e$, $4\mu+33e$ and $0\mu+66e$. The numbers correspond to the percentage of times an EAS of the type given by the first raw is identified as an EAS of the type given by the left colum.

EASType	$8\mu+0e$	$4\mu+33e$	$0\mu+66e$	Data
$8\mu+0e$	70%	36%	7%	70%
$4\mu+33e$	16%	25%	15%	19%
$0\mu+66e$	14%	39%	78%	11%

Table 3 shows the results for the classification into three distinct categories labeled $8+0e$, $4+33e$ and $0+66e$. In this case the neural network has a worse than average value for the classification of events of the intermediate category, $4+33e$, although it stills works reasonably well for the other extreme categories, i.e., 8+0e and 0+66e. Again the real EAS data run into the trained neural network are classified as containing mostly events of the muon-rich 8+0e category.

3. Conclusion

We have used simple cuts in the bi-dimensional parameter space of amplitude vs rise-time, see Fig. 1, to identify signals due to isolated electrons and isolated muons, as well as low energy EAS, see Fig.2. We have used these signals to form artificial traces for EAS by adding pre-determined numbers of real electron and muon traces to form shower traces with known values of μ/EM. These artificial signatures for EAS were in turn used to train Kohonen neural networks with the aim to classify EAS events according to their μ/EM value. We found out that the mean values of the amplitude, 10-50% and 10-90% rise-times for the artificial EAS formed were very sensitive to the arrival time distributions used to combine the electron and muon traces. The real data values were best reproduced by using exponential arrival times for electrons and muons with decay values narrowly distributed around 15 ns for muons and 25 ns electrons. We found that the network that gives the highest rate of correct classification has two dimensions with 8 neurons in the first dimension and 4 in the second. As shown in tables 1-3, neural networks classify in a way which is better than random and encourage further work along this direction.

References

1. Pierre Auger Project Design Report, http://www.auger.org/admin/ Design-Report/index.html, 230 p (revised 1997).
2. Y. Jeronimo, H. Salazar, C. Vargas and L. Villaseñor, Proc. of the Tenth Mexican School on Particles and Fields, U. Cotti et al. (eds.), AIP Conf. Proc. 670, 479-486 (2003).
3. M. Alarcón et al., Nucl. Instr. and Meth. in Phys. Res. A. 420 [1-2], 39-47 (1999).

Λ^0 POLARIZATION IN $pp \rightarrow p\Lambda^0 K^+(\pi^+\pi^-)^5$ AT 27.5 GeV[*]

J. FÉLIX[1], D. C. CHRISTIAN[2], M. D. CHURCH[3], M. FORBUSH[4,a],
E. E. GOTTSCHALK[3,b], G. GUTIERREZ[2], E. P. HARTOUNI[5,c], S. D. HOLMES[2],
F. R. HUSON[4], D. A. JENSEN[2], B. C. KNAPP[3], M. N. KREISLER[5], J. URIBE[5,d],
B. J. STERN[3,e], M. H. L. S. WANG[5,b], A. WEHMANN[2], L. R. WIENCKE[3,f]
and J. T. WHITE[4,a]

[1]Instituto de Fisica, Universidad de Guanajuato, León, Guanajuato, México; e-mail:
felix@fisica.ugto.mx. [2]Fermilab, Batavia, Illinois 60510, USA. [3]Columbia University,
Nevis Laboratories, Irvington, New York 10533, USA. [4]Department of Physics, Texas
A&M University, College Station, Texas 77843,USA. [5]University of Massachusetts,
Amherst, Massachusetts 01003, USA. Present affiliation: [A]University of California,
Davis, CA 95616. [B]Fermilab,Batavia, Illinois 60510, USA. [C]Lawrence Livermore
National Laboratory, Livermore CA 94550. [D]University of Texas, M.D. Anderson
Cancer Center, Houston, TX 77030. [E]AT&T Research Laboratories, Murray Hill, NJ
07974. [F]University of Utah, Salt Lake City, UT 84112.

The polarization of 1973 Λ^0's from the specific reaction $pp \rightarrow p\Lambda^0 K^+(\pi^+\pi^-)^5$
created from 27.5 GeV incident protons on a liquid Hydrogen target, as function of x_F, P_T, and $M_{\Lambda^0 K^+}$, is, inside statistics, consistent with the polarization of Λ^0's from $pp \rightarrow p_{fast}\Lambda^0 K^+$ at 800 GeV.

1. Introduction

This is an experimental true that Λ^0's from unpolarized pp inclusive and exclusive collisions, at different energies, are produced polarized[1]; that this polarization depends on x_F, P_T, and $\Lambda^0 K^+$ invariant mass[2]. Additionally,

[*]We acknowledge the assistance of the technical staff at the AGS at Brookhaven National Laboratories and the superb efforts by the staffs at the University of Massachusetts, Columbia University, and Fermilab. This work was supported in part by National Science Foundation Grants No. PHY90-14879 and No. PHY89-21320, by the Department of Energy Contracts No. DE-AC02-76CHO3000, No. DE-AS05-87ER40356 and No. W-7405-ENG-48, and by CONACYT of México under Grants 458100-5-4009PE and 2002-CO1-39941.

278

other baryons like Ξ, Σ, etc., are produced polarized in the same experimental circumstances[3].

Some authors have proposed many theoretical ideas trying to understand that Λ^0 polarization[4]. These models lack of predictive power, and the problem of Λ^0 polarization, and in general of baryon polarization, remains as an open problem. Some experiments have been conducted to measure Λ^0 polarization, in exclusive pp collisions, trying to unveil Λ^0 polarization origin studying specific final states where Λ^0 is produced. This paper reports the results of a study of Λ^0 polarization in the specific final state

$$pp \to p\Lambda^0 K^+\pi^+\pi^-\pi^+\pi^-\pi^+\pi^-\pi^+\pi^-\pi^+\pi^-. \tag{1}$$

In that reaction Λ^0 is produced polarized. And this polarization is function of x_F, P_T, and $M_{\Lambda^0 K^+}$.

This sample consists of fully reconstructed pp events produced at 27.5 GeV. All final-state particles are measured and identified. Two previous measurements have been published for polarization of Λ^0 from[5]

$$pp \to p\Lambda^0 K^+\pi^+\pi^-\pi^+\pi^-\pi^+\pi^- \tag{2}$$

$$pp \to p\Lambda^0 K^+(\pi^+\pi^-)^N; N = 1, 2, 3, 4. \tag{3}$$

at 27.5 GeV. An one measurement from[2]

$$pp \to p\Lambda^0 K^+. \tag{4}$$

at 800 GeV.

2. The BNL 766 Experiment and Λ^0 Data

The data for this study were recorded at the Alternating Gradient Synchrotron (AGS) at Brookhaven National Laboratories in experiment E766, described in detail elsewhere[6]. For this study, 1973 Λ^0's satisfied selection criterium cuts reported elsewhere[6].

3. Λ^0 Polarization and Results

This study of Λ^0 polarization explores the dependence of the polarization on the kinematic variables P_T, x_F and $M_{\Lambda^0 K^+}$ in the final states represented by Eq. (1). The way these variables are defined and the way Λ^0 polarization is measured are explained elsewhere[5].

Λ^0 polarization has the same sign in both hemispheres $x_F > 0$ and $x_F < 0$ -positive-, contrary to Λ^0 polarization from reactions type Eq. (2)

that has opposite sign -negative in the $x_F > 0$ hemisphere and positive in the $x_F < 0$ one[2]-.

To improve the statistical power of polarization measurements, Λ^0 polarization from both hemispheres are summed directly. The results of Λ^0 polarization are in Figure 1, as function of $M_{\Lambda^0 K^+}$. This Λ^0 polarization behavior has been observed previously in other reactions[2].

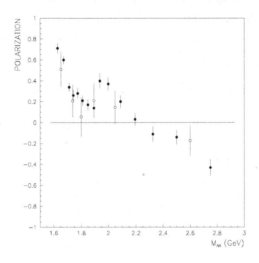

Fig. 1. Λ^0 Polarization as function of the $M_{\Lambda^0 K^+}$ invariant mass. Open circles, this experiment; filled ones, Reference 2 (J. Félix et al).

A Monte Carlo analysis is used to study possible systematic effects, caused by finite acceptance and finite resolution, that might bias the Λ^0 polarization measurements. The Monte Carlo sample is generated with unpolarized Λ^0's using a model for reactions (1) that faithfully reproduces all kinematic distributions. This sample of events is subjected to the same analysis programs and cuts used for the data. The measured polarization for this Monte Carlo sample, which is generated with zero polarization, is found to be consistent with zero as function of x_F, P_T, and $M_{\Lambda^0 K^+}$.

In a second analysis that makes a direct comparison to data, Monte Carlo events are weighted by $(1 + \alpha \wp cos\theta)$ with $\wp$ measured, for each bin of $M_{\Lambda^0 K^+}$. The Λ^0 polarization in the Monte Carlo is in good agreement with the data, all the χ^2/dof between data and Monte Carlo distributions are close to 1. The $cos\theta$ distributions are not quite linear, due to detector acceptance, and these acceptance induced variations in $cos\theta$ are reproduced by the Monte Carlo.

Λ^0 polarization in reactions (1) depends on x_F, P_T, $M_{\Lambda^0 K^+}$; this polarization is similar to that determined in $pp \to p\Lambda^0 K^+$, at 800 GeV^2. Therefore, provide that there is energy enough in the reaction to create Λ^0, its polarization is independent of the beam energy.

References

1. K. Heller *et al*, Phys. Lett. **B68**, 480(1977). G. Bunce *et al*, Phys. Lett. **B86**, 386(1979).
2. T. Henkes *et al*, Phys. Lett. B283, 155(1992). J. Félix *et al*, Phys. Rev. Lett. 88, 061801-4(2002).
3. J. Duryea *et al*, Phys. Rev. Lett. 67, 1193(1991). R. Rameika *et al*, Phys. Rev. **D33**, 3172(1986). C. Wilkinson *et al*, Phys. Rev. Lett. 58, 855(1987). B. Lundberg *et al*, Phys. Rev. **D40**, 39(1989). F. Lomanno *et al*, Phys. Rev. Lett. 43, 1905(1979). S. Erhan *et al*, Phys. Lett. **B82**, 301(1979). F. Abe *et al*, Phys. Rev. Lett. 50, 1102(1983). K. Raychaudhuri *et al*, Phys. Lett. **B90**, 319(1980). K. Heller *et al*, Phys. Rev. Lett. 41, 607(1978). F. Abe *et al*, J. of the Phys. S. of Japan. 52, 4107(1983). P. Aahlin *et al*, Lettere al Nuovo Cimento 21, 236(1978). A. M. Smith *et al*, Phys. Lett. **B185**, 209(1987). V. Blobel *et al*, Nuclear Physics **B122**, 429(1977).
4. T. A. DeGrand *et al*, Phys. Rev. **D24**, 2419(1981). B. Andersson *et al*, Phys. Lett. **B85**, 417(1979). J. Szweed *et al*, Phys. Lett. **B105**, 403(1981). K. J. M. Moriarty *et al*, Lett. Nuovo Cimento 17, 366(1976). S. M. Troshin and N. E. Tyurin, Sov. J. Nucl. Phys. 38, 4(1983). J. Soffer and N.E. Törnqvist, Phys. Rev. Lett. 68, 907(1992). Y. Hama and T. Kodama, Phys. Rev. **D48**, 3116(1993). R. Barni *et al*, Phys. Lett. **B296**, 251(1992). W. G. D. Dharmaratna and G. R. Goldstein, Phys. Rev. **D53**, 1073(1996). W. G. D. Dharmaratna and G. R. Goldstein, Phys. Rev. **D41**, 1731(1990). S. M. Troshin and N. E. Tyurin, Phys. Rev. **D55**, 1265(1997). L. Zuo-Tang and C. Boros, Phys. Rev. Lett. 79, 3608(1997).
5. J. Félix *et al*, Phys. Rev. Lett. 76, 22(1996). J. Félix, Ph.D. thesis, Universidad de Guanajuato, México, 1(1994). J. Félix *et al*, Phys. Rev. Lett. 82, 5213(1999).
6. J. Uribe *et al*, Phys. Rev. D49, 4373(1994). E. P. Hartouni *et al*, Nucl. Inst. Meth. A317, 161(1992). E. P. Hartouni *et al*, Phys. Rev. Lett. 72, 1322(1994). E. E. Gottschalk *et al*, Phys. Rev. D53, 4756(1996). D. C. Christian *et al*, Nucl. Instr. and Meth. A345, 62(1994). B. C. Knapp and W. Sippach, IEEE Trans. on Nucl. Sci. NS 27, 578(1980). E. P. Hartouni *et al*, IEEE Trans. on Nucl. Sci. NS 36, 1480(1989). B. C. Knapp, Nucl. Instrum. Methods **A289**, 561(1980).

DEFINITION OF THE POLARIZATION VECTOR*

VÍCTOR M. CASTILLO-VALLEJO

Departamento de Física Aplicada, CINVESTAV-IPN, Unidad Mérida
A.P. 73 Cordemex 97310, Mérida, Yucatán, México
victorcv@mda.cinvestav.mx

JULIÁN FÉLIX

Instituto de Física, Universidad de Guanajuato,
León, Guanajuato, 37150, México
felix@tutuli.ugto.mx

To study 1/2-spin particles polarization only requires a knowledge of a vectorial quantity. Particles with spin S> 1/2 can have polarization vector as well as polarization tensors of higher rank. This work clarifies the physical meaning of polarization vector definition ($\vec{P}$) for systems with half-integer spin. Prevalent definitions are introduced and examined. The behavior of $\vec{P}$ under symmetry operations (rotation, parity and time reversal) is analyzed. Polarization vector definition is extended to a polarization four-vector definition.

1. Polarization Vector Definitions

A particle is polarized if the mean value of its spin along some direction is different from zero. In general, the polarization vector $\vec{P}$ is the mean value of some polarization operator $\hat{P}$. Basically there are four definitions

---

*Supported in part by CONACyT Grant 2002-CO1-39941.

of $\vec{P}^{1-7}$ reported in the world literature as follows:

$$\vec{P} = \frac{\langle \vec{S} \rangle}{S}, \tag{1}$$

$$P = max\langle \vec{S} \cdot \hat{e} \rangle = \langle \vec{S} \cdot \hat{e}_0 \rangle; \quad \vec{P} = P\hat{e}_0, \tag{2}$$

$$P_M = \langle \hat{T}_{1M} \rangle, \quad M = -1, 0, +1; \quad \vec{P} = \sum_M P_M \hat{e}_M \quad, \tag{3}$$

$$\vec{P} = \langle \vec{\sigma} \rangle. \tag{4}$$

$\vec{P}$ is defined in the frame where the particle of interest is at rest. The expectation values for each operator are calculated using the spin functions of operators $\hat{S}$. In addition the operators $\hat{P}$ are Hermitian operators.

Equations (1)-(4) are independent of the representation of spin functions. An explicit form of the polarization operator depends on the representation of these functions. For simplicity, the base where $\hat{S}_z$ is diagonalized is used. In those equations the absolute value of $\vec{P}$ is called *the degree of polarization* [a], and its range depends on the normalization convention. Eq. (1) defines $\vec{P}$ using a cartesian coordinate system [1,3]; and Eq. (3) defines it using a spherical coordinate system[3]. In this way, for a fixed direction $\hat{n}$, $\vec{P} \cdot \hat{n}$ can take negative values: $-1 \leq \vec{P} \cdot \hat{n} \leq 1$ for Eq. (1) and $-S \leq \vec{P} \cdot \hat{n} \leq S$ for Eq. (3); in Eq. (2) $\vec{P}$ is defined using polar coordinates[5,6], and $\vec{P} \cdot \hat{n}$ only can take positive values: $0 \leq \vec{P} \cdot \hat{n} \leq S$.

The operators $\hat{T}_{LM}$, with $M = -L, -L+1 \cdots L$, are called *polarization tensors of rank* L. They are expressed using a spherical base. Eq. (3) corresponds with the case $L = 1$. One advantage of this expression is that a generalization to higher order polarization tensors is straightforward, despite that their physical meaning are neither obvious and clear. When $L = 1$, the operators $\hat{T}_{LM}$ are proportional to the spherical components of the spin operator. Eq. (4)[6] is defined only for spin 1/2 ($\vec{\sigma}$ are the Pauli matrices[7]). Its equivalence to Eq. (1) is clear.

In Eq. (2) the maximum is calculated[5] over all the directions $\hat{e}$. Hence the maximum of the expectation value is calculated for $\vec{S}$ fixed and by varying the direction $\hat{e}$. When $\vec{S}$ and $\hat{e}$ are parallel, $\hat{S} \cdot \hat{e}$ takes its maximum value and $\hat{e} = \hat{e}_0$. In this case the components of $\vec{P}$ are written down using a polar basis. The components of $\vec{P}$ in a cartesian basis are $P_i = \langle \vec{S}_i \cdot \hat{e}_i \rangle$, with $\hat{e}_i = \hat{i}, \hat{j}, \hat{k}$. Hence, from the last expression, definitions (1) and (2) are equivalent, regardless the spin values, up to the normalization factor S.

---

[a]Some authors use different definitions of degree of polarization. See for example Ref.[1].

Proof of the equivalence between definitions (1) and (3) runs in a similar way. Equations (1) and (4) are equivalent too because $\vec{\sigma} = 2\vec{S}$ (for $S = 1/2$).

Regardless the $\vec{P}$ value, its completely specifies the spin state χ [8]. If $|\vec{P}| < 1$, spin states are commonly described by the formalism of density matrix[9]. Equations (1)-(4) only describe spin-states with $|\vec{P}| = 1$ because for $1/2-$spin there are two possible projections of the spin along some fixed axis. When $S > 1/2$, the three components of $\vec{P}$ are insufficient to fix the parameters involved in the $(2s + 1)$-dimensional spinor χ and hence, the formalism of polarization vector does not fix the $|\vec{P}|$ value. Therefore, a polarization vector can describe S-greater-than-1/2 spin states with any value of $\vec{P}$.

2. Transformation of $\vec{P}$ Under Symmetry Operations

2.1. *Rotation*

Under rotations, spinors χ are transformed by the unitary operator $\hat{D}^s(\alpha, \beta, \gamma)$[1], where α, β, γ are the Euler angles characterizing the rotation. The polarization vector of the rotated spinor is

$$\begin{aligned}
\vec{P}' &= \chi'^\dagger \hat{P} \chi' \\
&= \chi^\dagger \hat{D}^{s\dagger} \hat{P} \hat{D}^s \chi \\
&= \chi^\dagger \hat{P}' \chi \\
&= \hat{R}(\alpha, \beta, \gamma) \vec{P},
\end{aligned}$$

with $\hat{P}' = \hat{D}^{s\dagger} \hat{P} \hat{D}^s$. $\hat{R}(\alpha, \beta, \gamma)$ is the rotation operator in the Euclidean space. Therefore $\vec{P}$-operator transforms as a vector under rotations, independently of the definition used for $\vec{P}$.

2.2. *Parity inversion*

The action of Hermitian parity operator $\wp$ on the canonical state $\chi(s)$ is achieved by $\chi' = \wp\chi = \eta\chi$, where η is the intrinsic parity of the particle represented by χ. Then, the polarization of the inverted spinor is

$$\begin{aligned}
\vec{P}' &= \chi'^\dagger \hat{P} \chi' \\
&= \chi^\dagger \wp^\dagger \hat{P} \wp \chi \\
&= \eta^2 \chi^\dagger \hat{P} \chi \\
&= \chi^\dagger \hat{P} \chi \\
\vec{P}' &= \vec{P}.
\end{aligned}$$

Polarization operator is invariant under inversion of the coordinate system.

2.3. *Time inversion*

$\vec{S}$ changes its sign under timer time reversal operation T. Therefore polarization vectors defined in terms of $\vec{S}$ transforms as

$$\vec{P}' \propto \langle \vec{S}' \rangle$$
$$= \langle T^{-1} \vec{S} T \rangle$$
$$= -\langle \vec{S} \rangle$$
$$\propto -\vec{P}.$$

Polarization operator changes its sign under time reversal operation.

3. Four-Vector of Polarization

Polarization is defined in the frame where the particle is at rest but this polarization can be obtained from the polarization calculated in any frame. To do that, it is necessary to define a covariant polarization operator. The four-vector of spin operator is defined in terms of Pauli-Lubanski operators $\hat{W}_\mu$. In the frame where the particle of interest is at rest the spatial components of this operator correspond to the well known spin operators. From Eq. (1) the polarization four-vector operator is $P_\mu = (0, \vec{P}) = \frac{1}{s}(0, \vec{S})$.

The contravariant components of the polarization operator, in the frame where the particle is moving with velocity $\vec{\beta}$, are

$$P'^\mu = \begin{pmatrix} P'^0 \\ P'^x \\ P'^y \\ P'^z \end{pmatrix} = \begin{pmatrix} \gamma \vec{\beta} \cdot \vec{P} \\ \alpha \beta x \vec{\beta} \cdot \vec{P} + \hat{P}_x \\ \alpha \beta y \vec{\beta} \cdot \vec{P} + \hat{P}_y \\ \alpha \beta z \vec{\beta} \cdot \vec{P} + \hat{P}_z \end{pmatrix}. \tag{5}$$

This operator acts over states of motion of the particle. Eq. (5) can be used to calculate polarization in the moving frame, and then a boost is done in order to obtain the polarization in the particle rest frame.

References

1. Elliot Leader, *Spin in particle physics*, Ed. Cambridge Univ. Press, (2001), London.
2. Kam Biu Luk, *Ph.D. Thesis*, Rutgers University, (1983), New Jersey.
3. D. A. Varshalovich *et al*, *Quantum theory of angular momentum*, Ed. World Scientific, (1988), Singapure.
4. P. M. Ho *et al*, *Phys. Rev.* **D44**, 3402 (1991).
5. R. Hagedorn, *Relativistic kinematics*, Ed. W.A. Benjamin, (1963), CERN.
6. J. Félix, *Mod. Phys. Lett.* **A12**, 363 (1997).

7. J. Sakurai, *Mod. Quantum Mechanics*, Ed. Addison-Wesley, (1985), U.S.A.
8. Claude Cohen *et al*, *Quantum Mechanics*, Ed. John Wiley @ Sons, (1977), Francia.
9. P. D. Collins, *Regge theory & high energy physics*, Ed. Cambridge Univ. Press, (1977), London.
10. Suh Urk Chung, *Spin formalisms*, Ed. CERN, (1971), Génova.

THE MINOS EXPERIMENT

M. SANCHEZ

Department of Physics,
Harvard University,
Cambridge, MA 02138 USA
msanchez@physics.harvard.edu

The MINOS experiment will use a long baseline ν_μ beam, measuring the neutrino signal 1 km downstream from production and 735 km later with similar Near and Far planar steel/scintillator detectors, located at Fermilab and at the Soudan mine in the United States. Its main physics goals consist in verifying dominant $\nu_\mu \to \nu_\tau$ oscillations and a precise measurement of the oscillation parameters. The Far Detector has been successfully taking atmospheric neutrino data for over a year. With the Near detector installation underway, MINOS is preparing for the neutrino beam turn on early next year. A general overview of the experiment along with latest results on cosmic neutrino data and proposed physics goals will be shown.

1. Introduction

MINOS (Main Injector Neutrino Oscillation Search) is a long baseline neutrino oscillation experiment that will probe the region of parameter space responsible for atmospheric neutrino oscillations predicted by the Super-Kamiokande experiment [1] and independently confirmed by MACRO [2] and Soudan 2 [3].

A neutrino beam (NUMI) produced at Fermilab will be detected first by the 980 ton Near Detector at Fermilab and then by the 5.4 kton Far Detector, 735 km away in the Soudan mine in Northern Minnesota. Comparing the observed neutrino energy spectrum at the two locations allows a precise measurement of the oscillations parameters. It is a high statistics experiment and the two detectors allow a reduction of systematic uncertainties yielding a 10 % measurement of Δm_{23}^2, the mass difference involved in atmospheric neutrino oscillations. This and other physics goals are further described in section 3. The beam construction at Fermilab is currently nearing completion and the MINOS Near Detector is currently being commissioned. The MINOS Far Detector has been recording high quality data since the beginning of August 2003. A preliminary study of atmospheric neutrino interactions shows good performance and understanding of the detector.

2. The MINOS Detectors and the NUMI Beam

The MINOS detectors are sampling iron calorimeters. The Far Detector has a total of 486 octagon steel planes, 2.54 cm thick, alternated with 1 cm plastic scintillator planes. The detectors are magnetized with a coil producing a field of 1.5 Tesla. Each scintillator plane is made of 4 cm wide strips and up to 8 m in length depending on the position on the plane. The strips in alternating planes are rotated at $\pm 45°$ to the vertical thereby providing two orthogonal coordinates. Signals from the scintillator are collected via wavelength shifting fibers and carried by clear optical fibers to be read out using photomultiplier tubes. The energy resolution has been measured to be $23\%/\sqrt{E}$ for electromagnetic shower and $55\%/\sqrt{E}$ for hadronic showers.

The Near Detector is designed as similar as possible to the Far Detector but smaller in all dimensions since the neutrino beam spot is only 50 cm wide. A higher rate of neutrino interactions ($\sim$5 with the low energy beam) will be observed in each 10 μs spill therefore faster readout electronics are used without multiplexing.

The NUMI neutrino beams is produced by accelerating 120 GeV protons at the Fermilab Main Injector (MI) towards a graphite target producing mesons (mostly $\pi^{\pm}$, $K^{\pm}$) which are focused by two neutrino horns with parabolic inner conductors. The pions and kaons then decay in a 675 m long decay pipe with 1m radius evacuated to <1.5 Torr. The relative positions of the two horns and the location of the target can be adjusted determining the spectrum of the focused mesons. The most recent results from Super-Kamiokande[1] favor $\Delta m^2 \sim 0.002\ eV^2$ therefore, given the 735 km baseline, the best sensitivity is provided using the "low energy" beam configuration.

3. Physics Goals

The MINOS physics analyses start by identifying the type of neutrino interaction that occurred in the detectors using topological criteria. The Near and Far spectra for the different neutrinos types are then compared. MINOS

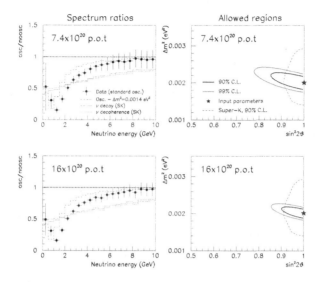

Fig. 1. The left hand plots show the expected ratio of reconstructed neutrino energy spectrum in the far detector to that observed in the near detector for different numbers of incident protons on the NUMI target, assuming $\Delta m^2_{23} = 0.002$ eV^2 and maximal mixing. The right hand plots show the corresponding expected 90% and 99% C.L. including expected systematic uncertainties.

provides sensitivity to neutrino oscillations in the full region of parameter space allowed by the results of the Super-Kamiokande experiment. The main goals of the MINOS experiment can be summarized as follows:

(1) Demonstrate the oscillatory behavior of neutrinos by observing the oscillation dip as seen in Figure 1 and provide a high statistic discrimination against alternative models *e.g.* decoherence, ν decay and extra dimensions.

(2) Provide a precise measurement of Δm^2_{23} of the order of 10%.

(3) Search for subdominant $\nu_\mu \to \nu_e$ oscillations. There is a possibility for a 3σ discovery beyond CHOOZ limits [5] allowing for a first measurement of θ_{13}.

(4) MINOS is the first large deep underground experiment with a magnetic field, this allows a direct measurement of neutrino versus antineutrino oscillations using atmospheric neutrino interactions.

4. Cosmic Neutrino Data Analysis

We present preliminary observations of atmospheric neutrino interactions in the MINOS Far Detector in two event categories: neutrino induced upward-going muons from neutrino interaction in the surrounding rock; and events where the neutrino interaction occurs within the detector volume. The dominant background in both cases is from cosmic-ray muons.

The upward going muon analysis uses data from 2.3 kty exposure. Charged current muon neutrino events are reconstructed as muon tracks and hadronic showers. Precise timing information allows to determine whether a reconstructed event corresponds to an upward or downward going neutrino. A total of 58 upward going events with muon tracks of length greater than 2m are observed.

In the MINOS detector, at a depth of 2070 mwe, the expected cosmic ray muon background is approximately 50K events per day, whereas the expected signal rate is of 0.30 ± 0.05 charged current $\nu_\mu / \bar{\nu}_\mu$ interactions per day. Fiducial and topological cuts combined with the use of the veto shield achieve a rejection rate of over 10^6.

From the 1.85 kty exposure considered here 37 contained events are selected. The final veto shield rejects 51 events in data, consistent within one standard deviation with the Monte Carlo expectation of 61 ± 6. The selected events are consistent with both the expectation of 40 ± 8 events assuming no neutrino oscillations, and with the expectation of 29 ± 6 assuming $\Delta m^2 = 0.0025\ eV^2$ and $\sin^2 2\theta_{23} = 1.0$.

5. Conclusions

The MINOS experiment already has high quality physics data from a year of running the Far Detector and shows the detector working well. As shown, the atmospheric neutrino analyses are going forward and these data will be used to make a first direct observation of separate ν_μ and $\bar{\nu}_\mu$ atmospheric neutrino oscillations. In the meantime, the Near Detector and NUMI beam installation are in an advanced state and first physics data is anticipated early 2005.

References

1. Super-Kamiokande Collaboration, Y. Fukada *et al*, *Phys. Lett.* **B433**, 9-18 (1998); *Phys. Rev. Lett.* **85**, 3999-4003 (2000).
2. MACRO Collaboration, M. Ambrosio *et al*, *Phys. Lett.* **B566**, 35-44 (2003).
3. Soudan 2 Collaboration, M. Sanchez *et al*, *Phys. Rev.* **D68**, 113004 (2003).
4. MINOS Collaboration, P. Adamson *et al*, MINOS technical design report. *NUMI-L-337*.
5. CHOOZ Collaboration, M. Apollonio *et al*, *Phys. Lett.* **B466**, 415-430 (1999).

STUDY OF SCINTILLATING COUNTERS AND THEIR APPLICATION TO V0 DETECTOR OF ALICE

JULIA E. RUIZ TABASCO

Departamento de Física Aplicada, CINVESTAV-IPN, Unidad Mérida
A.P. 73 Cordemex 97310, Mérida, Yucatán, México
julita@moni.mda.cinvestav.mx

This work is about the study of scintillating detectors and its importance and application like one of the subdetectors in the ALICE (A Large Ion Collider Experiment) proyect.

1. Introduction

The interest of the human being to discover the origin of the universe to guide him to extrapolate his theories to the experimental verification of the evolution and constitution of the matter.

It's know there are strong evidences that ordinary matter, which we are constitute (protons, electrons, neutrons, etc.), is only part of a expanded and colded universe that began from a initial explotion. But, something must to exist in this transition (from the big bang to the universe we observe actually) and it have to be precisely a still dense and hot state in which quarks and gluons were free. This state is call quark-gluon plasm, today, quarks and gluons are confinated inside protons and neutrons but people know this state exist in the neutron stars.

2. ALICE Experiment

With the purpose to recreate the quark-gluon plasm in the laboratory and to study its properties to arise the proyect of the ALICE experiment that will use the LHC acelerator in CERN. This device will race around lead nucleus to a center of mass energy of 5.5 TeV/nucleon and luminosity of 1027 cm-2 s-1. These ions will collide head on, the collision will compress and will heat up so much that will be reach the density and temperature conditions needed, for during a time of 8x10-23 sec. and like QCD predic, exists a transition of hadronic matter to a plasm of quarks and gluons desconfinated, immediatly they will cluster to form until 80000 primary particles who will move away of the interaction point. To determine the existence of this state matter and to study its characteristics must be analyzed a big number of observables.

These observables (the produced particles) give information about the initial conditions and the evolution of the spacetime, that is the reason because its not only important but also indispensable to detect them.

ALICE will consist of diferent types of subdetectors, every one will measure diferent parameters and they will aid to the reconstruction of the event and its subsequent phases. The central part will be embedded inside a solenoidal magnet (L3) and will measure hadrons, electrons and photons in every event. It is formed for:

(1) The trayectory system:

- The inner tracking system (ITS), whose purpose is the detection of secundary vertexs, hyperons and charms.
- Time Proyection Chamber whose task is find trayectories, measure momentum and to identify particles (dE/dx)

(2) The system dedicated exclusively to identification particles:

- Time of flying, will identify pions, kaons and protons produced in the central region with momentum less than 2.5 GeV/c.
- High Momentum Particle Identification Detector (HMPID) specialized in hadrons of high momentum.
- Transition Radiation Detector (TRD) will identify electrons with momentum over 1 GeV/c
- The electromagnetic calorimeter of one arm, The Photon Spectrometer (PHOS) will search photons and will measure π^0 and η of high momentum. The arm of the muons spectrometer located toward the interaction point consist of an arrangement of absorbers, a big dipolar magnet and fourteen stations of trigger and path chambers.

But ALICE also use a number of small subdetector systems located in small angles (Zero Degree Calorimeter, Photo Multiplicity Detector, CASTOR, V0) used to define and shoot the trigger when global characteristics of an event are satisfied, particulary the impact parameter; but in this work

I want to emphatize the V0 system that will be used like the principal device of trigger.

The construction of every one of these subdetectors is responsability of the different countries that shape the colaboration, to Mexico correspond the V0 system, specifically the V0L which will work in coordination with the V0R (responsability of the french team).

The V0 System have to provide the following information:

- Rough estimate of the interaction vertex.
- Multiplicity measurement.
- Luminosity determination.
- Beam gas rejection.
- Level 0 trigger for the ALICE detector

Therefore the kind of detector that we choose must have these properties:

- A clear and uniform signal from minimum ionization particles (MIP).
- Large dynamic range for charged particle multiplicity.
- Low noise.
- Fast response and good time resolution.

A kind of detector that satisfied the last requirements is the scintillating detector which takes advantage of its luminescent properties in presence of radiation to produce light. This light is converted in a electrical sign and then electronically analyzed to give information about the incident radiation.

A scintillating material is that who produce a short light pulse when it is exposed to some kind of radiation or energy (light, heat, etc.). When a particle pass across a scintillating media loss some of its energy, and this can excite some atoms in the media. The difference with the clasical explication is that the atoms desexcited emit only a part of the energy excitation (3%), the other part is lost in no radiative decadence (1 photon/100 eV).

There are different kind of scintillating materials: gaseous, glass, liquid, crystal, plastic. We chosse plastic, specifically the response, etc.BC408, because the emission wavelength, is manageable, has a quick response, etc.

The plastic is polished, then is cover with a reflecting media to collect more light, and with a insulator media to don't let the exterior light go inside.The tests done with different reflecting and insulator materials determined the teflon and insulator tape like better reflecting and insulator media respectively.

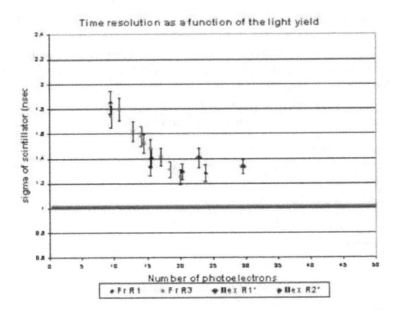

Fig. 1.   Results of the beam test in 2002.

Then the light produced (photon number produced are proportional to the absorbed energy) is collected and transported by optic fibers, specifically the BCF92 Wavelength Light Shift Fiber (WLS) to adjust the emitted wavelength with a better aceptance of the photomultiplicator tubes. Fifteen centimeters of WLS fiber is melt with 40 cm of no WLS fiber to reduced the attenuation. These fibers are also cover with teflon and insulator tape.

The light is transported by internal reflection inside the optic fiber to the photosensible device, the Photo Multiplier Tube (PMT). The photons collide in the cathode and for the photosensible effect photoelectrons are ejected (the number of photoelectrons ejected are proportional to photon incident number). These are directed by a potential to the dynodes where they are multiplicated by the secundary emission process, the final current produce a out signal, so in this way the absorbed energy in the scintillating material have been converted in a current sign with charge proportional to energy.

The later treatment of the signal consist in to eliminate noise and present the information like a charge distribution en the ADC o like a time distribution in the TDC and to estimate the photoelectrons number and the resolution time, respectively.

Like we know the physic of detection and the generation process of the sign, there is that to do general considerations like geometry and segmentation of the detector. Because space restriction, the last design (it is going to be change) choosen have five rings segmented in 12 independient sections with the optic fibers embedded radially in the material (this technique is know like megatile) but is important talk about this design because the results presented were obtained with prototypes of this.

The result of the beam tests in 2002 year are shown in the figure 1. We need that resolution time is less than one nanosecond and its possible to see that although we get good photoelectrons number, the limit of 1 nanosecond didn't be reach. So we needed to improve our design!.

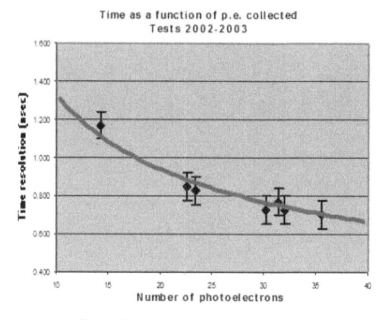

Fig. 2.   Results obtained in the beam test 2003.

3. Results and Conclusions

In the next beam test the prototype detector is with the optic fibers parallel to cover more area to collect light. And the results (figure 2) indicate that resolution time is less than 1 nsec. like we want with a good light collection too.

Finally, we can conclude that the scintillating detector to meet the requirements of detection that the V0 system must have and therefore the V0L will be a scintillating detector.

References

1. Journal Phys. G: Nucl. Part. Phys 30 (2004)1517
2. ALICE Technical Proposal, Ahmad et al. CERN/LHCC/ 95-71, 1995
3. S. Kim, et al. Nucl. Inst. Meth. Phys. Res. A306(1995)206
4. ALICE Technical Design Report, Forward Detectors FMD, T0, V0 CERN/LHCC-2004-025, 2004.
5. R.I. Dzhelyadin, et al., DELPHI Internal Note 86-108, TRACK 42 , CERN 1986
6. S. Vergara, private communication. ALICE Internal Note in preparation.

DATA SELECTION OF Ξ^0 MUON SEMILEPTONIC DECAY IN KTeV

R. A. GOMES

(for the KTeV Collaboration)

University of Campinas,
CP6165, 13083-970,
Campinas, São Paulo, Brazil
ragomes@ifi.unicamp.br

The Ξ^0 semileptonic decays have found great opportunity to be investigated in KTeV experiment at Fermilab. We report the procedure of data selection of the Ξ^0 muon semileptonic decay in 1999 KTeV data, which culminated in the observation of nine events of this decay mode.

1. Introduction

1.1. *Hyperon semileptonic decays*

The neutron semileptonic decay (β-decay) figures as the only decay mode of the neutron, $n \to pe^-\bar{\nu}_e$. However, the hyperons can experience more than one channel to decay. The dominant decay of hyperons is the hadronic two body decay of the type $A \to B\pi$, where A and B are the initial and final hyperon and π represents a neutral or charged pion. The exception is the Σ^0 that decays into a dominant radiative process, $\Sigma^0 \to \Lambda\gamma$.

The hadronic decays of hyperons represent always more than 99% of the total decay rate, resulting in very small probability to other hyperon decay modes. The semileptonic decays of hyperons are in the order of 10^{-3} to 10^{-6} of the total decay rate, leading to the difficulty of obtaining experimental data to investigate these processes. Table 1 shows the branching ratio of observed hyperon semileptonic decays, also with the number of observed events. The last column is the difference of mass ΔM between the final and initial baryon. The processes with ΔM higher than the mass of the muon (~ 106 MeV/c^2) are also allowed to decay into the muon mode.

Table 1. Hyperon Semileptonic Decays. All values from PDG, except $\Xi^0 \to \Sigma^+ \mu^- \bar\nu_\mu$ decay for which we used the preliminary KTeV measurement from ref. [2].

Process	Branching Ratio	Number of Events	ΔM (MeV/c^2)
$\Lambda \to pe\bar\nu_e$	8.3×10^{-4}	20k	177
$\Lambda \to p\mu^- \bar\nu_\mu$	1.6×10^{-4}	28	
$\Sigma^+ \to \Lambda e^+ \nu_e$	2.0×10^{-5}	21	74
$\Sigma^- \to ne^- \bar\nu_e$	1.0×10^{-3}	4.1k	258
$\Sigma^- \to n\mu^- \bar\nu_\mu$	4.5×10^{-4}	174	
$\Sigma^- \to \Lambda e^- \bar\nu_e$	5.7×10^{-5}	1.8k	82
$\Xi^0 \to \Sigma^+ e^- \bar\nu_e$	2.7×10^{-4}	176	126
$\Xi^0 \to \Sigma^+ \mu^- \bar\nu_\mu$	4.3×10^{-6}	9	
$\Xi^- \to \Lambda e^- \bar\nu_e$	5.6×10^{-6}	2.9k	205
$\Xi^- \to \Lambda \mu^- \bar\nu_\mu$	3.5×10^{-6}	1	
$\Xi^- \to \Sigma^0 e^- \bar\nu_e$	8.7×10^{-6}	154	128

1.2. *KTeV experiment*

The KTeV experiment is mainly known for the study of CP violation and rare decays in the neutral kaon system. However, the neutral beam of KTeV also contains a copious amount of hyperons, allowing the investigation of rare decays of them, in particular the Ξ^0 semileptonic decays.

The decay region of KTeV is located 94 m from the production target to eliminate most K_S and to ensure K_L in the fiducial volume. The consequence to hyperons is that only very high momentum Λ and Ξ^0 can reach the detectors – the Ξ^0 momentum is peaked at 290 MeV/c.

Basically, the KTeV detector apparatus consisted of a charged particle spectrometer (two pairs of drift chambers with an electromagnet between them), an electromagnetic calorimeter with 3100 independent CsI crystals and a muon system (lead and steel walls followed by hodoscope planes).

2. Event Selection

2.1. *Reconstruction analysis*

The data selection of Ξ^0 muon semileptonic decays, $\Xi^0 \to \Sigma^+ \mu^- \bar\nu_\mu$, will consider the subsequent decay of the Σ^+ into $p\pi^0$ (branching ratio of 51.6%) and the π^0 into $\gamma\gamma$ (98.8%). The final detectable state will be a proton, a muon and two photons, noting that we miss the detection of the neutrino.

As basic requirements for the analysis of this decay we demand at least two neutral clusters (not associated with charged tracks) in the calorimeter, generated by two photons coming from the π^0, two charged tracks in the spectrometer – a positive one with high momentum that remains in the beam and goes into the hole of the calorimeter and a negative one that hits the calorimeter with minimum ionizing energy – and an identified muon, detected in the muon system.

The reconstruction of the decay begins by using the energies E_1 and E_2 of the two neutral clusters and their separation d_{12} at the calorimeter to reconstruct the longitudinal position of the π^0 vertex (Z_{π^0}), using

$$Z_{\pi^0} \approx Z_{CsI} - \sqrt{E_1 E_2}\, \frac{d_{12}}{m_{\pi^0}}$$

where Z_{CsI} is the distance between the target and the CsI calorimeter and m_{π^0} is the π^0 invariant mass[a]. This reconstructed π^0 longitudinal position is then considered as the Σ^+ one, since the π^0 immediately decays after its production, due to a lifetime of $\sim 10^{-16}$ s.

We then reconstruct the proton track from its momentum and project it to the z position of the Σ^+ vertex calculating the other two coordinates (x, y) of it as well as the Σ^+ four-momentum. The Ξ^0 vertex is determined by the distance of closest approach between the Σ^+ and the muon tracks, allowing us to reconstruct also the Ξ^0 four-momentum.

2.2. *Background modes for $\Xi^0 \rightarrow \Sigma^+ \mu^- \bar{\nu}_\mu$ decay*

The Ξ^0 muon decay has three sources of possible background modes, (a) $K_L \rightarrow \pi^+\pi^-\pi^0$ and $K_L \rightarrow \pi^0\pi^+\mu^-\bar{\nu}_\mu$ decays with high momentum π^+ faking a proton, (b) $\Lambda \rightarrow p\mu^-\bar{\nu}_\mu$ and $\Lambda \rightarrow p\pi^-$ plus two accidental photons, and (c) $\Xi^0 \rightarrow \Lambda\pi^0$ followed by $\Lambda \rightarrow p\pi^-$, $\Xi^0 \rightarrow \Sigma^0\gamma$ followed by $\Sigma^0 \rightarrow \Lambda\gamma$ and $\Lambda \rightarrow p\pi^-$, $\Xi^0 \rightarrow \Lambda\gamma$ followed by $\Lambda \rightarrow p\pi^-$ plus an accidental photon, and $\Xi^0 \rightarrow \Sigma^+ e^- \bar{\nu}_e$ with an accidental muon and an electron lost. We considered the possibility of the π^- to decay into $\mu^-\bar{\nu}_\mu$ or to punch through the muon system and fake a muon.

We reconstructed some specific variables to distinguish these backgrounds to the signal mode, as the invariant mass of K_L, Λ, Σ^0 and Ξ^{0b}. The Ξ^0 beta decay is eliminated using the different response of the electron

---

[a]We need to assume the photons are proceeding from a π^0 to calculate its vertex.
[b]Four different Ξ^0 invariant masses were reconstructed, using $\Lambda\pi^0$, $\Sigma^0\gamma$, $\Lambda\gamma$ and $\Sigma^+ e^-$.

in the calorimeter and requirements demanding the hits in the muon system to match the extrapolated negative track. The Λ background decays are easy to distinguish due to trigger requirements and offline selection criteria implemented to avoid accidental activity. For instance, an event must deposit more than 18 GeV of energy in the calorimeter and each neutral cluster must have more than 3 GeV.

Most background modes were investigated using Monte Carlo simulation to generate at least 10 times their expected number of events. Due to enormous number of K_L in the decay region, it is not feasible to generate the required number of $K_L \to \pi^+\pi^-\pi^0$ events by Monte Carlo. We then used a wrong sign charge analysis consisted in changing the sign of the charged tracks in our signal final state, taking advantage of the symmetry between π^+ and π^- in $K_L \to \pi^+\pi^-\pi^0$ and of the anti-hyperon suppression in KTeV target production.

To improve the selection criteria already implemented, we observed that all possible background modes have a charged particle vertex giving rise to two observed charged tracks, except the $\Xi^0 \to \Sigma^+ e^- \bar{\nu}_e$ decay. We then implemented a crucial selection cut on the transverse momentum of the charged vertex squared ($> 0.018 \ (\text{GeV}/c)^2$), which is very efficient to eliminate those backgrounds, since the two observed charged tracks from the signal (proton and muon tracks) does not compose a real vertex.

3. Conclusions

We applied the selection criteria to 1999 KTeV data and among 300 million Ξ^0 decays we have observed nine events of the $\Xi^0 \to \Sigma^+ \mu^- \bar{\nu}_\mu$ decay with no background seen from Monte Carlo simulation or wrong sign charge analysis.

Acknowledgments

The author would like to thank Dr. C. O. Escobar and Dr. E. J. Ramberg for valuable discussions and the support from Fermi National Accelerator Laboratory, Ministry of Education of Brazil and National Science Foundation.

References

1. Review of Particle Physics, *Phys. Lett.* **B592**, 1 (2004).
2. R. A. Gomes, in Proc. of *'BEACH 04'*, *6th Int. Conf. on Hyperons, Charm & Beauty Hadrons*, Chicago, Illinois (2004); to be published in *Nucl. Phys.* **B** *(Proc. Suppl.)*.

CHAOTIC PROPERTIES OF HIGH ENERGY COSMIC RAYS

A. TICONA, R. TICONA, N. MARTINIC, I. POMA, R. GUTIERREZ,
R. CALLE and E. RODRIGUEZ

Instituto de Investigaciones Físicas
Universidad Mayor de San Andrés
Cota Cota, Calle 27 s/n, La Paz, Bolivia

Using the dimension of correlation analysis, we study the time series intervals of the high-energy cosmic ray data, obtained in the mount of Chacaltaya, in the experiment S.Y.S. Some of the analyzed groups present a fractal dimension close to 3. This fractal dimension seems to show that the transport system of cosmic ray would be chaotic under some circumstances. Results show that only data of charged particles present this fractal property.

1. Introduction

Cosmic rays arrive to earth before traveling very long distances and go through many magnetic fields, under these circumstances, the cosmic rays flow could behave sometimes (or all the time) as a chaotic flow. If it is so, we were able to detect some signal of chaos in the detection experiments.

The experiment of the Kinky University [1] reported the existence of a chaotic signal in the arrival time series of cosmic rays. Although, an Italian group refuted these results [2], other groups [3] found some chaotic signal too, using the same analysis, all of them based in experiments at the sea level.

In this work we analyze the data of the S.Y.S. experiment installed at the mount of Chacaltaya, using the dimension of correlation [4] analysis too. This experiment is at 5200 meters over the sea level, so we have a great quantity of data (around 1 every 2 seconds), compared with the Kinky experiment (1 every 5 minutes). This way, we are able to check better these kinds of effect.

2. The S.Y.S. Experiment

Working since 1977, the main objective of this experiment is to detect air showers produced as an effect of the interaction of primary radiation with the atmosphere. The threshold size of the events is $5x10^5$, this corresponds to an energy around to 10^{14} eV.

This experiment counts with 45 density detectors to detect the electromagnetic component and 32 burst detectors to detect the hadronic component of the showers. It also counts with 13 fast time detectors in order to determine the direction of events. The first analyze is made in order to get the main information of the primary cosmic ray, i.e. energy, direction and kind of radiation, also information of the shower is consider, i.e. time of arrival, size, age and center of the event.

As a second step, we select the time arrivals of the events that have energy higher than the threshold. We also consider only the events that count with an arrival direction. The zenith angle of the arrival direction must not be higher than 45 degrees, because for these angles we observe few events, due to the quantity of atmosphere that events must go through, before reaching detectors.

With the arrival time of the selected events we get the differences of time. These differences behave exponentially, as can be seen in the Figure 1. This behavior is a characteristic of chaotic systems.

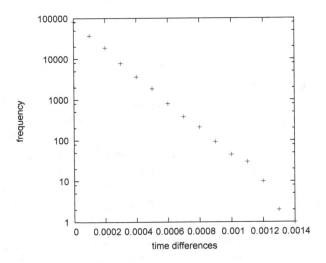

Fig. 1.   Distribution of the time differences.

3. Dimension of Correlation

Once we have the time differences, we build vectors which components are these time differences. We chose an embedded dimension m for these vectors. The possible fractal dimension of the system must be less than the half of $m - 1$ [4], so we chose $m = 11$ for our analysis, because we do not expect a fractal dimension higher than 5.

Then we find the $m-$dimensional distance between each pair of vectors. Using these distances, we build the function of correlation $C_m(r)$, that corresponds to the number of distances with a value less than r. Now, the dimension of correlation is given by:

$$D_m(r) = \frac{\Delta(log C_m(r))}{\Delta(log r)}. \tag{1}$$

If the function of correlation has a region with constant slope, the dimension of correlation shows a constant value in this region. This behavior shows that the system is chaotic. If the data analyzed does not have any region with a constant slope in the function of correlation, it means that chaos is not present in these data.

It is recommended to analyze groups with 500 to 1000 data, in order to be able to observe the possible chaotic signal [4]. We analyzed around 6 months of data, between October of 1993 to March of 1994. We got around 6000 groups and more than 100 of these groups show a region with constant slope in the function of correlation, as we can see in Figure 2.

Some special cases even of white noise, use to present a little region of constant slope in the function of correlation. In these cases, the value of the dimension of correlation changes very fast with the embedded dimension. Then, we analyze each group with constant slope, from different values of embedded dimension. Two examples are shown in Figure 3.

After this analysis just around 20 groups can be considered as candidates to have chaotic behavior, because they keep almost the same value of the dimension of correlation, for different values of the embedded dimension.

In order to check some special characteristic in the selected groups, first we study the arrival directions, which compared with normal groups do not present any special behavior. As a second step, we checked the accumulated flux per unit of time, and for all the cases did not show any special behavior either.

4. Discussion

For all the groups that present a fractal dimension, we did not find any signal that shows a local effect producing this behavior. This also leads to

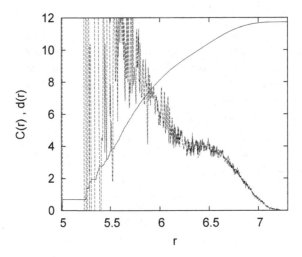

Fig. 2.   Function of Correlation and Dimension of Correlation, of a group of 512 data.

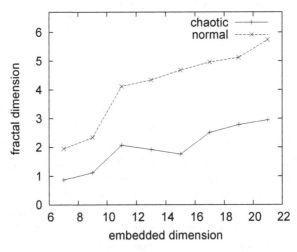

Fig. 3.   Variation of the fractal dimension as a function of the embedded dimension, of two groups of data.

think that the fractal signal belongs to a chaotic system that affects the cosmic ray flow. Many explanations have been proposed [1,3] for example, comets dust that interact with the primary radiation, but in this case we would expect that the number of events increases in the groups that present this kind of behavior and also, a direction in space, with a higher number of events.

Moreover, in February the 7_{th} we have a group that shows a fractal behavior, at the same time, the experiment of the Kinky University also reported a group with this behavior [4]. As we mention, the number of events per unit time, is completely different in these experiments, this made that the analyzed groups belongs to different periods of time. We also have to mention that the experiments are located almost at opposite places in the earth. This could indicate that this kind of signal belong to the complete system of the cosmic ray flow.

The fractal dimension found in all of our groups is between $2, 5$ and 3. The oscillations in the value could indicate that the chaotic signal is mixed with noise that belongs to a non-chaotic signal. This effect has also been studied mixing data belonging to a strange attractor with white noise. Then, the dimension of the strange attractor suffers some variations from its real value.

Using the hadron less analysis we studied each group taking out the data belonging to gamma rays, and the fractal behavior was kept. In the other hand, the signal that belongs to the gamma rays only did not present a fractal behavior in any case.

The better explanation to this effect, seems to belong to the whole transport system of the cosmic ray flow, which in some cases, maybe because some variations in galactic magnetic fields, behaves in a chaotic way, mixing the normal signal with some particles that are under the effect of this chaotic behavior [1,3].

Acknowledgments

We thanks to the organizer committee of the V-SILAFAE, for their dedication and the success of this event.

References

1. T. Kitamura et.al, *Astroparticle Physics*, **6**, 279 (1997);
2. M. Aglietta et.al, *Europhys. Lett.* **34(3)**, 231 (1996).
3. Ohara S. et al., *Proceedings of the 24th International Cosmic Ray Conference*, **1** 289 (1995); W. Unno et al., *Europhysics Letters*, **39**, 465 (1997).
4. JP Eckmann and D. Ruelle, *Reviews of Modern Physics* 57, 617 (1985); P. Grassberger and I. Procaccia, *Physical Review A*, **Vol. 28 N. 4**, (1983); G. Nicolis, *INTRODUCTION TO NONLINEAR SCIENCE*, **Cambridge, University Press**, (1995); M. Schroeder, *Fractals, Chaos, Power Laws. Minutes From An Infinite Paradise*, **New York: WH Freeman**, (1991).

ENERGY SPECTRUM OF SURVIVING PROTONS

R. CALLE, R. TICONA, N. MARTINIC, A. TICONA,
E. RODRIGUEZ and R. GUTIERREZ

Laboratorio de Rayos Cosmicos Chacaltaya, La Paz Bolivia

N. INOUE

Departamento de Física, Universidad de Saitama, Japon

Using the EAS technics we have obtained fluxes of primary particles that have
not interacted nucleary with the atmosphere of the earth. We show their energy
spectrum. The dependence of the cross section of the nuclear interaction on the
energy for the present range is also analyzed. Finally, we compare the results
with the ones obtained by simulation.

1. Introduction

The primary particles with high energy that come from our galaxy produce
extensive air showers (EAS) in the atmosphere of the earth. The experiment EAS-EXC (Extensive Air Shower plus Emulsion X Rays Chamber)
located in the mount of Chacaltaya (5200 meters over sea level, S 16°21',
O 68°8') use plastic scintillators to detect the secundary produced by the
interaction of primary particles with atoms of the atmosphere. The 45 detectors, denominated density detectors N, give the density of electrons per
detector. These detectors are distributed over an area of around 50 meters
of radio. There are also 32 detectors that made the emulsion chamber and
calorimeter of hadrons, they cober an area of 8 m^2, as can be seen in figure
1. Each one of the calorimeter detector called B detectors (Burst detector),
is under a lead layer of 0.15 m of thickness. The emulsion chamber consists
of sensible layers of X ray filmes, between the lead layers. Normally, the
calorimeter detects charged particles produced in the emulsion chamber.

By the interaction of the primary paerticles. The experiment was designed to observe at the same time the electromagnetic component (EM component), the nuclear component (hadrons mentioned before) and the families, i.e. the patterns in the X ray filmes. Although, there is not a complete agreement between the models, it is known that the nuclear component is related with the EM component, for most information read Aguirre et al. [ref. 1] and Kawasumi et al., [ref. 2].

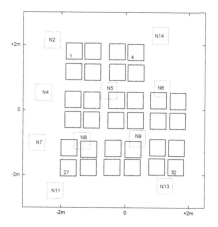

Fig. 1. The regular pattern of squares represent the calorimeter of hadrons located at the center of the experiment EAS. The rest of the squares correspond to the N detectors located around 1.5 m over the calorimeter.

The present work use the events generated by hadrons registered between 1992 and 1995 in the experiment. i.e. the data of the calorimeter converted to number of charged particles n_b (particles from the B detector). This density is in general of the order of 10 to 10^5 particles per 0.25 m^2 in each B unity. The convertion of energy in the calorimeter gives the mean lost of energy of the muon in the plastic scintillator. This part of the detector system begins when one of the detectors register an amount higher than the threshold $n_b \geq 10$. Just a part of the data is assumed to correspond to protons reaching detectors without interacting with the atmosphere. The process of selection will be explained in section 2. Aguirre et al. [ref. 3] and Kawasumi et al., [ref. 4], reported big bursts produced by charged particles in the plastic scintyllators of 60 m^2 area, without a shower. These authors assumed that these Burst were generated by surviving primary protons.

2. Data Analysis

From the original data base, have been selected groups of events that show more than $n_b \geq 10$ particles in the B detectors, following the next selection criterion:

(1) Data are taken without considering the number of particles detected in the N detectors.
(2) Primary particles are considered only if they go through the detectors N5, N6, N8 or N9.
(3) Data is taken considering the ones that go through detectors N5, N6, N8 or N9 without considering the data in the rest of the N detectors.

In all the cases only one of the B detectors must have data, but the rest of the B detectors must have no particle detected. All this process, corresponds to one kind of selection criterion of the Burst detectors.

The main intention is to select the most reliable data, representing primary particles reaching detectors without interacting with the atmosphere. After this selection, the energy spectrum is shown in figure 2.

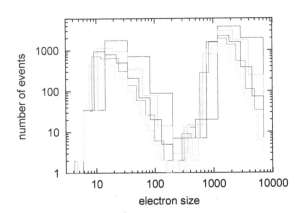

Fig. 2. Size spectrum of the primary surviving protons. Convertion from size to energy is done by using Ec. 1. All the histograms correpond to the same data base, differences is due to the different bins of size used.

We should stand out, that our software use all the selection criterion at the same time, in order to chose the events. Figure 2 was obtained doing the conversion from size to energy, by using the ecuation:

$$\frac{E}{eV} = 10^{10.33} \times n_b^{0.74} \tag{1}$$

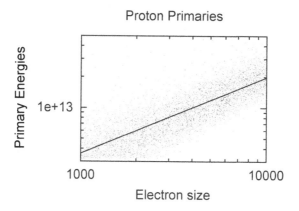

Fig. 3. Exponential fit for the simulated data using the CORSIKA algorithm.

Where, E is the energy and n_b is the size of the Burst.

Simulations were performed by using the CORSIKA algorithm adjusted to the experiment of Chacaltaya, generating around 200000 events. The correlation between size and energy of the simulated events for the range of energy used here, gives the parameters of the Ec1. (See figure 3). For all the three selection criterion used, we have almost the same result shown in figure 4. Numeric values for the fits of the two cases, are given in table 1.

Table 1. Parameters of the power law fit, it means: $10^{a_o} x^{a_1}$, for the number of events as a function of energy. These parameters corresponds to the fits in figure 4.

	a_o	a_1
high energy group	62.14	−4.7
	62.33	−46
low energy group	35.2	−3.12
	35.4	−2.94

3. Discussion

Aguirre et al., [ref. 1], analized the calorimeter using another criterion. They just payed attention to the high energy part of the spectrum. In the present work, we analized almost all the spectrum. The principal question to answer is if the two peaks belongs to surviving protons. Aguirre et al.,

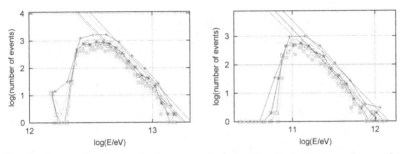

Fig. 4.   Energy distribution for the group selected with criterion 1. It can be seen that the exponent of the power law is higher for the first figure. The exponent for the first figure is equal to -3 and for the second is equal to -4. The different histograms corresponds to different bins in the horizontal axis.

[ref. 3], got an exponent equal to -3.19 ± 0.19 (it means -2.19 for the integral spectrum) with a different arrangement in the mount of Chacaltaya (60 m^2 of plastic scintyllator under around 380 gcm^{-2} of lead, plus some coincidence detectors). The energy range connsidered in [ref. 3] is between 3TeV to 10TeV, wich coincides with the first figure in figure 4. The time interval of observation with these detectors was of 130 days. In the other hand, the relation between energy and size was given by: $E/eV \approx 0.9 \times 10^9 n_b$ obtained by montecarlo methods. In the present work we do not only consider a greater period of time, but also the blindage in the calorimeter is around the order of the mean free path of the nuclear interaction, this gives us a higher quantity of data. The primary spectrum mesured by Grigorov et al., [ref. 5], gives an exponent for the power law, equal to -2.6 ± 0.01 for an energy range between 0.1 TeV and 10^3 TeV.

Considering the cross section and the mean free path of the energetic particle, Aguirre et al., [ref. 3] suggested a logarthimical dependence of the energy due to the higher exponent obtained. In the other hand, if the exponent was the same to the one obtained experimentally out of the atmosphere, it were not necessary to consider a logarithmical dependence for the range of the energy studied here, considering that the mean free path do not depend on the energy.

References

1. Agirre C., et al., Phys. Rev. D 62, 32003 (2000).
2. Kawasumi N., et al., Phys. Rev. D53.
3. Kamata K., et al., Canad. Journ. Phys., 46, S63 (1968).
4. Grigorov N. L., et al., 12ICRC, 5, 1746 (1971) Hobart.

CALCULUS OF THE RATIO ϵ'/ϵ IN THE K MESON DECAY THROUGH THE 331 MODEL

J. C. MONTERO, C. C. NISHI and V. PLEITEZ

Universidade Estadual Paulista,
Instituto de Física Teórica,
Rua Pamplona, 145, 01405-900, São Paulo, Brazil

O. RAVINEZ

Universidad Nacional de Ingenieria UNI,
Facultad de Ciencias
Avenida Tupac Amaru S/N apartado 31139 Lima, Peru

M. C. RODRIGUEZ

Fundacão Universidade Federal do Rio Grande/FURG,
Departamento de Física,
Av. Itália, km 8, Campus Carreiros 96201-900, Rio Grande, RS, Brazil

We consider a 331 model in which both hard (through the Kobayashi-Maskawa matrix) and soft (via a phase in a trilinear term of the scalar potential) CP violation mechanisms can be implemented. Assuming that only the soft CP violation is at work, we are presently studying the possibility that this mechanism explain the observed CP violation in neutral kaons and B-mesons. Here we discuss briefly the obtainment of ϵ'_K and ϵ_K. Detailed calculations will be given elsewhere.

1. Introduction

Recently several models for hard CP violation, through charged scalar [1], or soft violation [2,3] were considered in literature. In particular in Ref. [2] it was proposed a soft CP violating mechanism in the context of a 331 model [4] in which in order to give mass to all fermions it is necessary to introduce only three scalar triplets, i.e., the introduction of the sextet of the model in Ref. [5] is not mandatory. In that model, CP violation arises only through the doubly charged scalar and it is possible to calculate the electron and neutron dipole moment for a wide range of values of the parameters of the model. Presently, we are looking for the predictions of the model concerning the ϵ_K and ϵ'_K parameters.

Let us review briefly the doubly charged scalar sector. As expected, there are a doubly charged, would be Goldstone boson G^{++}, and a physical doubly charged scalar Y^{++}

$$\begin{pmatrix} \rho^{++} \\ \chi^{++} \end{pmatrix} = \frac{1}{N} \begin{pmatrix} |v_\rho| & -|v_\chi| e^{-i\theta_\chi} \\ |v_\chi| e^{i\theta_\chi} & |v_\rho| \end{pmatrix} \begin{pmatrix} G^{++} \\ Y^{++} \end{pmatrix}, \tag{1}$$

where $N = (|v_\rho|^2 + |v_\chi|^2)^{1/2}$; the mass square of the Y^{++} field is given by

$$M^2_{Y^{++}} = \frac{A}{\sqrt{2}} \left(\frac{1}{|v_\chi|^2} + \frac{1}{|v_\rho|^2} \right) - \frac{a_8}{2} \left(|v_\chi|^2 + |v_\rho|^2 \right), \tag{2}$$

where we have defined $A \equiv \mathrm{Re}(\alpha v_\eta v_\rho v_\chi)$ with α a complex parameter in the trilinear term of the scalar potential and a_8 is the coupling of the quartic term $(\chi^\dagger \rho)(\rho^\dagger \chi)$ in the scalar potential [2].

Notice that since $|v_\chi| \gg |v_\rho|$, we have that it is ρ^{++} which is almost Y^{++}.

It was shown there that only two phases survive the re-definition of the phases of all field in the model: a phase of the trilinear coupling constant in the scalar potential α and the phase of a vacuum expectation value, say v_χ. Moreover, only one of those phases survives because of the constraint equation

$$\mathrm{Im} \left(\alpha v_\chi v_\rho v_\eta \right) = 0. \tag{3}$$

which implies that $\theta_\chi = -\theta_\alpha$.

Here we consider the interactions involving doubly charged Higgs scalars which couple to the known quarks and the exotic ones. There is a contribution to ϵ that arises by the exchange of a Y^{++} but it seems negligible if both scalars are heavier than the $U^{\pm\pm}$. Hence we will only consider box diagram involving a vector boson U^{++} and a scalar Y^{++} as shown in Fig. 1.

2. CP Violation in the Neutral Kaons

Concerning the direct *CP* violating parameters ϵ' its contribution for the neutral kaons comes mainly from the penguin diagram shown in Fig. 2. The

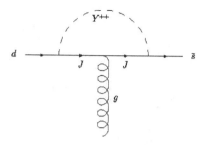

Fig. 1.   Penguim diagram.

vertices are given in Ref. [2] The recently experimental obtained value for this ratio is [6]

$$\text{Re}\left(\frac{\epsilon'}{\epsilon}\right) = 0.00200 \pm 0.00030 \,(\text{stat}) \pm 0.00026 \,(\text{syst}) \pm 0.00019 \,(\text{MC stat}) \quad (4)$$

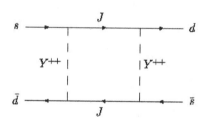

Fig. 2.   Box diagram with two scalars.

Hence in this model we have naturally that $0 \neq |\epsilon'| \ll |\epsilon|$.

The definitions for the relevant parameters in the neutral kaons system are the usual ones [7]:

$$\varepsilon' = \frac{e^{i(\delta_2 - \delta_0 + \frac{\pi}{2})}}{\sqrt{2}}\left[\frac{\text{Im}A_2}{\text{Re}A_0} - \frac{\text{Re}A_2 \text{Im}A_0}{\text{Re}^2 A_0}\right], \quad \varepsilon = \frac{e^{i\frac{\pi}{4}}}{\sqrt{2}}\left[\frac{\text{Im}A_0}{\text{Re}A_0} + \frac{\text{Im}M_{12}}{\Delta M}\right], \quad (5)$$

and taking into account that the CP violation is of the soft type, i.e, the penguin diagram with one Higgs scalar and one gluon gives the dominant contribution to the kaon decay, it results.

$$\frac{\varepsilon'}{\varepsilon} = -\frac{1}{22.2}\left[\frac{2\xi}{\epsilon_m + 2\xi}\right], \quad (6)$$

where, following Gilman and Wise [8], we have defined

$$\xi = \frac{\mathrm{Im}A_0}{\mathrm{Re}A_0} \quad \text{and} \quad \epsilon_m = \frac{\mathrm{Im}M_{12}}{\mathrm{Re}M_{12}}. \tag{7}$$

In Eq. (6) we have used the $\Delta I = 1/2$ rule for the nonleptonic decays, $\mathrm{Re}A_0/\mathrm{Re}A_2 \simeq 22.2$ and that the phase $\delta_2 - \delta_0 \simeq -\frac{\pi}{4}$ is determined by hadronic parameters following Ref. [9] and it is, therefore, model independent. The experimental status for the ε'/ε ratio has stressed the clear evidence for a non-zero value and, therefore, the existence of direct CP violation. The present world average is given by the Eq. (4)

Inserting this value in the Eq.(6) we obtain

$$\frac{\xi}{\epsilon_m} = -0.0192 \tag{8}$$

In the prediction of ε'/ε, $\mathrm{Re}A_0$ and $\mathrm{Re}M_{12}$ are taken from experiments, whereas $\mathrm{Im}A_0$ and $\mathrm{Im}M_{12}$ are computed quantities [10]. Taking into account the experimental values for $\mathrm{Re}A_0 = 3.3 \times 10^{-7}$ GeV and $\mathrm{Re}M_{12} = 1.75 \times 10^{-15}$ GeV we obtain the constraints that $\mathrm{Im}A_0$ and $\mathrm{Im}M_{12}$ must obey:

$$\frac{\mathrm{Im}A_0}{\mathrm{Im}M_{12}} = -0.0361 \times 10^8. \tag{9}$$

with,

$$\mathrm{Im}\, A_0 = \sqrt{2}\frac{g_s m_J^2 \, |v_\rho|}{16\pi^2 N^2} \, I_1 \, (B_{\tilde{g}})_{12}(B_J)_{11} \sin 2\theta_\chi \tag{10}$$

$$\langle \pi\pi(I=0)| \, \bar{s}(p_2)i\frac{\sigma^{\mu\nu}}{M_K^2}L \, q_\nu \, T^a d(p_1)\bar{Q}T_a\gamma_\mu Q \, |K^0\rangle, \tag{11}$$

where

$$I_1 = -i\{\frac{1}{2(m_Y^2 - m_J^2)} + \frac{m_Y^2}{(m_Y^2 - m_J^2)^2} + \frac{m_Y^4}{2(m_Y^2 - m_J^2)^3}ln(\frac{m_J^2}{m_Y^2})\}. \tag{12}$$

According to the bag model calculations [11] the term between brackets has de value

$$\langle \pi\pi(I=0)| \, \bar{s}(p_2)i\frac{\sigma^{\mu\nu}}{M_K^2}L \, q_\nu \, T^a d(p_1)\bar{Q}T_a\gamma_\mu Q \, |K^0\rangle = 2.0 \, Gev^2. \tag{13}$$

2.1. *Box diagrams*

The diagrams contributing to the ε parameter are of two types. One where we have the exchange of two Y^{--} and other with one U^{--} and one Y^{--}. They are shown in the Figs. 2 and 3, respectively. The imaginary part of these sort of diagrams has been derived in Refs. [12] and [13]. The scalar quark lagrangian interaction sector was given in Ref.[2] and the correspondent gauge quark lagrangian interaction sector is:

$$L_W = -\frac{g}{\sqrt{2}}\bar{J}(O_L^d)_{1\alpha}\gamma^\mu L d_\alpha U_\mu^{++} + H.c. \tag{14}$$

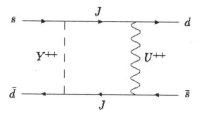

Fig. 3.   Box diagram with one scalar and one vectorial boson.

The contribution of Fig. 3 is given by

$$\mathrm{Im}H_\epsilon^{YU} = \frac{g^2 m_J^2 \sqrt{2}|v_\rho| \sin 2\theta_\chi}{2N^2}(O_L^d)_{11}(O_L^d)_{12} \tag{15}$$

$$\times \left[(B_{\tilde{g}})_{12}(B_J)_{11}I_l M_l - (B_{\tilde{g}})_{11}(B_J)_{21}I_j M_j\right], \tag{16}$$

where we have defined

$$\begin{aligned} M_l &= \left[\bar{s}(q_2)\gamma^l\gamma^\mu L d(q_1)\right]\left[\bar{s}(p_2)\gamma_\mu L d(p_1)\right] \\ M_j &= \left[\bar{s}(q_2)\gamma^\mu L d(q_1)\right]\left[\bar{s}(p_2)\gamma_\mu\gamma^j R d(p_1)\right] \end{aligned} \tag{17}$$

with the integrals

$$I_l = \int \frac{d^4 k}{2\pi^4}\frac{(k+q_2)_l}{D_2} \quad I_j = \int \frac{d^4 k}{2\pi^4}\frac{(k+p_1)_j}{D_2}, \tag{18}$$

The expression D_2 has deform

$$D_2 = \left[(k+p_1)^2 - m_j^2\right]\left[(k+q_2)^2 - m_j^2\right]\left[k^2 - m_Y^2\right]\left[(k+p_1-p_2)^2 - m_U^2\right]. \tag{19}$$

The terms I_l and I_j were calculated in the Ref. [12] assuming that $m_U, m_Y \gg m_J, m_s, m_d$. In the present model we consider the hierarchy $m_U, m_Y, m_J, \gg m_s, m_d$, obtaining the following results ($x = \frac{m_J^2}{m_Y^2}$)

$$I_l = \frac{i}{32\pi^2 m_y^2 m_U^2}\ln(\frac{m_J^2}{m_U^2})\left[h(x)(q_1-p_2)_l + 4g(x)p_{1l}\right], \tag{20}$$

and

$$I_j = \frac{-i}{32\pi^2 m_y^2 m_U^2}\ln(\frac{m_J^2}{m_U^2})\left[h(x)(q_2-p_1)_j + 4g(x)p_{2j}\right]. \tag{21}$$

The next step is to calculate terms like $\langle K^0 \mid M_j \mid \bar{K}^0 \rangle$ and $\langle K^0 \mid M_l \mid \bar{K}^0 \rangle$ which will be made in the global article

Acknowledgments

This work was partially supported by CNPq under the processes 305185/2003-9 (JCM) and 306087/88-0 (VP), and O. Ravinez, would like to thanks to FAPESP (BRAZIL) for financial support.

References

1. D. Bowser-Chao. D. Chang and W-Y. Keung, Phys. Rev. Lett. **79**, 1988 (1997).
2. J. C. Montero, V. Pleitez and O. Ravinez, Phys. Rev. D **60**, 076003 (1999), hep-ph/9811280.
3. H. Georgi and S. L. Glashow, Phys. Lett. **B451**, 372 (1999), hep-ph/9807399.
4. V. Pleitez and M. D. Tonasse, Phys. Rev. D **48**, 2353 (1993).
5. F. Pisano and V. Pleitez, Phys. Rev. D **46**, 410 (1992); P. Frampton, Phys. Rev. Lett. **69**, 2889 (1992); R. Foot, O. F. Hernandez, F. Pisano and V. Pleitez, Phys. Rev. D **47**, 4158 (1993).
6. A. Alavi-Harati *et al.* (KTEV Collaboration), Phys. Rev. Lett. **83**, 22 (1999); G. D. Barr, *et al.* (NA31 Collaboration), Phys. Lett. **B317**, 233 (1993); V. Fanti *et al.* (NA48 Collaboration), Phys. Lett. **B465**, 335 (1999); A. Lai *et al.* (NA48 Collaboration), Eur. Phys. J. **C22**, 231 (2001), hep-ex/0110019.
7. S. Eidelman, *et al.* (Particle Data Group), Phys. Lett. **B592**, 1 (2004).
8. F. Gilman, M. Wise, Phys. Rev. D **20**, 2392 (1979)
9. Y. Nir, CP Violation - A New Era, 55 Scottish Universities Summer School in Physics - Heavy Flavor Physics, 2001.
10. M. Ciuchini, G. Martinelli, Nucl. Phys. Proc. Suppl. **99B**, 27 (201); hep-ph/0006056.
11. J. F. Donoghue, J. S. Hagelin, B. R. Holstein, Phys. Rev. D **25**, 195 (1982).
12. J. Liu, L. Wolfenstein, Nucl. Phys. B289, 1 (1987)
13. D. Chang, Phys. Rev. D 25, 1318 (1982).

USE OF A SCINTILLATOR DETECTOR FOR LUMINOSITY MEASUREMENTS*

G. CONTRERAS

Departamento de Física Aplicada, CINVESTAV- Unidad Mérida
A.P. 73 Cordemex 97310 Mérida, México
jgcn@mda.cinvestav.mx

C. J. SOLANO SALINAS

Universidad Nacional de Ingenieria, Peru
and
CINVESTAV Unidad Merida, Mexico
jsolano@uni.edu.pe

A. M. GAGO

Sección Física, Departamento de Ciencias,
Pontificia Universidad Católica del Perú,
Apartado 1761, Lima, Perú
agago@fisica.pucp.edu.pe

One of the main goals of the V0 setup for the ALICE experiment at CERN will be the measurement of the instantaneous luminosity passing by the ALICE detector. We study here this possibility and also the efficiency of this setup for different processes of inelastic production. The utility of this setup to reject events not coming for the interaction vertex and to reconstruct this vertex position is discused.

---

*This work has been partially supported by CONACYT of México and for Pontificia Universidad Católica of Perú Project DAI-3108.

1. Luminosity

1.1. *Total cross section*

To obtain the the differential cross section $d\sigma/d\Omega$, we may consider a large number of identically prepared particles. We can the ask, what is the number of incident particles crossing a plane perpendicular to the incident direction per unit area per unit time? This is just proportional to the probability flux due to the incident beam. Likewise we may ask, what is the number of scattered particles going into a small area $d\sigma$ sub-tending a differential solid-angle element $d\Omega$?

$$\frac{d\sigma}{d\Omega}d\Omega = \frac{Number\ of\ particle\ scattered\ into\ d\Omega\ per\ unit\ of\ time}{Number\ of\ incident\ particles\ crossing\ unit\ area\ per\ unit\ of\ time} \tag{1}$$

$$\sigma_T = \int \frac{d\sigma}{d\Omega}d\Omega$$

$$= \frac{Total\ number\ of\ particle\ scattered\ per\ unit\ of\ time}{Number\ of\ incident\ particles\ crossing\ unit\ area\ per\ unit\ of\ time} \tag{2}$$

$$\sigma_T = \frac{Rate}{Luminosity} \tag{3}$$

At the LHC energies the TOTEM [7] experiment will measure the total, elastic and inelastic cross sections using this technique. It will measure as well the single diffractive cross section. TOTEM will measure the total cross section with the luminosity independent method [3,4] which requires instantaneous measurement of elastic scattering at low momentum transfer and of the total inelastic rate. CDF [2] and E710 [1] have measured the total proton-antiproton cross section at the Fermilab Tevatron Collider at c.m. system (c.m.s.) energies $\sqrt{s} = 1800$ GeV using this luminosity-independent method. The total cross section is the sum of the elastic and inelastic rates divided by the machine luminosity L:

$$\sigma_T = \frac{1}{L}(R_{el} + R_{in}) \tag{4}$$

The optical theorem relates the total cross section to the imaginary part of the forward elastic scattering rate:

$$\sigma_T^2 = \frac{16\pi(\hbar c)^2}{1+\rho^2}\frac{1}{L}dR_{el}/dt|_{t=0} \tag{5}$$

where ρ is the ratio of the real to imaginary part of the forward elastic scattering amplitude. Dividing 5 by 4 we obtain

$$\sigma_T = \frac{16\pi(\hbar c)^2}{1+\rho^2}\frac{dR_{el}/dt|_{t=0}}{R_{el} + R_{in}} \tag{6}$$

This method provides a precise measurement of the total cross section independently of the luminosity which frequently has a large uncertainty. TOTEM expects to measure the total cross section at an absolute error of about 1 mb.

TOTEM will detect the elastic scattering in the largest possible interval of momentum transfer from the Coulomb region $(-t \approx 5x10^{-4}GeV^2)$ up to at least $-t \approx 10GeV^2$. The extrapolation of the total cross section requires detection of elastic scattering in the low momentum region down to $-t \approx 10^{-2}GeV^2$. ρ is measured fitting this data to the elastic differential cross section:

$$\frac{1}{L}\frac{dN_{el}}{dt} = \frac{4\pi\alpha^2(\hbar c)^2G^4(t)}{|t|^2} + \frac{\alpha(\rho - \alpha\ \phi)\sigma_T G^2(t)}{|t|}exp(-B|t|/2)$$
$$+ \frac{\sigma_T^2(1+\rho^2)}{16\pi(\hbar c)^2}exp(-B|t|) \qquad (7)$$

The three terms in eq. 7 are due to, respectively, Coulomb scattering, Coulomb-nuclear interference, and nuclear scattering. L is the integrated accelerator luminosity, dN_{el}/dt is the observed elastic differential distribution, α is the fine structure constant, ϕ is the relative Coulomb-nuclear phase, given by [6] $ln(0.08|t|^{-1} - 0.577)$, and $G(t)$ is the nucleon electromagnetic form factor, which we parametrize in the usual way as $(1 + |t|/0.71)^{-2}$ [t is in $(GeV/c)^2$]. E710 [5] used this technique for $\sqrt{s} = 1800$ GeV at the Tevatron.

1.2. *ALICE luminosity*

ALICE experiment, through the V0 level 0 trigger, will calculate its luminosity using the visible cross section (luminosity monitor constant) for its Level 0 trigger, σ_{V0}, based on the TOTEM result for $\bar{p}p$ inelastic cross section at $\sqrt{s} = 8$ TeV.

Instantaneous luminosity based on V0 scaler rates:

The instantaneous luminosity L is related to the counting rate R_{V0} in the V0 counters by:

$$L = \frac{R_{V0}}{\sigma_{V0}} \qquad (8)$$

where σ_{V0} is the cross section subtended by these counters. The counting rate (and, thus, the instantaneous luminosity L) is measured for each of the bunch crossings.

This is **strictly true only if the instantaneous luminosity is low enough that the counting rate corresponds to the interaction rate**. As the luminosity increases there is the possibility for having multiple interactions in a single crossing. For this case, the counting rate is less than the interaction rate since multiple interactions get counted only once.

The multiple interaction correction may be calculated based on the Poisson statistics. The average number of interactions per crossing, $\bar{n}$, is given by:

$$\bar{n} = L\,\tau\,\sigma_{V0} \tag{9}$$

where τ is the crossing time . The multiple interaction correction factor is then:

$$\frac{L}{L_{measured}} = \frac{\bar{n}}{1 - e^{-\bar{n}}} \tag{10}$$

The V0 trigger efficiency:

The V0 trigger efficiency, ϵ_{V0}, can be determined by using samples of data collected while triggering on random beam crossings (zero bias). These unbiased data samples could be collected by triggering the data acquisition system solely on the crossing time of the pp beams at the collision region and not in the presence of an interaction. It can be used the TOF detectors for this.

They measure the pedestals using those zero bias data events with no interaction hits in-time with the beam crossing. This is used to correct the distributions when an in-time hit is seen in the arrays. After subtracting the pedestals the efficiency is determined from the fraction of events that have a good vertex interaction position, around the center of the Alice detector.

Inelastic scattering cross-sections:

The total, elastic and single diffractive cross sections will be obtained from TOTEM results. For the single diffractive cross section we will do a cross-check with our own measurement.

The double diffractive cross section, σ_{DD}, is not a direct measurement but it is calculated using the elastic and single diffractive cross sections,

σ_{el}, and σ_{SD} respectively. The calculation uses factorization (see Figure 1) to approximate the equivalence of the ratio:

$$\frac{\sigma_{DD}}{\sigma_{SD}} \approx \frac{\sigma_{SD}}{\sigma_{el}} \tag{11}$$

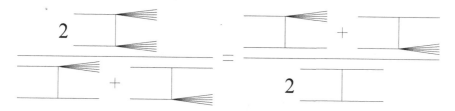

Fig. 1. Factorization to approximate the double diffractive cross section.

Since the single diffractive cross sections include either particle diffracting, then σ_{SD} is replaced by $\sigma_{DD}/2$ in the equation 11. A 10% systematic error is applied to this factorization method. Then the double diffractive cross section:

$$\sigma_{DD} \approx \frac{\sigma_{SD}^2}{4\,\sigma_{el}} \tag{12}$$

The QCD component of the inelastic cross section is simply the inelastic cross section with the single and double diffractive components subtracted off.

$$\sigma_{QCD} = \sigma_{inelastic} - \sigma_{SD} - \sigma_{DD} \tag{13}$$

Level 0 acceptances for inelastic scattering processes:

We are using the Pythia MC generator to simulate the inelastic processes and obtain their acceptances. This Monte Carlo has separate switches to allow generation of single diffractive, double diffractive, and QCD (inelastic non-diffractive) events separately.

The calculation of Luminosity Monitor Constant, σ_{V0}, would be given by:

$$\sigma_{V0} = \epsilon_{V0}(\epsilon_{SD}\sigma_{SD} + \epsilon_{DD}\sigma_{DD} + \epsilon_{QCD}\sigma_{QCD}) \tag{14}$$

First simulated results:

Table 1.   10000 fully simulated Pythia
pp-Minimum Bias events. Minimum
Bias (MB) events = QCD + diffractive
events.

Process	Generated
$q_i q_j \rightarrow q_i q_j$	17.28%
$qg \rightarrow qg$	13.64%
$gg \rightarrow q\bar{q}$	0.98%
$gg \rightarrow gg$	37.44%
other QCD processes	0.02%
QCD processes	69.36%
$AB \rightarrow AY$	8.79%
$AB \rightarrow XB$	9.52%
$AB \rightarrow XY$	12.33%
Diffractive	30.64%

Table 2.   Efficiencies of different kind of processes, by the detectors V0L, V0R, one of them, and both of them. Minimum Bias (MB) events = QCD + diffractive events.

Process	V0L	V0R	V0L or V0R	V0L and V0R
$q_i q_j \rightarrow q_i q_j$	97.62%	96.81%	99.88%	94.56%
$qg \rightarrow qg$	99.92%	99.85%	100 %	99.78%
$gg \rightarrow q\bar{q}$	100 %	100 %	100 %	100 %
$gg \rightarrow gg$	99.97%	99.97%	100 %	99.94%
other QCD proc.	100 %	100 %	100 %	100 %
QCD	99.38%	99.16%	99.97%	98.57%
$AB \rightarrow AY$	40.61%	64.73%	65.90%	39.30%
$AB \rightarrow XB$	76.99%	46.53%	78.10%	45.30%
$AB \rightarrow XY$	70.96%	65.53%	86.61%	49.87%
Diffractive	64.14%	59.39%	78.06%	45.46%
MB	88.58%	86.98%	93.26%	82.30%

Using a sample of 10000 fully simulated pp-Minimum Bias events ($\approx$ 13GB in disk) we obtained the percentages of the different kind of generated processes. The Minimum Bias (MB), according Pythia, is the sum of the QCD (not deep inelastic) and diffractive processes. See table 1. We obtained as well the acceptances for all these processes, see table 2. For these studies we used the Pythia version 6.150 that come with the Aliroot version V3.08 and the Root version v3.03.09.

2. Rejection of Events Not Coming from the Collision Vertex

We study the particles coming from the collision vertex (primaries) and for other interactions (secondaries) for the different rings of V0L and V0R detectors. We can see that the V0R detector has a lot of more secondaries hits than the V0L. This is due to the proximity to the center region of collision and to all the mechanical and electronic support. We observed that 97% of the primaries particles arrive within 0.5 ns which basically is the difference between the primaries coming from the closest and the farest interaction points $(-5.3 \leq Z_{vtx} \leq 5.3)$. We used this fact to reject the late hits and we could eliminate more than 80% of the secundaries hits. Even with this rejection we still observe a contamination of secundary hits, specially for the V0R detector, where is observed as well a strong dependence on the ring position. See figure 2.

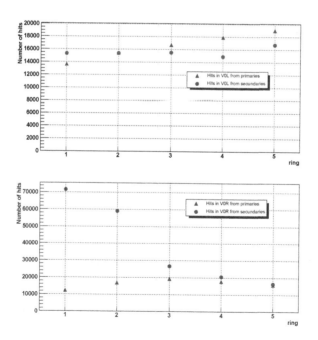

Fig. 2. Hits per ring for V0L and V0R.

3. Conclusions

These results are very preliminary. We are still working in the full simulation of the V0L detector (for example we are using the Litrani to study the propagation of light through the cintillator), to get the total efficiency and time resolution. We know that even with fully simulated events, still the full detector is not finished and that a real situation can be quite different from the simulated processes so we still need to work a lot on this.

References

1. E710 collaboration: N. Amos *et al*, Phys.Lett.B243 (1990) 158.
2. CDF collaboration: F. Abe *et al*, Phys.Rev.D50 (1994) 5550.
3. CERN-Pisa-Rome-Stony Brook Collaboration, Phys.Lett.B62 (1976) 460.; U. Amaldi *et al*, Nucl.Phys.B145 (1978) 367.
4. UA4 Collaboration, M. Bozzo *et al*, Phys.Lett.B147 (1984) 392.
5. E710 collaboration: N. Amos *et al*, Phys.Rev.Lett.68 (1992) 2433.
6. G.B. West and D.R. Yennie, Phys.Rev. 172 (1968) 1413.
7. http://totem.web.cern.ch/Totem/
8. J. Bantly *et al*, Fermilab-TM-1930, D0 Note 2544, Dec. 29, 1995.

SIMPLIFIED FORMS OF A LINEAR AND EXACT SET OF EQUATIONS FOR THE GRAVITATIONAL FIELD IN THE VERY EARLY UNIVERSE

ENER SALINAS

London South Bank University, FESBE
103 Borough Road, London SE1 0AA, UK
salinae@lsbu.ac.uk

The equivalence of three different methods to obtain the gravitational field of ultra-high relativistic particles is presented. Emphasis is placed on the linearity of the derived field equations. The structure of the regarded energy-momentum tensors helps simplifying these equations. The electromagnetic case is also considered introducing potentials with null and geodesic properties.

1. Introduction

There are some reasons to think that some solutions of Einstein equations having impulsive-wave type could represent some phenomena at very early stages of the universe [1], [2]. Moreover, there is presently an increasing interest in the topic from an unexpected source, namely the possibility of producing microscopic black holes in high energy collisions in a lab environment, such as the Large Hadron Collider (LHC) at CERN. Indications of such events are expected from particles whose mass is mainly dominated by kinetic energy (e.g. protons at 7 TeV) rather than rest mass [3]. Obtained signals may be in the form of copious particles e.g. photons, electrons, and quarks. Moreover, gravitational waves will also be emitted. In this paper, a linear with respect to a scalar field f and yet exact subset of the full non-linear Einstein field equations is presented, which yield solutions of the type that can be related to the phenomena just described. The set of field equations is an improvement the one given originally by A. H. Taub. Some properties of these equations are analyzed, focusing attention in vacuum and pure radiation energy-momentum tensors. In addition, the electromagnetic case is analyzed considering a special type of 4-potentials which are vector fields tangent to a null-geodesic congruence. This article is organized as follows: First there will be presented the two most known treatments of the problem, i.e. Aichelburg-Sexl (AS) boost and the Dray and t'Hooft (DH) shockwave approach. Subsequently a third equivalent approach will be study, namely the Generalized Kerr-Shield (GKS) transformation.

2. First Formulation: The AS Boost

The geometry of spacetime around highly relativistic particles can be described by metrics obtained by procedures such as the AS boost of the spacetime around a massive particle at rest [4]. The boost is made simultaneously as one lets the mass of the particle, $m = \gamma^{-1}p_m$ approach zero, being $\gamma = \sqrt{1 - (v/c)^2}$. The resulting effect is the metric of a gravitational shock wave

$$ds^2 = -dU\,dV + dy^2 + dz^2 - 4p_m\,\delta(U)\ln(y^2 + z^2)dU^2 \qquad (1)$$

Where $U = t - x$ and $V = t + y$ are the null coordinates.

In a similar way one can boost the metric of a charged particle (Reissner-Nordstrom), a rotating (Kerr) or one which contains rotation and charge (Kerr-Newman). Other backgrounds different from Minkowsky are also possible, such as de Sitter and anti-de Sitter, or Schwarzschild spacetimes.

3. Second Formulation: The DH Shockwave Approach

T.Dray and G.'t Hooft [2] introduced an equivalent approach to the AS boost, they assumed a metric of the following shape

$$ds^2 = 2A(U, V)dU\,dV + B(U, V)\eta_{ij}dx^i dx^j, \quad U < 0 \qquad (2)$$

It is also assumed that this is a solution of Einstein equations in Vacuum. A discontinuity is introduced

$$ds^2 = 2A(U,\ V + \theta C)\,dU(dV + \theta C, i\,dx^i) + B(U, V)\eta_{ij}dx^i dx^j \qquad (3)$$

Where $U > 0$, $V \to V + C(x^i)$ and $\theta = \theta(U)$ is the step function. Under a change of coordinates, $\hat{U} = U$, $\hat{V} = V + \theta C$, $\hat{x}^i = x^i$, resulting in

$$ds^2 = 2A(\hat{U},\ \hat{V})\,d\hat{U}(d\hat{V} - \delta(\hat{U})C\,d\hat{U}) + B(\hat{U}, \hat{V})\eta_{ij}d\hat{x}^i d\hat{x}^j \qquad (4)$$

Where $\delta = \delta(U)$ is the Dirac delta "function". This procedure reconstructs the AS metric (Eq. (1)).This approach was also useful to interpret the gravitational interaction between in-falling matter and Hawking radiation and the quantum properties of black holes.

4. Third Formulation: The GKS Approach

Trautman, in 1962, studied the possibility of propagating information by gravitational waves using metrics of the form $g_{ab} = \eta_{ab} + 2f l_a l_b$, where η_{ab} is the Minkowski Metric, f is a scalar function representing the information and l_a is a null vector. Kerr and Schild used this method to study algebraically special metrics (e.g. field of a spinning body). In 1981, A.H. Taub generalized this idea [5]. Starting from a seed spacetime (S, g), not necessarily restricted to being flat, one can generate another spacetime (S, g) with the transformation

$$\hat{g}_{ab} = g_{ab} + 2f l_a l_b \tag{5}$$

Where $\hat{g}_{ab}$ is a scalar field over S and l_a is a null and geodesic vector field with respect to g_{ab}, that is: $g_{ab} l^a l^b = 0$ and $g^{bc} l_{a:b} l_c = 0$ respectively. From the expression $\hat{g}^{aw} \hat{g}_{wb} = \delta_b^a$ we obtain the contravariant form of the metric $\hat{g}^{ab} = g^{ab} - 2f l^a l^b$.

It is possible to proof the following properties of the GKS transformation:

- l_a is null and geodesic with respect to $\hat{g}_{ab}$
- The covariant derivative of l_a with respect to $\hat{g}_{ab}$ is $l_{a;b} = l_{a:b} + f_{,d} l^d l_a l_b$
- l_a has the same expansion, twist and shear with respect to both metrics, $\hat{\Theta} = \Theta$, $\hat{\omega} = \omega$, $\hat{\sigma} = \sigma$
- The connection coefficients with respect to $\hat{g}_{ab}$ have the form $\hat{\Gamma}^c_{ab} = \Gamma^c_{ab} + \Omega^c_{ab}$, where $\Omega^c_{ab} = (f l^c l_a)_{:b} + (f l^c l_b)_{:a} - (f l_a l_b)^{:c} + 2f(f_{,d} l^d) l^c l_a l_b$.

Using these properties, we can obtain expressions for the Riemann tensor, the Ricci tensor, and the scalar curvature. Thus the Einstein tensor is

$$\hat{G}^u_d = G^u_d - 2f(l^w l^u R_{wd} + \tfrac{1}{2} l^a l^b R_{ab} \delta^u_d)$$
$$+[(f l^w l^u)_{:d} + (f l^w l_d)^{:u} - (f l^u l_d)^{:w} - (f l^w l^v)_{:v} \delta^u_d]_{:w} \tag{6}$$

Introducing Eq. (6) into the field equations $\hat{G}^u_d = 8\pi \hat{T}^u$ gives an equation which is linear in f we can give the following interpretation: knowing the seed metric g_{ab} and selecting a null-geodesic vector, then by a choice of the energy-momentum tensor it is possible to transform the Einstein equations into a second order linear system of equations with respect to the scalar field f. Solving the equations we obtain a metric that has chances to be physically relevant. We remark that Taub [5] obtained a similar equation; however his derivation contains an error, which is corrected in Eq. (6).

A difficulty arises with the covariant index form, $\hat{G}_{eu} = \hat{g}_{eu} \hat{G}^u_d = g_{eu} \hat{G}^u_d + 2f l_e l_u \hat{G}^u_d$, since the second term becomes quadratic in f. However this is resolved by imposing that the energy-momentum tensor fulfils the condition $l_u \hat{T}_d^u = 0$. Specialization to vacuum seed metric $(R_{ab} = 0)$ simplifies the field equations to:

$$[(fl^w l^u)_{:d} + (fl^w l_d)^{:u} - (fl^u l_d)^{:w}]_{:w} = 8\pi \hat{T}_d^u \tag{7}$$

The possible useful cases are vacuum to vacuum, vacuum to pure radiation and vacuum to electromagnetism. Einstein equations are even simpler:

(a) VACUUM → VACUUM

$$[(fl^w l^u)_{:d} + (fl^w l_d)^{:u} - (fl^u l_d)^{:w}]_{:w} = 0 \tag{8}$$

(b) VACUUM → PURE RADIATION

$$[(fl^w l^u)_{:d} + (fl^w l_d)^{:u} - (fl^u l_d)^{:w}]_{:w} = 8\pi\rho l^u l_d \tag{9}$$

The simplicity of these equations is evident since its few terms and its linearity. For example the gravitational shock-wave metric can be reconstructed from this procedure. The null vector $l^a = k^a = 2(0,1,0,0)$, and the energy-momentum tensor $\hat{T}_{ab} = p\delta(\sqrt{y^2 + z^2})\delta(U)l_a l_b$ are plugged into Eq. (9) which gives the solution $f(U,y) = -2p\ln(y^2 + z^2)\delta(U)$, therefore Eq. (1) is again reconstructed [6].

(c) VACUUM → ELECTROMAGNETISM

Here we consider the energy-momentum tensor $\hat{T}_{ab} = \hat{F}_{ac}\hat{F}_b^c - \frac{1}{4}\hat{g}_{ab}\hat{F}_{cd}\hat{F}^{cd}$, where we introduce electromagnetic 4-potentials with the characteristic of being null $A_a A^a = 0$ and geodesic $A_{a;b}A^b = 0$. These conditions imply that the potential can be expressed in terms of two new functions $A_a = \zeta\chi_{,a}$, where the $\chi_{,a}\chi^{,a} = 0$, $\zeta_{,a}\chi^{,a} = 0$, with this structure, the energy momentum tensor acquires the simple form $\hat{T}_{ab} = \alpha^2\chi_{,a}\chi_{,b}$ where $\alpha^2 = \zeta_{,a}\zeta^{,a}$ the new metric has the following shape

$$\hat{g}_{ab} = g_{ab}^{Vacuum} + 2\left(\frac{f}{\alpha^2}\right)\chi_{,a}\chi_{,b} \tag{10}$$

Eq. (10) can be used to describe a metric describing plane waves propagating in curved spacetime.

5. Conclusions

A set of equations that generates solutions representing the gravitational effects of highly ultra-relativistic moving particles was presented. Knowing a seed metric and selecting a null-geodesic vector field (e.g. by the study of the geometry of the seed background), then by a proper (physically relevant) choice of the energy momentum tensor, we can transform the Einstein equations to second order linear equations with respect to a scalar function f. Properties of null vectors, and constructions of energy momentum tensors based on null and geodesic properties, help simplifying the equations. Other generalizations of the formalism presented here can be devised, for example using two null directions $\hat{g}_{ab} = g_{ab} + Al_a l_b + Bk_a k_b + C(l_a k_b + l_a k_b)$. The method presented here can be fully formulated in the Newman-Penrose formalism, or in the Geroch-Held-Penrose (GHP) formalism [7], where further simplifications are expected.

References

1. Penrose, R (1972), in: *General Relativity,* edited by L. O'Raifeartaigh. Claredon, Oxford, p.101.
2. Dray, T and 't Hooft, G. (1985) "The gravitational shock wave of a massless particle". *Nucl. Phys.* **B 253**, p.173.
3. Giddins S. B., Thomas, S. (2002) High Energy Colliders as Black Hole Factory: The End of Short Distance Physics, *Physical Review D* **Vol. 65**.Paper No. 056010.
4. Aichelburg, P.C. and Sexl, R.U. (1971) *Gen. Rel. Grav.* **2**, p.303.
5. Taub, A. H. (1981), Ann. Phys., NY **134**, p.326.
6. Alonso, R. and Zamorano, N.(1987) *Phys. Rev. D* **Vol.35**, p.6.
7. Geroch, R. Held, A, and Penrose, R. (1973) *J. Mat. Phys.*, **14**, p.874.

POSTERS

DETERMINATION OF THE B-MASS USING RENORMALON CANCELLATION[*]

C. CONTRERAS

Dept. of Physics,
Universidad Tecnica Federico Santa Maria,
Casilla 110-V Valparaiso, CHILE
carlos.contreras@usm.cl

Methods of Borel integration to calculate the binding ground energies and mass of $b\bar{b}$ quarkonia are presented. The methods take into account the leading infrared renormalon (IR) structure of the "soft" binding energy $E(s)$ and of the quark pole masses m_q, and the property that the contributions of these singularities in $M(s) = 2m_q + E(s)$ cancel.

1. Introduction

The calculation of binding energies and masses of heavy quarkonia $q\bar{q}$ using renormalon method has attracted the attention recently. The calculations, based on perturbative expansions, are due to the knowledge of up to N^2LO term $(\sim \alpha_s^3)$ of the static quark-antiquark potential $V_{q\bar{q}}(r)$[1] and partial knowledge of the N^3LO term there, and the ultrasoft (us) gluon contributions to a corresponding effective theory N^3LO Hamiltonian[2]; and the knowledge of the pole mass m_q up to order $\sim \alpha_s^3$ [3].

The quarkonium mass is given by: $M_{q\bar{q}} = 2m_q + E_{q\bar{q}}(\mu)$, where $E_{q\bar{q}}(\mu)$ is the hard $(\mu_h \sim m_q)$ + soft $(\mu_s \sim m_q\alpha_s)$ part of the binding energy. However, in the resummation of the perturbative serie of the pole mass of $M_{q\bar{q}}$ we find an ambiguity due to the IR-renormalon[4] singularity which appears at the value $b = 1/2$ of the Borel Transform. The static potential has a related ambiguity, such that the contribution of these singularities in the $2m_q + E_{q\bar{q}}(\mu)$ cancel[5]. $E_{q\bar{q}}(\mu)$ contains the $V_{q\bar{q}}(\mu)$ and the kinetic

---

[*]Work supported by Project USM-110321 of the UTFSM.

energy. Finally, the binding energy has contribution $E_{q\bar{q}}(us)$ from the us momenta regime $\sim m_q \alpha_s^2$, which is not related to the $b = 1/2$ renormalon singularity. Therefore, we can separate the s and us part of the $E_{q\bar{q}}$ and apply the Borel resummation only to the s part.

2. Extraction of the $\overline{m}_b$ from the Mass of the $\Upsilon(1S)$

The Borel resummation of the pure perturbative "soft" mass $M_\Upsilon(1S; s) = 2m_b + E_{b\bar{b}}(s)$ and the extraction of the b-mass are performed in the following steps (For details, see reference[6]):

(i) The pole mass is calculated in term of the $\overline{MS}$ renormalon-free mass $m_b = m_b(\overline{m}_b)$ and evaluated by Borel integration using the Bilocal method and Principal Value (PV) prescription[7]. The residue parameter N_m is $N_m = 0.555 \pm 0.020$.

(ii) The separation in $E_{b\bar{b}}(s, \mu_f)$ and $E_{b\bar{b}}(us, \mu_f)$ is parametrized by a factorization scale μ_f. It can be fixed using the renormalon cancellation condition $M_\Upsilon(1S; s) = 2m_b + E_{b\bar{b}}(s)$, then $\mu_f = \kappa m_b \alpha_s^{3/2}(\mu_s)$ and $\kappa = 0.59 \pm 0.19$.

(iii) The soft binding energy $E_{b\bar{b}}(s, \mu_f)$ is evaluated via Borel integration for the $b = 1/2$ IR-renormalon. We used the PV-prescription and the "σ-regularized bilocal expansion"[8]. The requirement of the absence of the pole around $b = 1/2$ fixed $\sigma = 0.36 \pm 0.03$.

(iv) $E_{b\bar{b}}(us, \mu_f) = E_{b\bar{b}}(us)^{(\text{p+np})} \approx (-100 \pm 106)$ MeV is obtained from the perturbative and non-perturbative result (gluonic condensate).

(v) Adding together the Borel-resummed values $2m_b$, $E_{b\bar{b}}(s)$ and $E_{b\bar{b}}(us)$, requiring the reproduction of the measured mass value $M_\Upsilon(1S)$ with contribution of the nonzero mass of the charm quark, we obtained the value for the $\overline{m}_b(\overline{m}_b) = 4.241 \pm 0.068$ GeV when $\alpha_s(M_Z) = 0.1180 \pm 0.0015$.

Our result of $\overline{m}_b = 4.241 \pm 0.068$ GeV is close to the values obtained from the QCD spectral sum rules.

References

1. M. Peter, Phys. Rev. Lett. **78** (1997) 602
2. B. A. Kniehl and A. A. Penin, Nucl. Phys. B **563** (1999) 200
3. K. G. Chetyrkin and M. Steinhauser, Nucl. Phys. B **573** (2000) 617
4. M. Beneke, Phys. Rept. **317** (1999) 1.
5. A. Pineda, JHEP **0106** (2001) 022

6. C. Contreras, USM-TH 160, [hep-th/0411370]
7. G. Cvetic and T. Lee, Phys. Rev. D **64**, 014030 (2001)
8. C. Contreras, G. Cvetic and P. Gaete, Phys. Rev. D **70** (2004) 034008.

CP VIOLATION IN $B \to \phi K^*$ DECAYS: AMPLITUDES, FACTORIZATION AND NEW PHYSICS

D. GÓMEZ DUMM and A. SZYNKMAN

IFLP, CONICET – Depto. de Física, Univ. Nac. de La Plata,
C.C. 67, 1900 La Plata, Argentina
dumm@fisica.unlp.edu.ar

The phenomenology of B decays into two vector mesons offers the possibility of measuring many observables related with the angular distribution of the final outgoing particles. Here we study in particular the decays into ϕK^* states, which can be used to disentangle annihilation contributions to the amplitude and are potentially important channels to look for physics beyond the Standard Model.

The study of B physics offers a good opportunity to get new insight about the origin of CP violation. The rich variety of decays channels of the B mesons allows to look for many different observables, providing stringent tests for the CKM mechanism proposed by the Standard Model (SM) and allowing to get useful information about the physics involved in hadronic decay amplitudes. These analyses may definitely unveil the presence of New Physics (NP) beyond the SM, or provide hints for future searches.

We have concentrated in particular on $B \to \phi K^*$ decays. Among the various charmless $B \to VV$ channels, these are the first ones that have been experimentally observed. They have the property of being penguin-dominated, and, moreover, it is seen that sizable penguin contributions carry a common weak phase within the SM. Thus they are particularly interesting towards the search of NP effects. In addition, $B \to VV$ processes offer the possibility of measuring many different observables, taking into account the angular distributions of the final outgoing states. In this note we briefly mention the results of two research works[1,2]:

1. Analysis of the Size of Annihilation Amplitudes in $B^\pm \to \phi K^{*\pm}$

The magnitude of annihilation contributions to decay amplitudes is in general not known (predictions from different QCD-based approaches are contradictory), and their estimation from experiment is quite difficult due to the large theoretical uncertainties in the calculation of hadronic matrix elements. We point out that the decay $B^{\pm} \to \phi K^{*\pm}$ offers an interesting opportunity in this sense, since annihilation amplitudes may be relatively large, and can be disentangled by looking at certain CP-odd observables. In particular, in the framework of factorization, the experimental information can be used to measure both annihilation form factors and strong final state interaction (FSI) phases. This analysis also serves as a test of the consistency of the factorization approach itself, taking into account the theoretical estimation of the effective $\Delta B = 1$ Hamiltonian within the SM and the experimental information on the angle γ of the unitarity triangle.

2. Signals of New Physics in Neutral $B \to \phi K^*$ Channels

In the case of $B^0 \to \phi K^{*0}$ and $\bar{B}^0 \to \phi \bar{K}^{*0}$ decays, annihilation amplitudes can be neglected and CP violation effects should lie below the level of 1% within the SM. Thus these channels appear to be quite promising to find signals of NP. In addition, the angular distributions of final outgoing particles offer the possibility of getting even more information from an eventual nonzero experimental result: comparative analysis of adequate observables can be used to distinguish between different sets of effective operators, and therefore between different NP scenarios. In our analysis we have considered an effective Hamiltonian which includes, besides SM physics, possible NP contributions to SM-like operators as well as NP contributions driven by new, nonstandard operators. We have worked within the factorization approach, focusing mainly on Left-Right symmetric models, i.e. looking for observables that could distinguish between left-handed and right-handed currents. We have shown that in some cases nonstandard contributions are subject to kinematic cancellations, allowing to obtain a sort of chirality filter for the underlying NP. Indeed, we have found that if both the absorptive and FSI strong phases, as well as the ratio between SM and NP scales, are assumed to be small quantities, then the so-called "triple products" show a different behavior upon the nature of the involved operators, working effectively as filters of right-handed NP currents.

Acknowledgments

This work has been partially supported by Fundación Antorchas (Argentina), and ANPCyT (Argentina) under grant PICT02-03-10718.

References

1. L.N. Epele, D. Gómez Dumm, A. Szynkman, Eur. J. Phys. C **29** (2003) 207.
2. E. Álvarez, L.N. Epele, D. Gómez Dumm, A. Szynkman, hep-ph/0410096.

DEGENERACIES IN THE MEASUREMENT OF NEUTRINO OSCILLATION PARAMETERS: PROBLEM AND SOLUTION IN NEUTRINO FACTORIES

J. JONES* and A. M. GAGO†

Seccion Fisica, Departamento de Ciencias,
Pontificia Universidad Catolica del Peru,
Apartado 1761, Lima, Peru
*jjones@rcp.net.pe
†agago@fisica.pucp.edu.pe

In this work we review the degeneracies in neutrino oscillation parameters, making a quantitative analysis to find out how to solve these degeneracies in future Neutrino Factories.

1. Introduction

Given a neutrino oscillation probability function $P = P(\theta_{ij}, \delta, \Delta m_{ij}^2; L, E_\nu)$, we get a degenerate solution if

$$P(\ldots, \alpha, \beta, \ldots; L, E_\nu) = P(\ldots, \alpha', \beta', \ldots; L, E_\nu), \tag{1}$$

where (α, β) and (α', β') are different sets for the same oscillation parameters. There are three types of degeneracies involving neutrino oscillation parameters: (θ_{13}, δ), $sgn(\Delta m_{31}^2)$ and $(\theta_{23}, \frac{\pi}{2} - \theta_{23})$. Taking into account all cases, we can get an eightfold degeneracy [1].

2. Neutrino Factories

Neutrinos produced at Neutrino Factories arise from the decay of muons. This type of decay has the advantage of producing almost and equal number of neutrinos and antineutrinos, as well as having a very well determined flux.

In order to simulate number of events at Neutrino Factories we use [2]:

$$\frac{\partial n}{\partial E_\nu'} = N \int_{4GeV}^{E_\mu} dE_\nu \phi_{\nu_i}(E_\nu) \times P_{\nu_i \to \nu_f}(E_\nu) \times \sigma_{\nu_f}(E_\nu) \times R(E_\nu, E_\nu') \times \epsilon(E_\nu')(2)$$

where E_ν and E'_ν are the real and measured neutrino energies, ϕ_ν is the neutrino flux, $P_{\nu_i \to \nu_f}$ is the oscillation probability, σ_ν is the cross-section, $R(E_\nu, E'_\nu)$ is the neutrino energy resolution and ϵ is the detection efficiency. The measured oscillation channels are $\nu_e \to \nu_\mu$ and $\nu_e \to \nu_\tau$.

We have considered a Neutrino Factory with 2.5×10^{20} useful muons, a muon energy of 50 GeV, and 5 years of data taking. We consider standard 50 Kt. (5 Kt.) muon (tau) detectors, both at $L = 3000$ Km.

3. Results and Conclusions

We present results for 2σ and 3σ confidence levels, considering the following "true" parameters: $\sin^2 2\theta_{12} = 0.8$, $\tan^2 \theta_{23} = 0.82$, $\tan^2 \theta_{13} = 2.5 \times 10^{-3}$, $\delta = 0°$, $\Delta m_{21}^2 = 8 \times 10^{-5}$ and $\Delta m_{31}^2 = 3 \times 10^{-3}$.

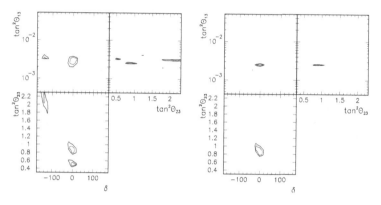

Fig. 1. 2σ and 3σ confidence levels, considering the real tau detection efficiency as reported in [3] (left), and an ideal efficiency, equal to the real muon detecion efficiency (right).

We can appreciate that the degeneracy problem can not be solved by Neutrino Factories using current technology, being the θ_{23} degeneracy responsible for this difficulty. In order to solve this problem, a better τ detection efficiency is needed.

References

1. V. Barger, D. Marfatia and K. Whisnant, *Phys. Rev.* **D65**, 073023 (2002).
2. P. Huber, M. Lindner and W. Winter, *Nucl. Phys.* **B645**, 3-48 (2002).
3. D. Autiero, G. De Lellis, A. Donini, M. Komatsu, D. Meloni, P. Migliozzi, R. Petti, L. Scotto Lavina and F. Terranova, *Eur. Phys.* **J. C33**, 243-260 (2004).

REVISION OF THE NEUTRINO OSCILLATION PROBABILITY IN THE SUPERNOVAE

L. ALIAGA* and A. M. GAGO†

Seccion Física, Departamento de Ciencias,
Pontificia Universidad Católica del Perú,
Apartado 1761, Lima, Perú
*leoaliaga@gmail.com
†agago@fisica.pucp.edu.pe

In this paper we review the comparison between the analytical and numerical solutions for the oscillation probability within the context of a Supernovae.

1. Theoretical Framework

The calculation of the survival probability in a medium of variable density can be done using two approaches. One is solving numerically propagation equation of neutrinos inside the Supernovae, which is given by[1]:

$$i\frac{d}{dx}\begin{pmatrix} \nu_e \\ \nu_\mu \end{pmatrix} = \frac{1}{2E}\left[U\begin{pmatrix} m_1^2 & 0 \\ 0 & m_2^2 \end{pmatrix}U^+ + \begin{pmatrix} 2V(x)E & 0 \\ 0 & 0 \end{pmatrix}\right]\begin{pmatrix} \nu_e \\ \nu_\mu \end{pmatrix} \quad (1)$$

where $m_{1,2}$ is the mass in vacuum, U is mixing matrix, E is the energy of the neutrinos and V(x) is the interaction potential of neutrino with the matter. The second approach is evaluating directly the survival probability formula, which is described by:

$$P_{ee} = \begin{pmatrix} 1 & 0 \end{pmatrix}\begin{pmatrix} \cos^2\theta & \sin^2\theta \\ \sin^2\theta & \cos^2\theta \end{pmatrix}\begin{pmatrix} 1-P_c & P_c \\ P_c & 1-P_c \end{pmatrix}\begin{pmatrix} \cos^2\theta_m & \sin^2\theta_m \\ \sin^2\theta_m & \cos^2\theta_m \end{pmatrix}\begin{pmatrix} 1 \\ 0 \end{pmatrix}(2)$$

where P_c is the well known crossing probability.

2. Inside the Supernovae: Results and Conclusions

In the Fig. 1., we present the neutrino probabilities. In the left top side, we show the isolines for P_c in $(\Delta m^2/E, tan^2\theta)$, obtained from the numer-

340

ical integration of Eq.(1) assuming a power law matter potential for the supernova. Additionally, we display the change of survival probability P_{ee} on the rigth top side for $E = 15 MeV$. In the same way, we show the P_c on the bottom side, calculated using the analytical prescription suggested in [3] with the realistic supernova matter density profile and the P_{ee} in $(\Delta m^2, tan^2\theta)$ for $E = 15 MeV$. From the comparison of both figures, we can see small differences between the two methods, which, certainly, would be smeared out in a convolution with the neutrino flux and cross section.

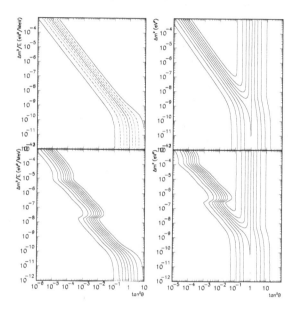

Fig. 1.   The P_c and P_{ee} for the realistic profile density analytical prescription suggested in [3] account (3).

References

1. T. K. Kuo and Pantaleone, *Rev. Mod. Phys.* **61**, 973-979 (1989).
2. Shigeyama and Nomoto. *Astrophys. J.* **360**, 242.
3. M. Kachelriess, A. Strumia, R. Tomas and J. W. F. Valle *hep-ph/* **0108100**.
4. Fogli, Lisi, Montanino and Palazzo. *Phys. Rev.* **D 65**, 073008.

CONSEQUENCES ON THE NEUTRINO MIXING MATRIX FROM TWO ZERO TEXTURES IN THE NEUTRINO MASS MATRIX

L. STUCCHI*, A. M. GAGO[†]

Sección Física, Departamento de Ciencias,
Pontificia Universidad Católica del Perú,
Apartado 1761, Lima, Perú
*luchano@universia.edu.pe
[†]agago@fisica.pucp.edu.pe

V. GUPTA

Departamento de Física Aplicada, CINVESTAV- Unidad Mérida
A.P. 73 Cordemex 97310 Mérida, México

The main purpose of this paper is to determine which two-zero textures are allowed by experimental data and what limits they impose on the elements of the neutrino mixing matrix that are not known yet.

1. Theoretical Framework

Neutrino oscillations are described by the three generation neutrino mixing matrix U known as the *Pontecorvo-Maki-Nakagawa-Sakata* [1] matrix which connects the flavor basis to the mass eigenstate basis. We use this matrix to relate the diagonal mass matrix $\hat{M}$ with the flavor basis one M:

$$M = U\hat{M}U^{\dagger} \tag{1}$$

We are interested in which zero textures of M are permitted by the experimental data [2].

2. Analysis

We solve the one-zero textures putting $\sin^2 \theta_{13}$ as a function of m_1 and we found that only five of those are allowed by experimental data (only $M_{\mu\tau}$ can not be zero). In order to obtain the valid two-zero textures, we cross the different combinations of the one-zero texture curves and we found only two of them are valid from the data, $M_{ee} = M_{e\mu} = 0$ and $M_{ee} = M_{e\tau} = 0$. When at least one off-diagonal element of M vanishes, one can deduce by studying the Jarlskog invariant that the δ phase is automatically zero, so there are no CP-violating effects. This situation is exactly what we find for Dirac neutrinos.

3. Results and Conclusions

We have found that only two textures of two zeros are allowed, those corresponding to type A [3,4], which happen to be very similar between them:

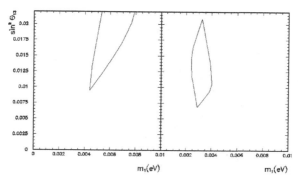

Fig. 1. Different zones for the valid sign combination of m_1, m_2 and m_3 for texture A1 ($M_{ee} = M_{e\mu} = 0$). Texture A2 ($M_{ee} = M_{e\tau} = 0$) has almost the same shape.

We can establish that theories beyond the Standard Model are limited to have only those textures indicated here, for Dirac neutrinos. Since we are obtaining limits to the values of the masses, $\sin^2 \theta_{13}$, and the δ phase each of those textures remains as a possibility as long as experiments agree with the restrictions.

References

1. Z. Maki, M. Nakagawa and S. Sakata, *Prog. Theo. Phys.* **28**, 27 (1962)
2. S.F. King *arXiv:hep-ph/0310204*
3. P.H. Frampton, S.L. Glashow and D. Marfatia *arXiv:hep-ph/0201008*
4. Z. Xing *arXiv:hep-ph/02011051*

EXPECTED FLUX OF HIGH ENERGY NEUTRINOS FROM OBSERVED ACTIVE GALACTIC NUCLEI

J. L. BAZO* and A. M. GAGO†

Sección Física, Departamento de Ciencias,
Pontificia Universidad Católica del Perú,
Apartado 1761, Lima, Perú
*jlbazo@fisica.pucp.edu.pe
†agago@fisica.pucp.edu.pe

Since neutrinos interact weakly, they represent a powerful tool to explore distant astrophysical objects, such as Active Galactic Nuclei (AGN). With a sample of 120 AGNs, we use the astrophysical parameters describing the neutrino flux to calculate the number of observable neutrinos in IceCube. We consider various hypotheses in the parameters involved in neutrino production and quantify their effect on the neutrino events.

1. Theoretical Framework

Active Galactic Nuclei are cosmic rays sources which are supposed to produce copiously neutrinos. Within the AGN jets, protons (PeV-EeV) are highly accelerated by a shock wave and should interact with ambient photons from the thermal radiation (UV bump), leading to the $\Delta^+(1232)$ resonance, followed by the cascade $n(\pi^+ \to \mu^+(\to \overline{\nu_\mu}e^+\nu_e)\nu_\mu)$. The ν_μ and $\overline{\nu}_\mu$ flux can be obtained with[1]:

$$\Phi_{\nu_\mu+\overline{\nu}_\mu} \approx \frac{1}{4\pi d_L{}^2\epsilon_\nu}\frac{1}{20}L_{p_{obs}}[(1-e^{-\tau})e^{(1-e^{-\tau})}] \tag{1}$$

Then the number of background free ν's induced events in IceCube is:

$$N_{\nu Events} = 4\pi N_{Av}.T.V.\rho \int_{\delta_{min}}^{\delta_{max}} pdf[\delta^{-exp}] \int_{10^{15}eV}^{10^{18}eV} \frac{\Phi_{\nu_\mu+\overline{\nu}_\mu}[\delta,\epsilon_\nu]}{\epsilon_\nu}\sigma_{\nu tot}d\epsilon_\nu d\delta \tag{2}$$

344

2. AGN Parameters

We focus are analysis on the following AGN parameters:

2.1. *Doppler boosting factor*

Because of the effect of the Doppler Factor δ in flux calculations, we focus on its estimation models: *Inverse Compton, Equipartition, Variability*[2] and *Minimum*. Furthermore, from proper motion observations[3] we obtain a power law distribution for the jet δ, with negative slope.

2.2. *Emission region geometry*

We consider one cylindrical and two spherical types of emission regions (models based on Δt_{obs})[4] and change the common formula $R' = \delta c \Delta t_{obs}$ by considering the angle to the line of sight and shock velocity.

3. Results and Conclusions

The emission region geometry could be determined from the total number of events (Fig.1). However after 3 years of observations it will not be conclusive as to decide in favor of one model. In the spherical case, we could use the angular resolution to distinguish between δ models.

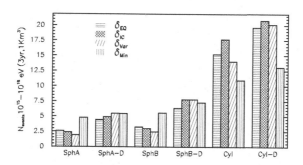

Fig. 1.   Expected total AGN ν_{μ} events after 3 years of observation in IceCube.

References

1. F. Halzen and D. Hooper, *Rept.Prog.Phys.* **65**, 1025 (2002).
2. A. Lähteenmäki and E. Valtaoja,E., *ApJ*, **521**, 493 (1999).
3. K. I. Kellermann, *et al. ApJ* **609**, 539 (2004).
4. R. J. Protheroe, accepted by *PASA* (2002).

MASPERI'S QUASISPIN MODEL OF THE SCALAR FIELD φ^6 THEORY WITH SOLITON SOLUTIONS*

M. AGÜERO[†], G. FRIAS and F. ONGAY[‡]

Facultad de Ciencias. Universidad Autonoma del Estado de Mexico,
Departamento de Fisica, [‡]Departamento de Matematica
Instituto Literario 100, Toluca, CP 5000, Mexico
[†]mag@uaemex.mx

The nonlinear lattice equation of the φ^6 theory is studied by using the technique of generalized coherent states associated to a $SU(2)$ Lie group. We analyze the discrete nonlinear equation with weak interaction between sites. The existence of saddles and centers is shown. The qualitative parametric domains which contain kinks, bubbles and plane waves were obtained. The specific implications of saddles and centers to the parametric first and second order phase transitions are identified and analyzed.

1. Hamiltonian Model

The lattice hamiltonian of the ϕ^6 theory obtained by Masperi and coworkers using the mean field approximation is given as [1]

$$
H = \sum_m \left\{ \begin{pmatrix} 0 & 0 & 0 \\ 0 & -\varepsilon & 0 \\ 0 & 0 & -\kappa \end{pmatrix}_m - \delta \begin{pmatrix} 0 & 1 & 0 \\ 1 & 0 & \alpha \\ 0 & \alpha & 0 \end{pmatrix}_m \begin{pmatrix} 0 & 1 & 0 \\ 1 & 0 & \alpha \\ 0 & \alpha & 0 \end{pmatrix}_{m+1} \right\}. \tag{1}
$$

where $\varepsilon, \kappa, \delta$ are parameters of the model, $\alpha = (\frac{\varepsilon}{\kappa-\varepsilon})^{1/2}$. Next, we analyze the quantum Hamiltonian (1) via the spin coherent states [3] and obtain the hamiltonian and the lattice nonlinear quasiclassical equation of motion for two interesting cases: $\alpha = 1$, and $\alpha \neq 1$.

*This work was supported by the UAEM Research Project 1758/2003.

a) Case $\alpha = 1$ or $2\varepsilon = \kappa$. We obtain the hamiltonian that is well known in lattice theories that supports analytic soliton solutions.

$$H_2 = \sum_m \varepsilon \frac{1 - |\psi_m|^2}{1 + |\psi_m|^2} - 2\delta \left(\frac{\bar{\psi}_m + \psi_m}{1 + |\psi_m|^2} \right) \left(\frac{\bar{\psi}_{m+1} + \psi_{m+1}}{1 + |\psi_{m+1}|^2} \right) \qquad (2)$$

b) Case $\alpha \neq 1$, the equation of motion is

$$i\,\dot{\psi}_m = \frac{2\psi_m (\kappa - \varepsilon) + 2\varepsilon \psi_m |\psi_m|^2}{(1 + |\psi_m|^2)} - $$

$$G \left(\frac{(\psi_{m+1} + c.c.) \left(\alpha + |\psi_{m+1}|^2 \right)}{\left(1 + |\psi_{m+1}|^2 \right)^2} + \frac{(\psi_{m-1} + c.c.) \left(\alpha + |\psi_{m-1}|^2 \right)}{\left(1 + |\psi_{m-1}|^2 \right)^2} \right) \qquad (3)$$

Where

$$G = 2\delta \frac{\left(\alpha + (1 - 2\alpha)\,\psi_m^2 + (2 - \alpha)\,|\psi_m|^2 - \psi_m^{\,2} |\psi_m|^2 \right)}{(1 + |\psi_m|^2)} \qquad (4)$$

2. Bifurcations and Phase Transitions

In the first approximation the last equation for $\psi_m = \zeta$, is a typical nonlinear Schrödinger equation:

$$i\,\dot{\zeta} = 2\,(k - \varepsilon)\,\zeta + 2\,(2\varepsilon - k)\,\zeta\,|\zeta|^2 - $$

$$4\delta \left(\zeta + \bar{\zeta} \right) \left[\alpha^2 + \alpha\,(1 - 2\alpha)\,\zeta^2 + 3\alpha\,(1 - \alpha)\,|\zeta|^2 \right] \qquad (5)$$

Next, we have obtained six regions in the parametric phase space (σ, Δ) with $\sigma = \frac{\varepsilon}{\kappa}$ and $\Delta = \frac{\delta}{\kappa}$. It is possible also to find different phase portraits of each sector in this parametric phase space.

The saddle points obtained here underlie the existence of both bubble and kink solitons and as is obvious, the centers are responsibles for linear waves. The first order phase transition occurs when the saddles and centers transform to another configuration with saddle and center singular points. The second order phase transition appears when a center bifurcates to saddle points and to other centers or these points transform to only one center.

The regions which could contain bubbles and kinks in the present paper converge to the regions of the parametric space of those results obtained by Masperi [2] and coworkers and in some sense they complement each other.

References

1. D. Boyanovsky and L. Masperi, *Phys. Rev*, **D 21**, n6,1550(1980).
2. L. Masperi, *Phys. Rev.*, **D 41**, 3263 (1990).
3. V. Makhankov, M. Aguero, A. Makhanov, *Journal of Physics A: Math. and Gen.*, **29**, 3005 (1996).

NONSTANDARD CP VIOLATION IN $B \to D_s D^0 \pi^0$ DECAYS[*]

A. SZYNKMAN

IFLP, Depto. de Física, Universidad Nacional de La Plata,
C.C. 67, 1900 La Plata, Argentina
szynkman@fisica.unlp.edu.ar

We study the possibility of measuring nonstandard CP violation effects through Dalitz plot analysis in $B \to D_s D^0 \pi^0$ decays. The accuracy in the extraction of CP violating phases is analyzed by performing a Monte Carlo simulation of the decays, and the magnitude of possible new physics effects is discussed. It is found that this represents a hopeful scenario for the search of new physics.

The origin of CP violation in nature is presently one of the most important open questions in particle physics. Indeed, the main goal of the experiments devoted to the study of B meson decays is either to confirm the picture offered by the Standard Model (SM) or to provide evidences of CP violation mechanisms originated from new physics. In fact, the common belief is that the SM is nothing but an effective manifestation of some underlying fundamental theory. In this way, all tests of the standard mechanism of CP violation, as well as the exploration of signatures of nonstandard physics, become relevant.

We discuss the possible measurement of nonstandard CP violation in $B \to D_s D^0 \pi^0$ [1], exploiting the fact that for these processes the asymmetry between B^+ and B^- decays is expected to be negligibly small in the Standard Model. The presence of two resonant channels provides the necessary interference to allow for CP asymmetries in the differential decay width, even in the limit of vanishing strong rescattering phases. From the experimental point of view, the usage of charged B mesons has the advantage of avoiding flavor–tagging difficulties. In addition, the processes $B \to D_s D^0 \pi^0$

[*]This work has been partially supported by CONICET Argentina

appear to be statistically favored, in view of their relatively high branching ratios of about 1%.

In order to measure the CP-odd phases entering the interfering contributions to the total decay amplitude, we propose to use the Dalitz plot fit technique. In general, three body decays of mesons proceed through intermediate resonant channels, and the Dalitz plot fit analysis provides a direct experimental access to the amplitudes and phases of the main contributions. In particular, this fit technique allows a clean disentanglement of relative phases, independent of theoretical uncertainties arising from final state interaction effects. The expected quality of the experimental measurements is estimated by means of a Monte Carlo simulation of the decays, from which we conclude that the phases can be extracted with a statistical error not larger than a couple of degrees, provided that the widths of the intermediate D^{*0} and D_s^* resonances are at least of the order of a hundred keV. On the theoretical side, within the framework of generalized factorization we perform a rough estimation of possible nonstandard CP violation effects on the interfering amplitudes. We take as an example the typical case of a multihiggs model, showing that the level of accuracy of the Dalitz plot fit measurements can be sufficient to reveal effects of new physics.

Let us finally stress that tree-dominated decays like $B \to D_s D^0 \pi^0$ are usually not regarded as good candidates to reveal new physics, since the effects on branching ratios are not expected to be strong enough to be separated from the theoretical errors. Our proposal represents a possible way of detecting these effects by means of CP asymmetries, which can allow the disentanglement of new physics contributions to penguin-like operators in a theoretically simple way.

Acknowledgements

A.S. acknowledges financial aid from Fundación Antorchas (Argentina).

References

1. L.N. Epele, D. Gómez Dumm, R. Méndez-Galain, A. Szynkman, Phys. Rev. D **69**, 055001 (2003).

SPINOR REALIZATION OF THE SKYRME MODEL

ROSENDO OCHOA JIMENEZ

Facultad de Ciencias de la Universidad Nacional de Ingeniería
Av. Tupac Amaru 210,
Lima, Perú
rochoaj@uni.edu.pe

YU. P. RYBAKOV

Department of Theoretical Physics, Peoples' Friendship University of Russia,
6, Miklukho-Maklaya str.,
117198 Moscow, Russia
ypybakov@sci.pfu.edu.ru

Skyrme proposed a mathematical model of the atomic nucleus that considers that nucleons is solitons is to say are dense disturbances of energy within the pions. To such solitons is called topological to them so that its stability is deduced of purely topological considerations. Within model of Skyrme it is managed to describe many statical properties of the baryons, but it is desired to have something but exact, that considers the degrees of freedom of quarks. In the spinorial accomplishment the field function is a 8 – spinor Ψ. with this is managed to increase the number of degrees of freedom.

In contemporary physics of particles for describing the baryons adapts the apparatus for quantum chromodynamics (QCD). However, many properties of baryons within the framework QCD cannot be explained, since is required the output beyond the framework of perturbation theory, which is well-off only with the high energies, in connection with this for the explanation the mass spectrum and structure of baryons is used the low-energy approximation, which is not rested on the perturbation theory. In this case in the theory the effective structures, which correspond to the bound states of quarks and gluons, appear. In the very the rough approximation, when according to the quark degrees of freedom it is carried out averaging and are considered only light quarks the appropriate fields takes values in the group of $SU(2)$ it appears the effective meson theory, which was called the Skyrme model [1,2]. Within the framework of the Skyrme model is succeeded in satisfactorily describing many static of the property of baryons; however, is desirable its refinement, which more fully considers quark degrees of freedom.

As first step sets out a spinorial version of the Skyrme model, which is analyzed in the article [3].

From the following currents [4]:

$$l_\mu = \partial_\mu \Psi \bar{\Psi}, \quad \bar{l}_\mu = \gamma_0 l_\mu^+ \gamma_0 = \Psi \partial_\mu \bar{\Psi} \tag{1}$$

with the help of these currents the Lagrangian density sets out [5]

$$L = \frac{1}{4\lambda^2} Sp(\bar{l}_\mu \gamma^\nu l^\mu \gamma_\nu) + \frac{\varepsilon^2}{16} Sp(\bar{l}^{[\mu} \gamma^\alpha l_{[\mu} \gamma_\alpha \bar{l}^{\nu]} \gamma^\beta l_{\nu]} \gamma_\beta). \tag{2}$$

The following step consists of finding the field with the Maximum symmetry. According to Coleman-Palais [2], if there exists an invariant semi-simple group of the Lagrangian density and the field is invariant under the action of this group, then this field is extremal of the model. The field must be invariant under rotations in real and isotopic spaces, and also must be invariant to the phase transformations of the group $U(1)$ and to the translation by the time. The invariant group has the form:

$$diag[SU(2)_S \otimes SU(2)_I] \otimes diag[U(1) \otimes T(1)]. \tag{3}$$

The invariant field is:

$$\Psi = (A + B\gamma_0 + C\gamma_5 + D\gamma_0\gamma_5) \begin{pmatrix} \Omega_{\frac{1}{2}, -\frac{1}{2}} \\ -\Omega_{\frac{1}{2}, \frac{1}{2}} \end{pmatrix} e^{-i\omega t}. \tag{4}$$

where A, B, C, D are constants that depend on r.

References

1. Skyrme T.H.R. *Nucle.Phys.* **31**, 556 (1962).
2. V.G. Makhankov, Y.P. Rybakov and V.I. Sanyuk, The Skyrme model: Fundamentals, Methods and Applications, Springer-Verlag (1993).
3. Yu. P. Rybakov, Farraj N. Abi *Bulletin of Peoples' Friendship University of Russia* **4** (1993).
4. Rybakov Yu. P. *VINITI series " Classical Field Theory and Gravitational Theory"* **2** Moscow (1991).
5. Yu. P. Rybakov, R. Ochoa Jimenez *Bulletin of Peoples' Friendship University of Russia* **7** (1999).

AUTHOR INDEX